Y0-BIY-306

Lecture Notes in Earth Sciences

52

Peter Ergenzinger Karl-Heinz Schmidt (Eds.)

Dynamics and Geomorphology of Mountain Rivers

Springer-Verlag

Lecture Notes in Earth Sciences

52

Editors:
S. Bhattacharji, Brooklyn
G. M. Friedman, Brooklyn and Troy
H. J. Neugebauer, Bonn
A. Seilacher, Tuebingen and Yale

Peter Ergenzinger   Karl-Heinz Schmidt (Eds.)

Dynamics and Geomorphology of Mountain Rivers

Springer-Verlag
Berlin Heidelberg New York
London Paris Tokyo
Hong Kong Barcelona
Budapest

Editors

Prof. Dr. Peter Ergenzinger
Prof. Dr. Karl-Heinz Schmidt
Institut für Geographische Wissenschaften
Freie Universität Berlin
Altensteinstr. 19, D-14195 Berlin, Germany

Geology
GB
1201.2
.D96
1994

"For all Lecture Notes in Earth Sciences published till now please see final pages of the book"

ISBN 3-540-57569-3 Springer-Verlag Berlin Heidelberg New York
ISBN 0-387-57569-3 Springer-Verlag New York Berlin Heidelberg

CIP data applied for

This work is subject to copyright. All rights are reserved, whether the whole or part of the material is concerned, specifically the rights of translation, reprinting, re-use of illustrations, recitation, broadcasting, reproduction on microfilms or in any other way, and storage in data banks. Duplication of this publication or parts thereof is permitted only under the provisions of the German Copyright Law of September 9, 1965, in its current version, and permission for use must always be obtained from Springer-Verlag. Violations are liable for prosecution under the German Copyright Law.

© Springer-Verlag Berlin Heidelberg 1994
Printed in Germany

Typesetting: Camera ready by author
SPIN: 10424086   32/3140-543210 - Printed on acid-free paper

PREFACE

This volume contains a selection of papers presented and discussed at the COMTAG-Workshop on "Dynamics and Geomorphology of Mountain Rivers". COMTAG (Commission on Theory, Measurement and Application in Geomorphology) is a commission of the International Geographical Union (IGU). The meeting was held in the monastery of Benediktbeuern in the Bavarian Alps in June 1992. The main objective of the meeting was to review the most recent developments in research on river bed dynamics and bedload transport in mountain rivers. Questions of mountain torrent control and environmental protection were also addressed. The general theme of the meeting finds its appropriate scientific and spatial location in the long tradition of bedload transport studies carried out in the fluvially active German Alps, which are often affected by flood and mass movement hazards. The conference provided an impulse for discussions between researchers in the fields of mountain torrent hydrology, water resources management and bedload transport modelling.

In the five years preceding the meeting the editors of this volume had headed a DFG (Deutsche Forschungsgemeinschaft) project on "Bedload transport and river bed adjustment in the Lainbach catchment" within the priority programme "Fluvial Geomorphodynamics in the late Quaternary". Results of the investigations and newly developed measurement techniques were introduced to the participants during the meeting and an excursion to the nearby Lainbach River. The meeting was attended by sixty four scientists from fifteen countries. Thirty four papers were presented in sessions on bedload transport in mountain torrents, measurement techniques of solid material transport, mass movements and sediment supply, river bed adjustment and roughness characteristics of steep mountain torrents, models of bedload transport, and catastrophic flooding. From a regional perspective the majority of the contributions dealt with the Alps with a special focus on investigations carried out at the northern fringe of the Alps. Most of the papers presented were submitted for publication, and selected papers have been included in this volume.

The workshop was financially supported by the Deutsche Forschungsgemeinschaft, the Commission of the European Communities (Directorate General for Science, Research and Development), the Freistaat Bayern (Ministerium für Unterricht, Kultur, Wissenschaft und Kunst) and the US-Army Research and Development Standardization Group. The participants and the organizers are grateful for these grants . We thank the president of COMTAG, Asher Schick, for his friendly support during the preparation and organization of the workshop. We are also very much indebted to the Katholische Stiftungsfachhochschule München and the Salesianer Don Boscos, Benediktbeuern, who opened the rooms of the monastery of Benedikbeuern for scientific sessions and social events during the conference. The organization of the meeting would not have been possible without the help of the local and regional administration, water and forest authorities. We highly appreciate this assistance. In addition, the editors thank the Springer-Verlag for the inclusion of the conference proceedings in this series and the colleagues F. Ahnert, J. Bathurst, W. Bechteler, I. Campbell, P. Carling, N.J. Clifford, S. Custer, T. Davies, A. Dittrich, R. Ferguson, K. Garleff, M. Hassan, R. Hey, H. Ibbeken, J. Karte, H. Keller, D. Knighton, J. Laronne, M. Meunier, M.D. Newson, D. Oostwoud-Wijdenes, I. Reed, K.S.Richards, A. Scheidegger and W. Symader for their valuable contributions as reviewers of the manuscripts that were submitted for this volume.

Berlin, October 1993

Karl-Heinz Schmidt   Peter Ergenzinger

CONTENTS

Introduction

SCHMIDT, K.-H. & P. ERGENZINGER: Recent developments and perspectives in mountain river research 3

Examples from the Alps

WILHELM, F.: Human impact and exploitation of water resources in the Northern Alps (Tyrol and Bavaria) 15

BECHTELER, W., H.-J. VOLLMERS & S. WIEPRECHT: Model investigations into the influence of renaturalization on sediment transport 37

RICKENMANN, D.: Bedload transport and discharge in the Erlenbach stream 53

D'AGOSTINO, V., M.A. LENZI & L. MARCHI: Sediment transport and water discharge during high flows in an instrumented watershed 67

WEISS, F.H.: Luminophor experiments in the Saalach and Salzach rivers 83

MIKOS, M.: The downstream fining of gravel-bed sediments in the Alpine Rhine River 93

SCHMIDT, K.-H.: River channel adjustment and sediment budget in response to a catastrophic flood event (Lainbach catchment, Southern Bavaria) 109

BUSSKAMP, R.: The influence of channel steps on coarse bed load transport in mountain torrents: Case study using the radio tracer technique 'PETSY' 129

WETZEL, K.: The significance of fluvial erosion, channel storage and gravitational processes in sediment production in a small mountainous catchment area 141

BANASIK, K. & D. BLEY: An attempt at modelling suspended sediment concentration after storm events in an Alpine torrent 161

BECHT, M.: Investigations of slope erosion in the Northern Limestone Alps 171

Examples from areas outside the Alps

BILLI, P.: Streambed dynamics and grain-size characteristics of two gravel rivers of the Northern Apennines, Italy 197

BARSCH, D., H. HAPPOLDT, R. MÄUSBACHER, L. SCHROTT & G. SCHUKRAFT: Discharge and fluvial sediment transport in a semi-arid high mountain catchment, Agua Negra, San Juan, Argentina 213

BARSCH, D., M. GUDE, R. MÄUSBACHER, G. SCHUKRAFT & A. SCHULTE: Sediment transport and discharge in a high arctic catchment (Liefdefjorden, NW Spitsbergen) 225

General investigations on sediment transport dynamics

GRANT, G.E.: Hydraulics and sediment transport dynamics controlling step-pool formation in high gradient streams: a flume experiment 241

ERGENZINGER, P., C. DE JONG, J. LARONNE & I. REID: Short term temporal variations in bedload transport rates: Squaw Creek, Montana, USA, and Nahal Yatir and Nahal Estemoa, Israel 251

HAMMANN, K. & A. DITTRICH: Measurement systems to determine the velocity field in and close to the roughness sublayer 265

MICHALIK, A. & W. BARTNIK: An attempt at determination of incipient bed load motion in mountain streams 289

DIEPENBROEK, M. & C. DE JONG: Quantification of textural particle characteristics by image analysis of sediment surfaces - examples from active and paleo-surfaces in steep, coarse grained mountain environments 301

COUSSOT, P. & J.M. PIAU: Some considerations on debris flow rheology 315

Introduction

RECENT DEVELOPMENTS AND PERSPECTIVES IN MOUNTAIN RIVER RESEARCH

Karl-Heinz Schmidt & Peter Ergenzinger
Fachrichtung Physische Geographie, Freie Universität Berlin
Altensteinstr. 19, 14195 Berlin

INTRODUCTION

Mountain rivers and especially mountain torrents are characterized by highly variable discharges and sediment transport volumes with bedload constituting a substantial part of the total load. Solid load transport is primarily controlled by the availability of material and the unsteady activity of sediment sources. The material supplied to sediment transport is highly irregular in grain size and angularity. A mountain torrent ("Wildbach" in the German terminology) has a steep average gradient, but the longitudinal profile consists of sections of different degrees of bed inclination. The term step-pool system has recently been applied for this series of reaches with steep to very steep slopes (steps) and intervening segments with moderately steep to gentle slopes, sometimes even with reverse gradients (WHITTAKER 1987, CHIN 1989; GRANT et al. 1990). For all these reasons mountain rivers are fundamentally different from lowland channels.

The unsystematic and irregular arrangement of the attributes of mountain torrents makes them highly incalculable and poses many problems for water management and flood control as well as for discharge, sediment yield and hazard monitoring and prediction, especially when considering increasing human activities in high mountain regions. The consequences of human impact on high mountain water resources are outlined by WILHELM (this volume). In recent years public and scientific interest in mountain rivers has intensified as a result of the proliferation of conflicts over water resources and energy development, reservoir construction, depletion of sediment sources, tourism, preservation of wilderness areas, timber harvesting, and natural hazards. Many of the processes and characteristics of mountain torrents and related environmental problems are addressed by the authors of this volume and will be briefly highlighted in the following introductory sections.

SEDIMENT SOURCES

Suspended sediment and bedload transport as well as river bed adjustment are closely linked to the availability and temporal behaviour of sediment sources. Major sediment sources are rill and wash erosion on bare slopes, landslides and debris flows from unstable slopes and erosion from the river bed and banks. The related processes are highly variable in both space and time. Identification of sediment sources is often difficult, but, besides field inspection, air photograph surveys or fingerprinting methods (PEART & WALLING 1986) and tracer techniques may give some valuable information. Sediment sources may be activated or de-activated by human interference. Forestry and tourism, especially ski tourism, result in an increase of bare slope areas with reduced infiltration capacities. Reservoir construction and artificial bed armouring result in the depletion of instream sediment sources.

The steep slopes of mountain rivers are prone to a variety of mass movement processes, which sometimes supply enormous amounts of solid material to the river bed. A recent review of mass movement processes from rapid rock avalanches and debris flows to slow earthflows has been given by BOVIS (1993). EISBACHER & CLAGUE (1984) present a report on destructive mass movements in high mountains and discuss appropriate countermeasures. There is a close interaction between the slope system and the channel system, but few studies and models take an integrated view of both systems. Mass inputs from the slopes are either directly taken up by river flow or they become a part of the storage term in the sediment budget equation (SLAYMAKER 1993). Sediment mobilized in upslope areas may be stored in the channel system for hundreds of years and influence sediment yield at the basin outlet during its residence time. Residence times are particularly long when high magnitude/low frequency events have occurred in the system.

In this volume the contributions of slope and channel sediment sources to the material accumulated in the river bed after a catastrophic event is analyzed by SCHMIDT. WETZEL gives an account of the relative significance of different types of sediment sources in a small Alpine catchment (Lainbach, Bavaria), and BECHT investigates the sediment production on steep slopes in several basins of the Northern Limestone Alps. Finally, COUSSOT & PIAU contribute a study related to the conveyance of material from sediment sources to the channel, when they present a review of the fundamental rheological attributes of debris flows.

SUSPENDED SEDIMENT TRANSPORT MEASUREMENT AND MODELLING

Information on suspended sediment transport by mountain rivers is important for the design and management of reservoirs, the operation of conduits for water power plants and for evaluating the impact of tourism and timber harvesting. High temporal and spatial variations are typical for the suspended sediment concentration at a given mountain river cross section. The temporal variations in concentration are often independent of discharge variations. Peak concentrations frequently occur prior to peak discharge; concentrations on the falling limb of the hydrograph are sometimes smaller for the same discharge value than on the rising limb, which results in hysteresis loops in the discharge/concentration graph. For these reasons it is very difficult to establish valid correlations between discharge and suspended sediment concentration in rating curve approaches. Temporal interpolation and extrapolation methods may lead to erroneous results (WALLING & WEBB 1981, OLIVE & RIEGER 1988). The variable delivery of individual sediment sources in the catchment is generally much more important than the hydraulic attributes of channel flow.

The difficulties of obtaining reliable time series of suspended sediment concentration values can be overcome by frequent sampling or indirect methods of continuous monitoring. Turbidity measurement has often been used for the indirect determination of the concentration of suspended solids (REINEMANN et al. 1982). Results of suspended sediment yield calculations can be much improved (BLEY & SCHMIDT 1991), though the turbidity signal is influenced not only by sediment concentration, but also by its grain size distribution, colour and content of organic material. The practical and theoretical aspects of the turbidity measurement technique have been reviewed by GIPPEL (1989) in great detail. Very little information exists on the spatial variation of suspended sediment concentration in the cross section. There is a striking discrepancy between the number of theoretical and laboratory models and empirical field measurements. Multi-point measurements in a mountain torrent (Lainbach, Bavaria) have shown that even in highly turbulent flow there are remarkable deviations in the cross profile (up to 100 % in the case study) (BLEY & SCHMIDT 1991). This uneven spatial distribution refers both to the suspended bedload and the wash load (cf. HOROWITZ et al. 1989). VETTER (1987) compared models of vertical concentration distributions with existing empirical data sets. The better understanding of the processes and mechanisms controlling the temporal and spatial variability of suspended sediment concentrations is a major challenge for future research.

Papers in this volume investigate suspended sediment transport in the extreme high mountain terrains of the Andes (BARSCH, HAPPOLDT et al.) and Spitsbergen (BARSCH, GUDE et al.). D'AGOSTINO et al. included suspended sediment in their investigations in an instrumented basin in the southern Alps. They also report on their experience with turbidity measurements. BANASIK & BLEY present an attempt at modelling suspended sediment concentration graphs for an Alpine catchment (Lainbach) being well aware of the difficulties imposed by the non-uniform distribution of sediment sources.

BEDLOAD TRANSPORT MEASUREMENT AND MODELLING

Environmental and water management implications of bedload transport are much the same as those sketched for suspended sediment transport. In addition, there are serious problems connected with the depletion of instream bedload sources. The entire bedload of mountain rivers is trapped when reservoirs are constructed. This human interference in the bedload budget leads to serious problems of degradation downstream of dams. Possibilities of rehabilitation for sections of Alpine rivers are discussed by MANGELSDORF et al. (1990). In bedload research there is an urgent need for more information on probabilities and conditions of entrainment, the travel lengths of particles, the influence of particle characteristics and bed topography on bedload transport, the determination of bedload yields, the simulation of particle distributions from point sources, the construction of efficient armouring devices etc. The increasing interest in gravel bed river research has been emphasized by the recent publication of three high-standard volumes on this topic (HEY et al. 1982; THORNE et al. 1987, BILLI et al. 1992).

But transport mechanisms of coarse bedload material are still poorly understood, especially in steep mountain rivers with irregular beds, a wide, badly sorted range of particle sizes and a highly variable pattern of flow and sediment transport. Sediment is supply-limited and no flow capacity oriented deterministic models can be applied. In many mountain rivers an armouring layer has developed (ANDREWS & PARKER 1987). Bedload is often transported in waves of different frequencies and amplitudes, which poses great problems for sampling and yield calculations (ERGENZINGER 1988, GOMEZ et al. 1989, BUNTE 1990, ERGENZINGER et al., this volume). Direct measurements of bedload volumes on a sufficient level of temporal resolution are only possible under very favourable conditions and with great monetary investments (e.g. HOFER 1987). LARONNE et al. (1992) use open slots in bedload channels to weigh the material infilled. D'AGOSTINO et al. (this volume) employ a large storage area for the direct determination of bedload yields. Indirect methods imply "hydrophone" techniques (RICKENMANN, this volume) or the signals of natural magnetic particles crossing coils (SPIEKER & ERGENZINGER 1990, ERGENZINGER et al., this volume). Bedload formulae (for a review of bedload formulae see GOMEZ & CHURCH 1989), which, under certain circumstances, may successfully be used in streams that transport sand and fine gravel, are not applicable to coarse material torrents. Critical conditions for particle entrainment are often expressed in terms of shear stress (MICHALIK & BARTNIK, this volume) or stream power. In the extremely irregular beds of mountain torrents it is virtually impossible to determine critical values of point shear stress or stream power for individual particles, since there is no information on local slopes or water depths in the course of a flood. Using a new tracer technique, the radio tracer system "PETSY" (ERGENZINGER et al. 1989), field measurements demonstrated that critical conditions can generally not be described by a single value. Following the ideas of EINSTEIN (1937), probability approaches are regarded as much more suitable for analyzing the process (BUSSKAMP & ERGENZINGER 1991, SCHMIDT & ERGENZINGER 1992). For measuring fluid forces acting on a single particle, a new automatic tracer system was developed (COSSY = Cobble Satellite System), which enables the shear stress at the point of entrainment to be calculated (ERGENZINGER & JÜPNER 1992).

In studies investigating the influence of particle weight and shape on transport lengths large samples of natural and artificial iron and magnetic tracers were used (GINTZ & SCHMIDT 1991, SCHMIDT & ERGENZINGER 1992). The results showed that shape exerts a decisive influence. The compact and more elongated particles travel longer distances than the platy ones. In this volume DIEPENBROEK & DE JONG provide a detailed account of textural particle characteristics. Tracer experiments also yielded important results concerning the distribution of particles from point sources (HASSAN et al. 1991, HASSAN & CHURCH 1992, SCHMIDT et al. 1992). The distributions can be best approximated by Gamma-functions.

Tracer systems are used by a number of contributors to this volume. BARSCH, GUDE et al. apply magnetic tracers to acquire information on transport by snowmelt runoff on Spitsbergen. The transit of natural magnetic particles across a sill with installed coil systems was registered by ERGENZINGER et al. SCHMIDT relocated magnetic tracers after a catastrophic flood event to identify areas of prevalent deposition, depths of burial and particle shape control in high magnitude events. BUSSKAMP investigates the influence of channel steps on the travel velocity of coarse particles with the help of radio tracers. WEISS used luminophor tracers to examine the spread of artificially introduced bedload material downstream of a dam.

RIVER BED ADJUSTMENT

Especially in step-pool rivers the process interactions between discharge, geometry, roughness and bedload transport are extremely complex (cf. LISLE 1987, WHITTAKER 1987) and very much a matter of speculation. River bed adjustment research in rivers with movable beds is directly related to this complex of mutual interrelations. Field investigations of bed adjustments in steep coarse-grained mountain torrents under non-uniform and unsteady flow conditions are extremely rare (cf. ERGENZINGER 1992). This also applies to studies of the effects of changing flow regimes. Discussions of flow resistance in gravel-bed rivers are usually confined to steady uniform flow in fixed boundary channels (HEY 1979). For describing the roughness in open channels the Darcy-Weisbach friction factor with its sound theoretical basis is most frequently used. The total roughness of a channel is determined by roughness components of different scales, i.e. the particle grain (and shape) roughness and the form roughness produced by small scale bedforms and larger scale riffle-pool sequences (ROBERT 1990, 1993), or step-pool sequences in steep mountain torrents.

Pebble clusters belong to the micro-scale roughness elements, they usually are composed of a large cobble forming an obstacle with accumulation of large particles on its stoss side and fine grains in its wake. There has been some discussion on the geometric and related processual attributes of clusters in recent years (BILLI 1998, REID et al. 1992). In step-pool systems the steps act like small weirs and the water plunges over them into the pools in a kind of tumbling flow, dissipating much energy. The step-pool sequence is stable during small and moderate floods, but may be destroyed during high magnitude/low frequency events (cf. SCHMIDT, this volume). The great importance of the pools as bedload source areas was demonstrated by magnetic tracer experiments, they are also favoured sites of deposition, which makes them the most active parts of the river bed in the bedload budget (GINTZ & SCHMIDT 1991, SCHMIDT & ERGENZINGER 1992).

As the Darcy-Weisbach friction factor is a compound expression of total channel roughness, it is necessarily much higher than that calculated for grain roughness alone. The numerical influence of form roughness on total roughness is still a matter of much debate, and the determination of grain roughness is far from being satisfactorily solved, especially when changing intra-flood roughnesses are concerned. A number of characteristic grain size parameters (D_i) have been proposed for describing grain roughness in flow resistance equations (BRAY 1982). Empirical evidence indicates that the roughness height of coarse material mixtures is best defined by 3.5 D_{84} (HEY 1979). The disadvantage of this approach is that the size distribution of the bed surface material is characterized by a single grain-size index (ROBERT 1990). Moreover, this index can only be determined before and after a flood, and no intra-flood information is available, though there are remarkable changes in bed geometry and roughness in the course of an event (ERGENZINGER 1992, SCHMIDT et al. 1992). It is therefore of much interest to observe changes in river bed configuration with the help of micro-relief measurements. A new sounding technique ("Tausendfüssler", see ERGENZINGER 1992) was developed, and the vertical differences between measuring points were used to calculate an average roughness value (k3) for the entire river bed or segments of the river bed.

In this volume BECHTELER et al. deal with river bed adjustment processes after renaturalization from a water engineering perspective. MIKOS investigates bedload adjustments by fining and sorting in the Alpine Rhine river. In SCHMIDT's paper the effects of a catastrophic flood on an equilibrium step-pool system and associated self-regulating processes are described. BILLI stresses the high variability of grain sizes on exposed surfaces and discusses his results against the background of equal mobility and downstream sorting. GRANT used flume experiments to determine the domain of flow conditions under which step-pool sequences are formed. HAMMANN & DIETRICH develop a physical model that allows the stability or instability of river beds to be determined more precisely than before. The model is based on exact measurements of turbulence characteristics and the velocity fields close to the liquid/solid interphase with Laser Doppler Anemometry.

Much more precise and short-interval field measurements are needed to elucidate the intricate river bed adjustment interactions between cross sectional geometry, roughness and velocity profiles. This could lead to a better physical understanding of the effects of bed micromorphology on the form drag in gravel bed rivers. Clearly, in future research the problem of the interrelations between flow hydraulics, bedload transport, roughness and bed micromorphology must be tackled with greater intensity.

REFERENCES

ANDREWS, E.D. & PARKER, G. (1987): Formation of a coarse surface layer as a response to gravel mobility. - In: THORNE, C.R., BATHURST, J.C. & HEY, R.D. (eds): Sediment Transport in Gravel-Bed Rivers, 269-300, Chichester (Wiley).

BILLI, P. (1988): A note on cluster bedform behaviour in a gravel-bed river. - Catena 15: 473-481.

BILLI, P., HEY, R.D., THORNE, C.R. & TACCONI, P. (eds) (1992): Dynamics of Gravel-Bed Rivers. - Chichester (Wiley).

BLEY, D. & SCHMIDT, K.-H. (1991): Die Bestimmung von repräsentativen Schwebstoff-Konzentrationsgängen - Erfahrungen aus dem Lainbachgebiet/Oberbayern. - Freiburger Geogr. Hefte 33: 121-129, Freiburg.

BOVIS, M.J. (1993): Hillslope geomorphology and geotechnique. - Progress in Physical Geography 17: 173-189.

BRAY, D.I. (1982): Flow resistance in gravel-bed rivers. - In: HEY, R.D., BATHURST, J.C. & THORNE, C.R. (eds): Gravel-Bed Rivers, 109-133, Chichester (Wiley).

BUNTE, K. (1990): Experiences and results from using a big-frame bed load sampler for coarse material bed load. - IAHS Publication 193: 223-231.

BUSSKAMP, R. & ERGENZINGER, P. (1991): Neue Analysen zum Transport von Grobgeschiebe. Messung Lagrangescher Parameter mit der Radiotracertechnik (PETSY). - Deutsche Gewässerkundliche Mitteilungen 35: 57-63, Koblenz.

CHIN, A. (1989): Step pools in stream channels. - Progress in Physical Geography 13: 391-407, London.

EINSTEIN, H.A. (1937): Der Geschiebetrieb als Wahrscheinlichkeitsproblem. - Mitteilungen der Versuchsanstalt für Wasserbau an der ETH Zürich: 3-112.

EISBACHER, G.H. & CLAGUE, J.J. (1984): Destructive mass movements in high mountains: hazard and management. - Geological Survey of Canada Paper 84-16.

ERGENZINGER, P. (1988): The nature of coarse material bedload transport. - IAHS Publication 174: 207-216.

ERGENZINGER, P. (1992): River bed adjustment in a step-pool system: Lainbach, Upper Bavaria. - In: BILLI, P., HEY, R.D., THORNE, C.R. & TACCONI, P. (eds): Dynamics of Gravel-Bed Rivers, 415-430, Chichester (Wiley).

ERGENZINGER, P. & JÜPNER, R. (1992): Using COSSY (COssy Satellite SYstem) for measuring the effects of lift and drag forces. - IAHS Publication 210: 41-49, Wallingford.

ERGENZINGER, P., SCHMIDT, K.-H. & BUSSKAMP, R. (1989): The Pebble Transmitter System (PETS): first results of a technique for studying coarse material erosion, transport and deposition. - Zeitschrift für Geomorphologie N.F. 33: 503-508, Berlin, Stuttgart.

GINTZ, D. & SCHMIDT, K.-H. (1991): Grobgeschiebetransport in einem Gebirgsbach als Funktion von Gerinnebettform und Geschiebemorphometrie. - Z. Geomorph. N.F. Suppl. Bd. 89: 63-72, Berlin, Stuttgart.

GIPPEL, C.J. (1989): The use of turbidity instruments to measure stream water suspended sediment concentration. - Monogr. Series 4, Department of Geogr. and Oceanogr., Univ. NSW, Australia.

GOMEZ.B. & CHURCH, M. (1989): An assessment of bed load sediment transport formulae for gravel bed rivers. - Water Resources Research 25 (6): 1161-1186.

GOMEZ, B., NAFF, R.L. & HUBBELL, D.W. (1989): Temporal variation in bedload transport rates associated with the migration of bedforms. - Earth Surface Processes and Landforms 14: 135-156.

GRANT, G.E., SWANSON, F.J. & WOLMAN, M.G. (1990): Pattern and origin of stepped morphology in high-gradient streams, Western Cascades, Oregon. - Geol. Soc. Am. Bull. 102: 340-352, Boulder.

HASSAN, M.A. & CHURCH, M. (1992): The movement of individual grains on the streambed. - In: P. BILLI, R.D. HEY, C.R. THORNE, P. TACCONI, (eds.): Dynamics of Gravel-Bed Rivers: 159-173, Chichester (Wiley).

HASSAN, M.A., CHURCH, M. & SCHICK, A.P. (1991): Distance of movement of coarse particles in gravel bed streams - Water Resources Research 27: 503-511.

HEY, R.D. (1979): Flow resistance in gravel-bed rivers. - Journal of the Hydraulics Division, A. Soc. Civ. Eng. 105: 365-379.

HEY, R.D., BATHURST, J.C. & THORNE, C.R. (eds) (1982): Gravel-Bed Rivers. - Chichester (Wiley).

HOFER, B. (1987): Der Feststofftransport von Hochgebirgsbächen am Beispiel des Pitzbaches. - Österreichische Wasserwirtschaft 39: 30-38.

HOROWITZ, A.J., RINELLA F.A., LAMOTHE, P., MILLER, T.L., EDWARDS, T.K., ROCHE, R.L. & RICKERT, D.A. (1989): Cross-sectional variability in suspended sediment and associated trace element concentrations in selected rivers in the US. - IAHS Publication 181: 57-66.

LARONNE, J.B., REID, I., YITSCHAK, Y. & FROSTICK, L.E. (1992): Recording bedload discharge in a semiarid channel, Nahal Yatir, Israel. - IAHS Publication 210: 79-86.

LISLE, T.E. (1987): Overview: Channel morphology and sediment transport in steepland rivers. - IAHS Publication 165: 287-297.

MANGELSDORF, J., SCHEUERMANN, K. & WEISS, F.H. (1990): River Morphology. A Guide for Geoscientists and Engineers. - Berlin (Springer).

OLIVE, L.J. & RIEGER, W.A. (1988): An examination of the role of sampling strategies in the study of suspended sediment transport. - IAHS Publ. 174: 259-267.

PEART, M.R. & WALLING, D.E. (1986): Fingerprinting sediment source: The example of a drainage basin in Devon, UK. - IAHS Publication 159: 41-55.

REID, I., FROSTICK, L.E. & BRAYSHAW, A. C. (1992): Microform roughness elements and the selective entrainment and entrapment of particles in gravel-bed rivers. - In: BILLI, P., HEY, R.D., THORNE, C.R. & TACCONI, P. (eds): Dynamics of Gravel-Bed Rivers, 253-266, Chichester (Wiley).

REINEMANN, L., SCHEMMER, H. & TIPPNER, M. (1982): Trübungsmessungen zur Bestimmung des Schwebstoffgehalts. - DGM 26: 167-174, Koblenz.

ROBERT, A. (1990): Boundary roughness in coarse-grained channels. - Progress in Physical Geography 14: 43-70.

ROBERT, A. (1993): Bed configuration and microscale processes in alluvial channels. - Progress in Physical Geography 17: 123-136.

SCHMIDT, K.-H., BLEY, D., BUSSKAMP, R., ERGENZINGER, P. & GINTZ, D. (1992): Feststofftransport und Flußbettdynamik in Wildbachsystemen.- Das Beispiel des Lainbachs in Oberbayern. - Die Erde 123: 17-28, Berlin.

SCHMIDT, K.-H. & ERGENZINGER, P. (1992): Bedload entrainment, travel lengths, step lengths, rest periods studied with passive (iron, magnetic) and active (radio) tracer techniques. - Earth Surface Processes and Landforms 17: 147-165, Chichester.

SLAYMAKER, O. (1993): The sediment budget of the Lillooet river basin, British Columbia. - Physical Geography 14: 304-320.

SPIEKER, R. & ERGENZINGER, P. (1990): New developments in measuring bed load by the magnetic tracer technique. IAHS Publication 189: 169-178.

THORNE, C.R., BATHURST, J.C. & HEY, R.D. (eds) (1987): Sediment Transport in Gravel-Bed Rivers - Chichester (Wiley).

VETTER, M. (1987): Der Transport suspendierter Feststoffe in offenen Gerinnen. - Mitteilungen des Instituts für Wasserwesen 19: 1-182, München-Neubiberg.

WALLING, D.E. & WEBB, B.W. (1981): The reliability of suspended sediment load data. - IAHS Publ. 133: 177-194.

WHITTAKER, J.G. (1987): Sediment transport in step-pool streams. - In: THORNE, C.R., BATHURST, J.C. & HEY, R.D. (eds): Sediment Transport in Gravel-Bed Rivers, 545-579, Chichester (Wiley).

(Papers that are cited in this introduction and are part of this volume are not included in the reference list)

Examples from the Alps

HUMAN IMPACT AND EXPLOITATION OF WATER RESOURCES IN THE NORTHERN ALPS (TYROL AND BAVARIA)

Friedrich Wilhelm
Institut für Geographie, Universität München
Luisenstraße 37/II, D-80333 München

Abstract

Extending 1200 km from east to west, with a surface area of 200.000 km", the Alps can be seen as the main climatic divide between Central and Southern Europe. About 2800 km" of this area show extensive glaciation. During the winter months snow covers the ground for a long time.

Owing to these facts and the high amount of precipitation on both the northern and the southern Alpine margins, the Alps may be called the "hydrophylatium principale" of Europe, releasing their abundant water into the rivers.

Human influence can be seen in different water storages, such as clouds, surface and sub-surface water. At this point, some remarks on the impact on individual storages will be given. It has not been proved that the amount of precipitation has actually increased. But there is no doubt that the emission of SO_2 and NO_x into the atmosphere has led to the phenomenon of acid rain.

Man uses water and snow in many different ways, e.g. to generate hydroelectric power, for irrigation and, especially during the winter, for recreation purposes. Soil compaction on ski-runs leads to a very high velocity of discharge, which results in an increasing rate of erosion in these areas.

Furthermore, factors such as artificial snow and pollution contaminate both surface water and sub-surface water.

Constructions for controlling stream flow are certainly one of the most obvious forms of human impact. Dams and reservoirs were built to prevent further floods, to raise the base level of the streams and to generate electricity.
As a result, the runoff system has changed rapidly. This concerns high water waves resulting from heavy rainfall and

the melting of snow during spring and summer times, as well as the runoff regime during winter, when a lower water level can be measured.

These impacts and their results do not only affect the hydrological system, they also influence the morphology of river beds.

1. Introduction

Water plays a decisive role in the formation of river beds and other geomorphological processes. Owing to the abundance of water, mountain streams are very active in the forming of geomorphological elements. Therefore, interferences with the water/regolith system become especially important. According to Athanasius Kirchner, a hydrologist of the Baroque period (SCHEUERMANN 1977), the Alps constitute the "hydrophylatium principale", the sponge that supplies the surrounding areas with its rich water resources. BAUMGARTNER ET AL. (1983) quantified this statement for the Alps (tab.1).

According to tab. 1 both precipitation and especially discharge rates of the Alpine portion of the catchment areas are higher than those of the total catchment areas. Especially discharge shows DA values up to about 400%. Whilst in the case of the Rhein (Rhine) and Donau (Danube) the Alpine parts of the catchments comprise only 6% of the total catchment areas, discharge in these Alpine catchments areas amounts to as much as 19% and 24% respectively, i.e. it is over-proportionally high.

Human impact on the hydrological system of the mountains has focused both on exploitation and prevention of damage. Thus man changes the natural system and, at the same time, the controlling mechanisms for dependent processes. Possibilities of changing the hydrological system are provided either by the storage elements: clouds, snow cover or glaciers, surface waters (lakes and rivers), and soil water especially ground water, and the transferring processes: precipitation, melting of snow/ice, surface and ground water runoff.

Tab.1: Hydrometeorological data for selected large drainages of the Alps (BAUMGARTNER et al. 1983, p.222)

River	F $10^3 km^2$	FA %	P cm	PA %	D cm	DA %
Rhein	224	6	110	158	48	273
Donau	817	6	86	174	26	400
Rhône	99	32	122	111	57	139
Po	75	32	120	135	61	182

(F = total catchment area in $10^3 km^2$; FA = Alpine portion of the catchment area (%); P = mean precipitation of the catchment area in cm; PA = precipitation falling in the Alpine region (%); D = mean discharge of the total area in cm; DA = the discharge coming from the Alpine region (%) in relation to mean discharge).

2. The Natural Environment

The distribution of moisture in the Alps is determined by their geographical position, surface area and altitude, and their geological and geomorphological structures which control discharge and storage. The southernmost parts of the Alps extend as far as the zone of Mediterranean climate, although their major part lies in the zone of westerly winds. With a total area of 240 000 km⁻ (BIRKENHAUER 1980) the Alps extend for about 1200 km in a W-E direction and 200 km in the meridional direction. They form a marked climatic boundary between the Mediterranean climate in the south and the climate of the temperate zone in the north. Owing to the provenance of the winds that bring precipitation the peripheral parts of the Alps are wetter than the central areas.

Furthermore, it is of great importance for the abundance of water of the Alps that they extend up to the glaciated zone and that, with rising altitude, mass and duration of the snow cover increase in winter. Because of the low saturation vapour pressure over a melting ice surface of only 6.11 hPa snow and ice show reduced evaporation. Owing to the high energy consumption of the melting process of 335 Jg^{-1}, there is only a slow decomposition of the snow cover. Meltwater infiltrating into the cold snow cover can freeze in lower layers and be stored again. Melted snow and ice contribute mainly to the supply of soil water and ground water (HERRMANN 1978; WILHELM & VOGT 1988).

Tab. 2: Water balances for the Northern and Central Alps (according to BAUMGARTNER et al. 1983, p. 193)					
Area	H (m)	P (cm)	D (cm)	E (cm)	100D/P %
Northern Alps	1270	182	134	48	73
Central Alps	1780	131	88	43	67

(H = mean elevation of area in m; P = precipitation in cm; D = discharge in cm; E = evaporation in cm).

Not only the short-term but also the seasonal discharge rates are decisively influenced by snow and ice. According to ASCHWANDEN & WEINGARTNER (1983/84) complex hydrographs showing two peaks can only be found in catchment areas with a mean altitude of less than about 1500 m. In higher regions the retention of the snow dominates the seasonal discharge pattern.

Utilization of the Alpine water resources, especially since the 19th century, has been positively affected by the rich water supply of the Northern Limestone Alps. This fact is underlined by the water balances according to BAUMGARTNER et al. (1983) (tab. 2).

3. Impact on Water Storages

Impact on hydrological systems can be defined in both quantitative and qualitative ways. Quantitative impacts are usually easily visible in the form of water management constructions and their effects can be forecasted, whereas qualitative impacts alter the physical and chemical characteristics of hydrological systems. Often they cannot directly be traced by human observation, which makes them even more dangerous. In the following, man-made changes in the atmosphere, snow cover, and surface water storage systems will be briefly described.

3.1. Impact on Atmospheric Water

Significant changes in the amount of precipitation as documented by the long-term precipitation records (tab. 3) for München (elevation of observation site 520 m), Innsbruck (579 m), and Marienberg (1335 m, catchment area of the Etsch) of the past 80 to 100 years, are within the range of natural variability of climate and cannot necessarily be traced back to human influence (greenhouse effect).

Tab. 3: Changes of precipitation with time. (Data sources: (FLIRI 1974; DEUTSCHER WETTERDIENST several years)

Locality	München	Innsbruck	Marienberg
Period of observation	1881-1985	1891-1985	1859-1985
Mean precipitation [mm]	939	920	684
Slope of regression [1]	1.23±0.18	-0.92±0.21	-0.54±0.13
confidence level	99.9%	99.9%	99.9%

[1] slope of regression (alteration of precipitation within time) and limit of 95% confidence-level.

During this time the quality of "hydrometeors" has changed as well because of increased energy consumption, a fact already indicated by SMITH (1872). Numerous examples of this development exist in Central, Western, and Northern Europe.

In the Alps, too, forest damage provides a warning testimony of the effects of accumulating pollutants in the atmosphere. In the Salzburg basin the SO_2 emission reached a maximum of 17,000 t/yr of SO_2 in the early 1980s according to WITTMANN & TÜRK (1988). Technical measures (filters, fuels containing less pollutants, etc.) reduced the emission to 10,500 t/yr of SO_2 by 1985 and to only 6500 t/yr of SO_2 by 1987. The alterations in the chemical composition of precipitation have been confirmed by measurements (fig. 1) for some years. According to these measurements the substances in the air (NO_x, SO_2, CO_2) at the observation site Wank (1800 m asl, near Garmisch-Partenkirchen) have decreased during the period (1982/91), with the exception of CO_2 which is less than at Mauna Loa (Hawaii) but rising at a steeper rate.

WANK, NO_x, CO_2
2 x 7 – weighted mean

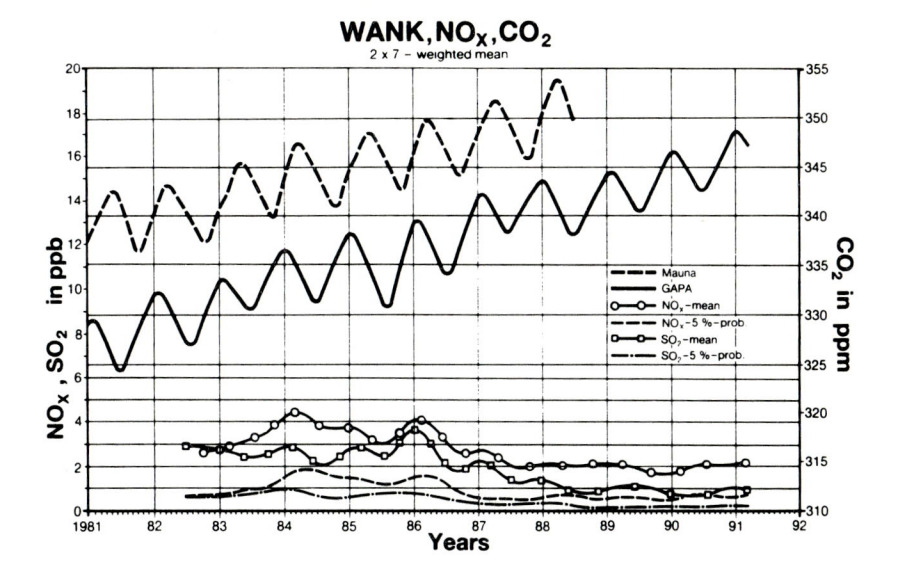

Fig.1: Alteration of NO_x, SO_2 and CO_2 measured at Wank (1800 m asl) near Garmisch Partenkirchen. (Data source: FRAUENHOFER-INSTITUT FÜR ATMOSPHÄRISCHE UMWELTFORSCHUNG, Garmisch-Partenkirchen).

3.2. The Snow Cover as Storage

Human alteration of storage within the snow cover can be for utilization purposes, e.g. for winter sports or for water supply by means of irrigation, and to prevent damage, e.g. constructions against avalanches. Here, the winter sports aspect will be emphasized.

Skiing has had a radical impact on the snow cover, both visible and invisible. Small farming villages have become famous winter sports resorts. In Switzerland skiing tourism has increased elevenfold in the last 30 years (MOSIMANN 1986). In some skiing resorts the number of guest beds greatly exceeds the number of inhabitants. Water consumption in the villages has risen dramatically. Whereas in former times the daily consumption of water was about 20 to 30 l per head, it amounts to 200-300 l for a much greater number of people today. In addition to that, slowly- or non-degradable substances in the sewage have increased. This load and the additional intake of salt from the skiing grounds impair the water quality of streams and rivers (fig. 2).

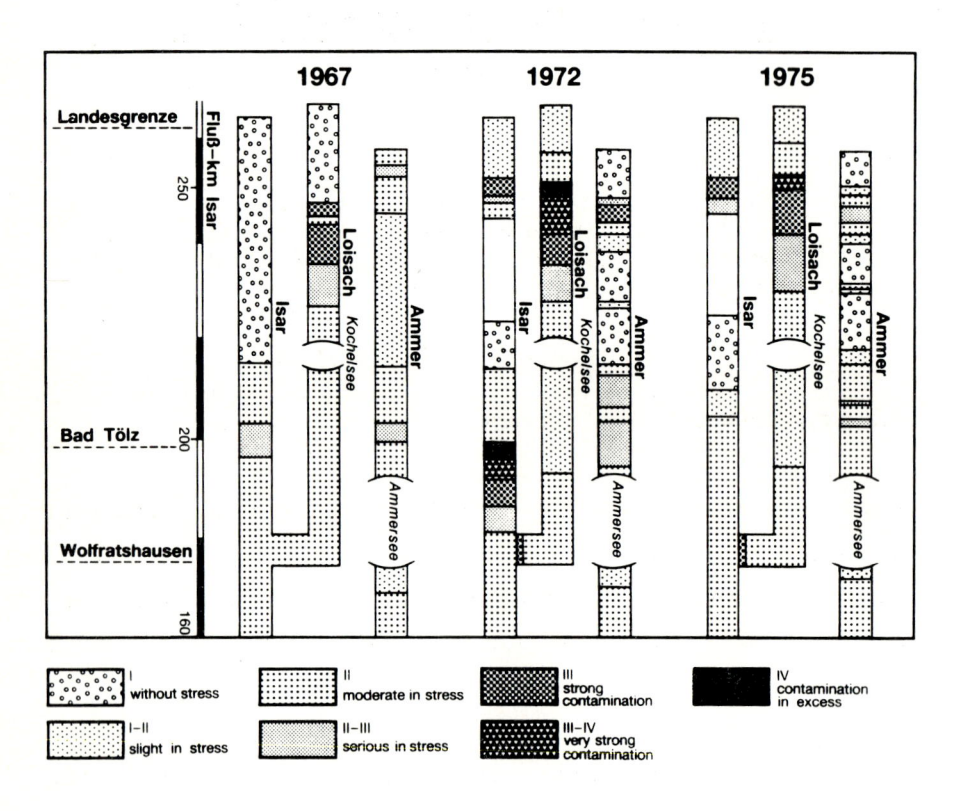

Fig. 2: Water quality of Isar, Loisach and Ammer from the sources to south of München and to the Amper respectively (from: BAYERISCHES STAATSMINISTERIUM FÜR LANDESENTWICKLUNG UND UMWELTFRAGEN 1980, adapted).

The growing expansion of winter sports facilities is drastically changing the natural environment. In addition to extensive levelling for downhill skiing, aisles have been cut into the mountain forests. In the winter months more snow accumulates on these areas and on the ski-runs than on the surrounding grounds.

The high investments involved made it necessary to guarantee skiing for winter guests also in years with little snow. Therefore, snow is produced artificially. This consumes about 200 to 300 l of water per m^2 in one winter (on the average 100,000 m^3 of water for each winter sports resort). By adding salt and chemicals in the production of artificial snow the snow cover gains stability and duration (NAUMANN 1989). The added preservatives contaminate the meltwater runoff (fig. 2).

For the construction of ski-runs the subsoil is cleared of trees and levelled. During the skiing season fresh snow is regularly removed from the runs and they are worked by caterpillar tractors. Thus the vegetation cover and the natural structure of the soil are destroyed. These impacts cause not only a 10-15% reduction of yields on pastoral land, but also an increased surface runoff with subsequent sheet flows and erosion in gullies and ditches (MOSIMANN 1986; MOSIMANN & LUDER 1980; DIETMANN 1985).

3.3. Surface Storages

In the Alps reservoirs are important for the supply of power because of the abundance of water and the high available fall-energy. In the winter of 1972/73 reservoirs fed by glacial meltwater covered 24% of Switzerland's power consumption (MÜLLER et al. 1976). Some natural and man-made reservoirs in the vicinity of the site of Benediktbeuern (Walchen-, Achen-, and Forggensee, as well as Sylvenstein reservoir) will serve as examples to discuss changes in the environment. The purpose of these reservoirs is to regulate the runoff of the rivers downstream (flood prevention, regulation of low water levels) and power production.

Tab. 4: Data on Walchensee, Achensee, Sylvensteinspeicher und Forggensee (according to BUNDESMINISTERIUM FÜR VERKEHR UND VERSTAATLICHTE BETRIEBE 1956; LAUFFER 1969; H. LINK 1970; BAYERNWERK A.G., undated)

Item	Achensee	Walchensee	Sylvenst. Speicher	Forggensee
Basin area km^2	218	770	426	1708
Lake surface [km^2]	6,8	16,5	5,5	14,5
Maximum depth [m]	133	192	29	35
Volume [10^{6}m^3]	580	1357	105	149
Eff.volume [10^{6}m^3]	71,4	110	100	149
Lakelevel osc.[m]	11,5	7	28	17
Energy cont.[10^{6}MW]	80,4	101,8	22,7	80,4
Year of constr.	1926/29	1921/23	1956/59	1950/53

In 1969/70 LINK (1970) counted a total of 322 reservoirs in the Alps with a surface of 275.8 km^2, an effective volume of about 10.4 km^3, and an energy content of 22.14 TWh. Only 13 of these reservoirs are situated in the Northern Limestone Alps; they cover an area of 58 km^2, their effective volume is 0.5 km^3, their energy content 331.1*10^6 MWh. According to calculations by HENSELMANN (1970) for lakes in southern Bavaria the reduction of the flood amplitude by lake retention depends mainly on the rate of inflow to the lake. In Walchensee this natural balance is also affected by reservoir management. In summer a volume of 110*10^6 m^3, mainly from meltwater of snow, is retained in the lake to be released during the shortage of water caused by snow retention in winter. From summer to autumn the lake level is kept constant.

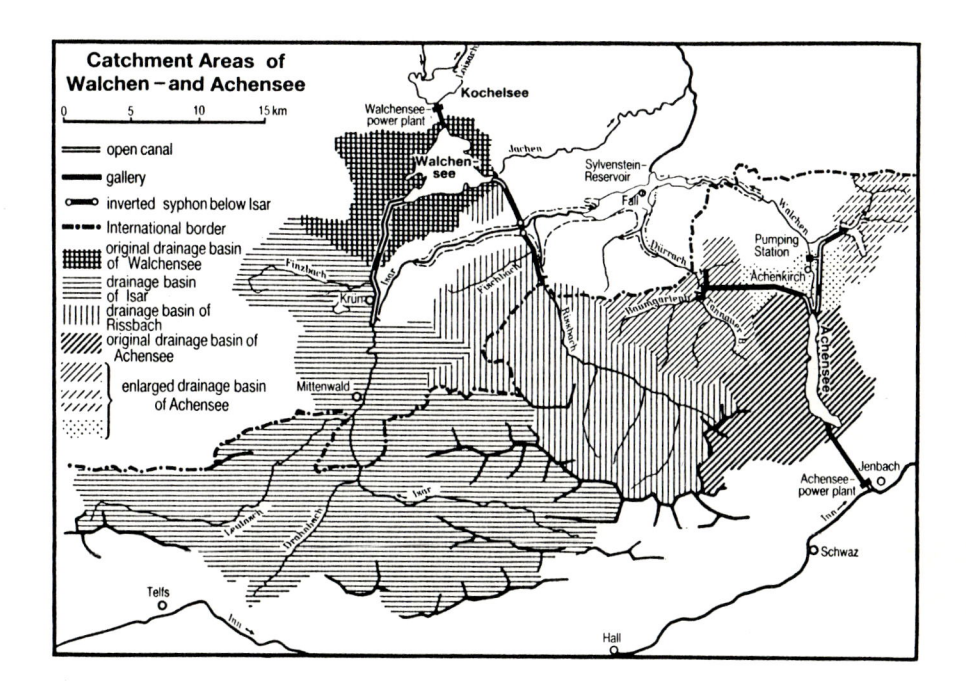

Fig. 3: Changes of the catchment areas of Walchen- and Achensee. (After FELS 1950/51; PENZ & RUPPERT 1975).

Owing to construction work for water resources management in connection with Achensee and Sylvenstein reservoir, Walchensee is no longer a natural hydrological system (fig. 3). Its original catchment area measured 74 km^2, and its outlet, the Jachen, provided an average discharge of 2.3 m^{3}s^{-1} to the Isar. Because of the tapping of the Isar at Krün weir (since 1924, MQ=13.7 m^{3}s^{-1}), of the Flinzbach (since 1943, MQ=0.5 m^{3}s^{-1}), and of the Rißbach area (since 1950,

MQ=7.6 m^3s^{-1}) the Walchensee catchment area has grown to 770 km^2. Thus the average discharge has risen to 24 m^3s^{-1}. Today it passes through the turbines of the Walchensee hydroelectric power plant to Kochelsee and so into the river basin of the Loisach. Furthermore 9.6 m^3s^{-1} are transferred via Achensee to the Jenbach power plant of the Tiroler Wasserkraftwerke (TIWAG) and thus to the Inn system.

These changes have reduced the natural annual discharge of the Isar at Bad Tölz from 1.9 km^3 to only 0.9 km^3. More than half of this volume flows off as flood discharge; the maximum discharge at Bad Tölz being 897 m^3s^{-1} in the period 1931/1966. On the other hand the minimum discharge is merely 2 m^3s^{-1}. Therefore, Sylvenstein reservoir has been planned as both a flood release measure and to regulate low water levels. The total volume of Sylvenstein reservoir has been designed to provide 30-40*10^6 m^3 as additional input during periods of low-level discharge, 50*10^6 m^3 as storage to cut flood peaks, and 20*10^6 m^3 as intermediate storage for higher discharge immediately after floods. Thus a low-level discharge of 10 m^3s^{-1} in winter and 20 m^3s^{-1} in summer should be guaranteed for the Isar at Bad Tölz.

The discharge at Bad Tölz has fundamentally changed owing to the impact of water resources management. Comparing the daily discharge of March and of August 1984 it can be seen that the minimum requirements of the authorities have been fulfilled. The hydrograph has become much smoother owing to the damping effect (retention) of Sylvenstein reservoir (fig. 4). Compared to 1950 the diversion of the Rißbach to Walchensee had further reduced discharge in 1984.

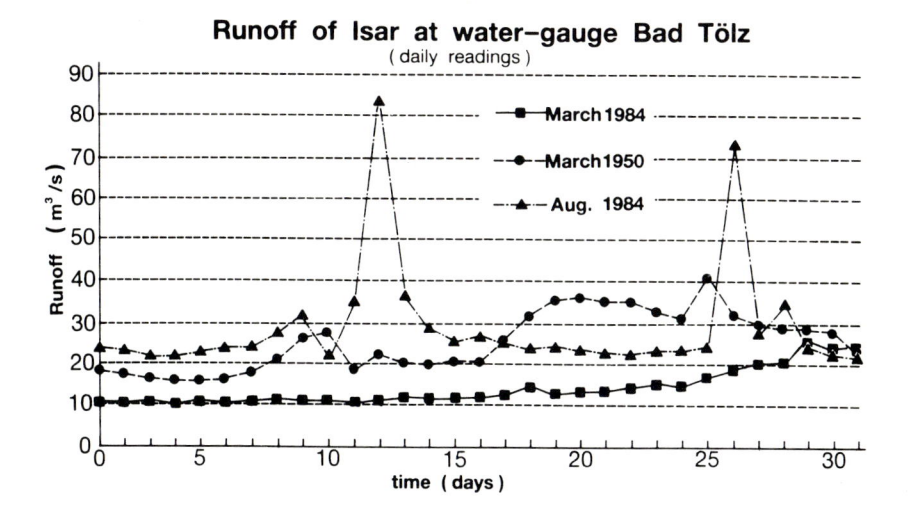

Runoff of Isar at water-gauge Bad Tölz
(daily readings)

Legend: March 1984; March 1950; Aug. 1984

Fig: 4: Impact of Silvenstein reservoir on the discharge of the Isar. 1950 prior to the construction of the reservoir. (Data source: LANDESAMT FÜR WASSERWIRTSCHAFT, several years).

The example of the discharge hydrographs of the Lech at Steeg and Lechbruck clearly shows the influence of Forggensee (fig. 5). Leaving the area of the Lechquellengebirge (local name) at Steeg (catchment area = 247 km") Lech shows two flood peaks between November and the end of March. The continuous low-level discharge indicates snow retention. During this period precipitation falls mainly as snow or sleet, even at the Schwangau-Horn observation site situated at the margin of the Alps. For this period the hydrograph of the discharge at Lechbruck downstream of Forggensee, which is exploited by the Roßhaupten power plant, (catchment area = 1708 km^2) shows a slightly oscillating plateau. The oscillations result from the influence of affluents from the part of the catchment area close to the Alpine margin in combination with the discharge controlled by the requirements of the power plant. The flood peaks at Steeg during the period of meltwater discharge from April to early June again form plateaus at Lechbruck. At this time storage capacity is provided in the reservoir for the remaining meltwater and floods through heavy rainfall during the filling of the reservoir until mid-June. Only from mid-June onwards, when the reservoir has to be full according to the directives of the supervising authority, do all gauges record prominent, although damped, and nearly synchronous flood peaks.

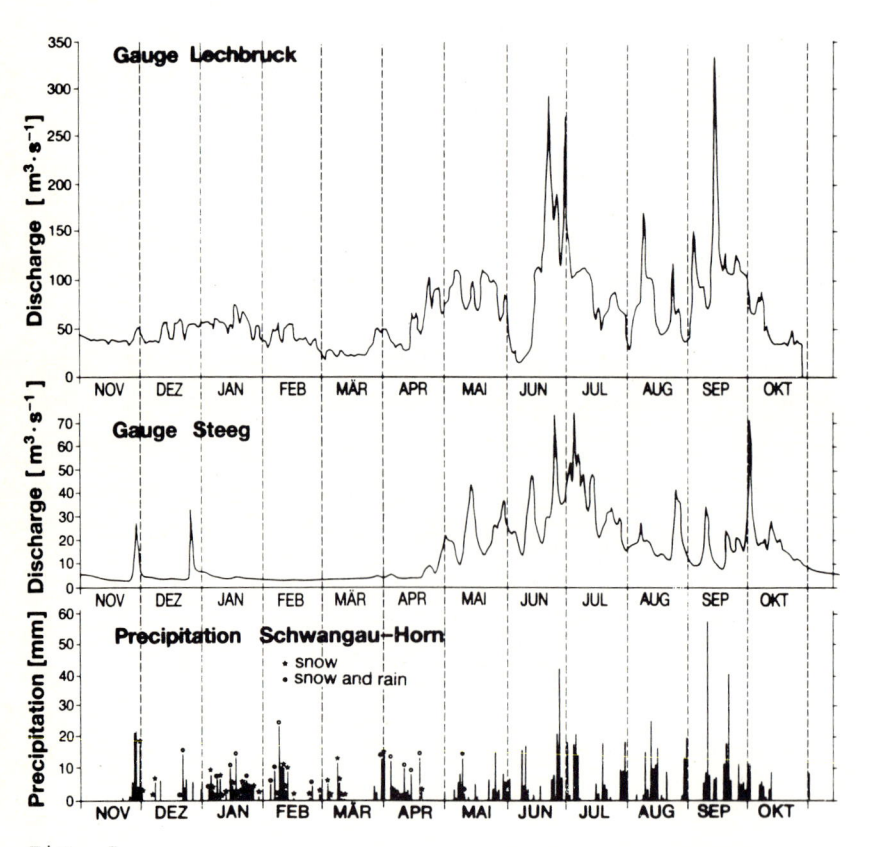

Fig. 5:
Impact of Forggensee (Lech reservoir) on discharge pattern shown at the gauges in Steeg and Lechbruck. Asterix on Precipitation means snow, dots snow and rain.(Data source: DEUTSCHER WETTERDIENST, several years; BAYERISCHES LANDESAMT FÜR WASSERWIRTSCHAFT 1984).

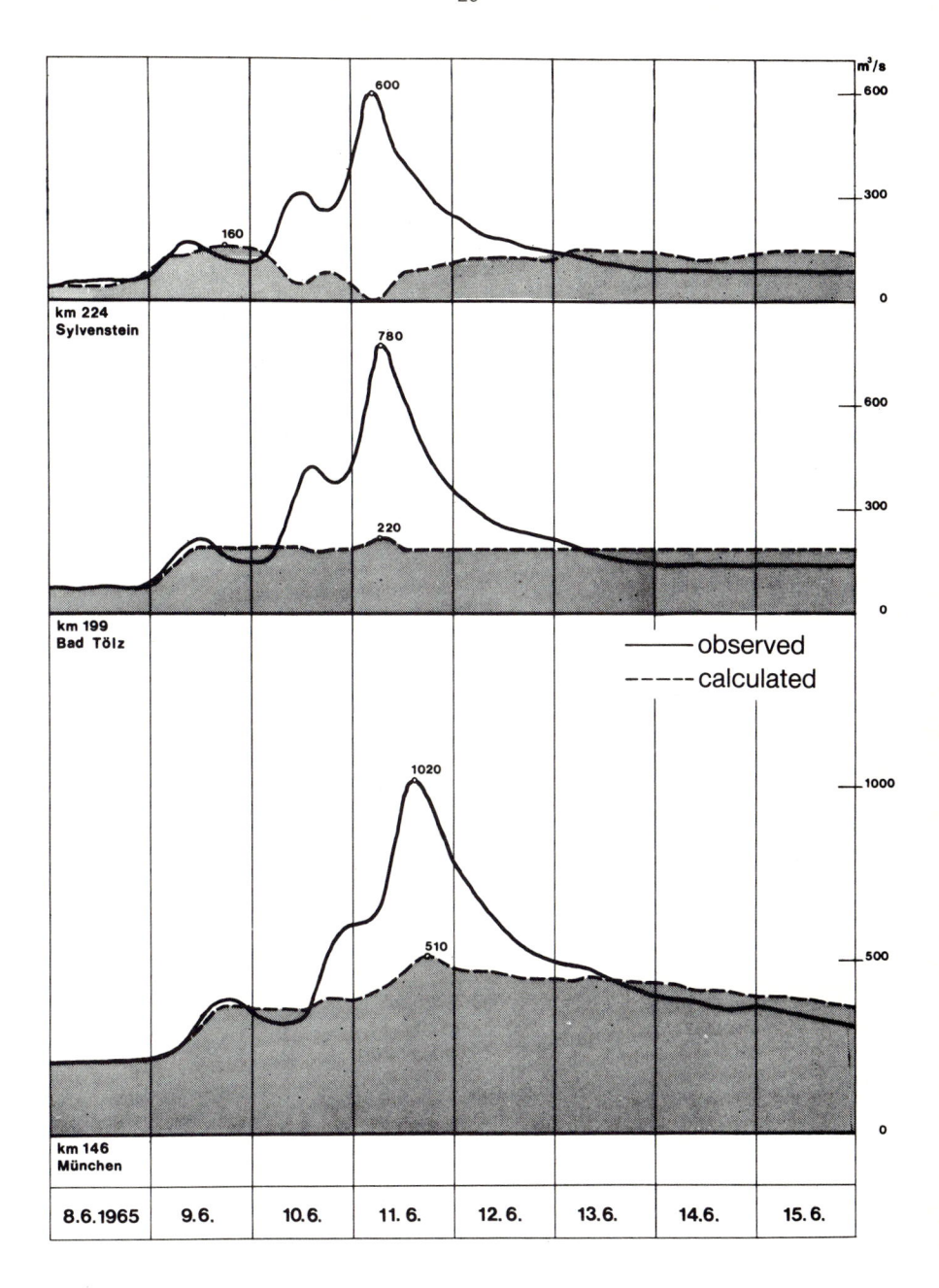

Fig. 6: Impact of Sylvenstein reservoir on the flood discharge of Isar. (From: BAYERISCHES STAATS-MINISTERIUM FÜR LANDESENTWICKLUNG UND UMWELT-FORSCHUNG 1980).

The hydrographs of the flood of June 1965 (fig. 6) for Sylvenstein, Bad Tölz, and München Prinzregentenbrücke exemplify the damping effect of the Walchensee/Sylvenstein system. This effect diminishes with increasing distance from the reservoir and with growing discharge of affluents flowing into the Isar downstream of the head; still it is pronounced even in München. At Sylvenstein the entire flood peak of 600 m^3s^{-1} was absorbed. At Bad Tölz it was reduced from 780 m^3s^{-1} to 220 m^3s^{-1}, i.e. to less than a third, and in München with 510 m^3s^{-1} to exactly half of 1020 m^3s^{-1} (BAYERISCHES LANDESAMT FÜR UMWELTSCHUTZ 1980).

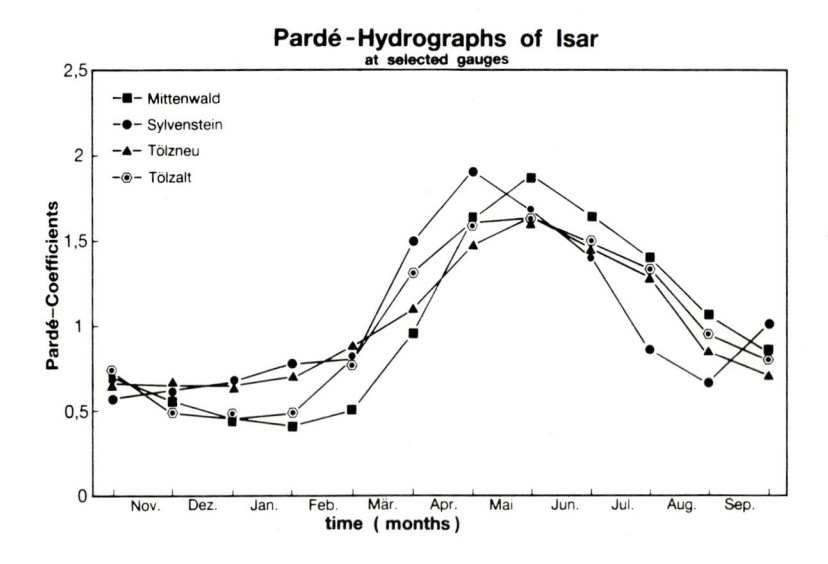

Fig. 7: Comparison of Pardé hydrographs of the Isar at the gauges in Mittenwald, Sylvenstein, and Bad Tölz. For Bad Tölz the hydrographs show long-term means for the time before and after construction of Sylvenstein reservoir. (Data source: BAYERISCHES LANDESAMT FÜR WASSERWIRTSCHAFT, several years).

In conclusion I should like to emphasize that the storage of water by the hydro-electric power plants of the Lech causes a very specific change in floods (UNBEHAUEN 1971; WILHELM 1992). Owing to diminishing retention capacities the flood discharge often increases compared to natural conditions.

Although discharge has been changed significantly by the impacts of water resources management, the Pardé hydrographs of these river systems show the same temporal distributions of maxima and minima for the Isar and Lech (fig. 7). The levelling effect of the storage, however, is reflected in the amplitude of the Pardé coefficients. Before the construction of the Sylvenstein reservoir (from 1929 to 1949) the ratio of maximum/minimum of the Pardé coefficients was 3.60. Afterwards, in the period from 1959 to 1984, it decreased to 2.54, the change of discharge due to regulation of low water levels from December to March being more pronounced than changes in summer.

4. Conclusion

These few examples show that environmental changes due to human impact can be decisive, though not always persistent. This holds especially true for hydrological systems, which can be manipulated more easily than the petrosphere and atmosphere. Quantitative and qualitative impacts on the hydrological cycle inevitably cause changes in the formation of river beds. Thus hydrology is linked with hydraulic engineering and river bed morphology.

The measures were undertaken under the guiding principles of the tasks of water resources management (WILHELM 1987) to provide water for human benefit while preserving the natural ecological system. Inadequate realisation of this vital task may cause environmental damage. This kind of action should not be considered irresponsible, as is often the case. There have been many shortcomings - and hence much damage - because we are still far from perceiving, let alone understanding, the correlations within the extreme complexity of the environmental system, even within the hydrological subsystem.

References

ASCHWANDEN, H. &. WEINGARTNER, R. (1983/84): Die Abflußregime der Schweiz. Teil 1: Alpine Abflußregime. Teil 2: Mittelländische und jurassische Abflußsysteme. Geogr. Inst. d. Univ. Bern, Abt. Phys. Geogr. Gewässerkunde. Bern.

BAUMGARTNER, A.; REICHEL, E. &. WEBER, G. (1983): Der Wasserhaushalt der Alpen. München.

BAYERISCHES LANDESAMT FÜR WASSERWIRTSCHAFT (1984): 100 Jahre Wasserbau am Lech zwischen Landsberg und Augsburg. Schriftenreihe des Bayer. LA. f. Wasserwirtschaft H. 19, München.

BAYERISCHES LANDESAMT FÜR WASSERWIRTSCHAFT (1988): Wasserwirtschaft in Bayern als Zukunftsauftrag und Herausforderung. Schriftenreihe Bayer. LA. f. Wasserwirtaschaft, H. 22, München.

BAYERISCHES LANDESAMT FÜR WASSERWIRTSCHAFT (Ed) (several years): Deutsches Gewässerkundliches Jahrbuch. München.

BAYERISCHES STAATSMINISTERIUM FÜR LANDESENTWICKLUNG UND UMWELTFRAGEN (Ed) (1980): Wasserwirtschaftlicher Rahmenplan Isar. München.

BAYERISCHE WASSERKRAFTWERKE AG (1988) Der Lech und der Lechausbau. München.

BAYERISCHE WASSERKRAFTWERKE AG (Ed) (undated): Walchensee-Kraftwerk mit Obernach- und Niedernach-Kraftwerk. München.

BIRKENHAUER, J. (1980): Die Alpen. = Uni-Taschenbücher 955, Paderborn, München.

DEUTSCHER SKIVERBAND (DSV) (Edsg) (1992): DSV-Atlas Skiwinter 1992. Fink-Kümmerly+Frey, Zürich.

DEUTSCHER WETTERDIENST (Ed) (several years): Deutsches Meteorologisches Jahrbuch BUNDESREPUBLIK Deutschland. Offenbach.

DIETMANN, T. (1985): Ökologische Schäden durch Massenskisport. Entwicklung und Veränderungen des Skigebietes am Fellhorn bei Oberstdorf/Allgäu von 1953 - 1982 durch seine Erschließung für den Massenskisport. In: Jb. Ver. z. Schutz d. Bergwelt 50, S. 107 - 158.

FLIRI, F. (1974): Niederschlag und Lufttemperatur im Alpenraum. Wiss AV-Hefte, H. 24, 111 S.

HENSELMANN, R. (1970): Der Abflußausgleich durch die Wasserstandschwankungen eines natürlichen Sees. Gezeigt am Beispiel bayerischer Seen. Bayerische Landesstelle f. Gewässerkd., H. 4, München.

HERRMANN, A. (1978): Schneehydrologische Untersuchungen in einem randalpinen Niederschlagsgebiet (Lainbachtal bei Benediktbeuern/Oberbayern).=Münchener Geogr. Abh., Bd. 22, München.

KARL, J.; MANGELSDORF, J. &. SCHEUERMANN, K. (1977): Die Isar, ein Gebirgsfluß im Spannungsfeld zwischen Natur und Zivilisation. In: Jb. d. Vereins u. Schutze der Bergwelt, 42, Jg. S. 175 - 224, München.

LAUFFER, H. (1969): Wasserkraft im Bundesland Tirol. In: Österreichische Wasserwirtschaft, Jg. 21, H.9/10.

LINK, H. (1970): Speicherseen der Alpen. In: Wasser- und Energiewirtschaft, S. 241 -358.

MICHLER, G. (1992): Exkursion in den Isarwinkel. In: Exkursionsführer zur COMTAG-Tagung vom 9. - 12. Juni 1992 in Benediktbeuern. In: Münchener Geogr. Abh. Bd. B 15, München.

MOSIMANN, TH. (1986): Skitourismus und Umweltbelastung im Hochgebirge. Geogr. Rdsch. 38/6, S. 303 - 311.

MOSIMANN, TH. &. LUDER, P. (1980): Landschaftsökologischer Einfluß von Anlagen für den Massenskisport. I. Gesamtaufnahme des Pistenzustandes (Relief, Boden, Vegetation, rezente Morphodynamik) im Skigebiet Crap Sogn Gion/Laax GR. Materialien zur Physiogeographie H.1, Basel.

MÜLLER, F; CAFLISCH, T.&. MÜLLER, G. (1976): Firn und Eis der Schweizer Alpen. Gletscherinventar. ETH Zürich, Geographisches Institut, Publ. Nr. 57, Zürich.

NAUMANN, E. (1989): Beurteilung von Beschneiungsanlagen durch Skifahrer und Einheimische - Eine empirische Untersuchung in drei gut ausgestatteten Wintersportorten. Diplomarbeit Lehrstuhl für Landschaftstechnologie (Fakultät für Forstwirtschaft) und für Geographie (Fakultät für Geowissenschaften) der Universität München. München.

OBERSTE BAUBEHÖRDE IM BAYERISCHEN STAATSMINISTERIUM DES INNERN (1990): Flüsse und Seen in Bayern, Wasserbeschaffenheit, Gewässergüte 1989. Schriftenreihe Wasserwirtschaft in Bayern, H. 23, München.

SCHEUERMANN, K. (1977): Der Wasserkreislauf im Weltbild eines Gelehrten der Barockzeit. In: DGM, 21/2, S. 21-27.

SCHEUERMANN, K. (1983): Die Isar im Wandel der Zeiten. In: Plessen, M.L.(Hrsg.): Die Isar. Ein Lebenslauf. München.

SCHEUERMANN, K. (1990): Der obere Lech im Wandel der Zeiten. In: Jb. d. Vereins zum Schutze der Bergwelt, 55. Jg., S. 105-121, München.

SMITH, R. A. (1872): Air and rain: the beginnings of a chemical climatology. London.

UNBEHAUEN, W. (1971): Die Hochwasserabflußverhältnisse der Bayerischen Donau. Hochwasser der Jahresreihe 1845/1965. Bes. Mitt. z. Dt. Gewässerkdl. Jb. Nr. 30, München.

WASSERWIRTSCHAFTSAMT WEILHEIM (1984): Sylvensteinspeicher. München.

WILHELM, F. (1987): Hydrogeographie. Das Geographische Seminar. Braunschweig.

WILHELM, F. (1992): Stellung der Flüsse im Wasserkreislauf. In: Geographie und Schule, 14. Jg., H. 75, S. 2 - 13.

WILHELM, F. &. VOGT, H. (1988): Entwicklung der Schneedecke. In: Abfluß in Wildbächen. Untersuchungen im Einzugsgebiet des Lainbaches bei Benediktbeuern/Oberbayern. = Münchener Geogr. Abh. Bd. B 6, S. 281 - 420; München.

WITTMANN, H. &. TÜRK, R. (1988): Immissionsbedingte Flechtenzonen im Bundesland Salzburg und ihre Beziehungen zum Problemkreis "Waldsterben".In: Berichte ANL 12, S. 247 - 258.

MODEL INVESTIGATIONS INTO THE INFLUENCE OF RENATURALIZATION ON SEDIMENT TRANSPORT

W. Bechteler, H.-J. Vollmers, S. Wieprecht
Institut für Wasserwesen
Universität der Bundeswehr München
D - 85577 Neubiberg

Abstract

Efforts have recently been made to renaturalize the Weißach River which had previously been regulated in a schematic, monotonous manner. During the last flood period an agglomeration of sediment occurred, which may provoke an overtopping of the dams. In a hydraulic model with movable bed (scale 1:20) the existing conditions and possible improvements were studied in order to prevent agglomerations. Furthermore, fundamental investigations were made with regard to the influence of constructional steps on sediment transport and water levels.

1. Introduction
1.1 General Remarks

Most natural rivers are in a sensitive state of hydraulic-sedimentological balance, which is defined by mutual influences of water and solids. Lowland rivers show in general a "continuous" transport behaviour throughout the year. Depending on the discharge, more or less material is being transported.

In the case of alpine rivers the transport behaviour is characterized by different factors of influence. As the bed material consists of rather coarse grain diameters bed load transport will only start at great bed shear velocities. In order to set the bed material into motion high discharges are necessary, i.e. only flood events are actual transport events. Up to 80 or 90 % of the total annual transport volume may be transported in the course of a single flood wave.

1.2 Problem Description

Recently the attempt has been made to renaturalize the last section of the Weissach River before it flows into Lake Tegernsee. This part had been previously regulated in a schematic monotonous way. During the last flood period an agglomeration of sediment has occurred, which may provoke an overtopping of the dams. In a hydraulic model with movable bed (scale 1:20) the existing conditions and possible improvements were studied in order to prevent agglomerations. Furthermore fundamental investigations were made about the influence of obstructions on sediment transport and water levels.

2. Similarity Considerations
2.1 General Remarks

The knowledge of the physical basis of sediment transport is still so fragmentary that it is not possible to develop a generally acceptable calculation approach. Owing to the practical importance of the phenomenon, however, there are quite a number of more or less empirical formulas, whose bases and ranges of application differ from each other. Therefore their calculation results may also differ considerably.

Tab. 1 presents a list of the total annual sediment transport in the Loisach River (Upper Bavaria) calculated according to several authors. The actual total annual sediment transport has been found to be about 30 000 m^3 by averaging the dredged material of several years.

Transport Model	water depth and discharge for start of motion		days per year with transport	annual total sediment transport
	[m]	[m^3/s]	[days]	[m^3]
Meyer-Peter, Müller, 1949	0.87	20.05	173	29 648
Laursen, 1958	0.99	25.65	109	32 331
Bishop, e.a., 1965	0.00	0.00	365	92 430
Pernecker, Vollmers, 1965	0.49	6.04	363	39 222
Engelund, Hansen, 1967	0.00	0.00	365	36 973
Graf, Acaroglu, 1968	0.00	0.00	365	32 626
Ackers, White, 1973	0.72	13.60	264	23 718
Ranga Raju, e.a., 1981	1.03	28.20	94	9 206
Karim, Kennedy, TLTM, 1983	0.69	12.40	279	8 666

Tab. 1: Annual Total Sediment Transport Loisach, Gauge Schlehdorf, VOLLMERS, 1992

In contrast to other branches of science the answering of practical questions is the main object in the engineering sciences. This is true especially in the case of hydraulic constructions! About 100 years ago hydraulic engineers "invented" the physical model, as already GALILEI realised the difficulties in calculating the course of waters (contrary to the movement of planets!). Up to the present day such models (in the meantime extended to mathematical ones) are part of the hydraulic engineer's standard tools in answering practical questions. Hydraulic models with movable beds belong to the highest category of difficulties and can only be successfully operated in laboratories with excellent equipment and great experience. The greatest problem is the conversion of model test results into natural values and vice versa (VOLLMERS, 1989). The creation of similarity between nature and model is time-consuming and physically not exactly feasible.

Finally it has to be pointed out that the major part of the present knowledge about sediment transport comes from investigations performed on simplified physical models.

2.2 Model Similarity

The movable bed physical model of the Weissach River was built at a geometric scale of 1 : 20 in the hydromechanics laboratory of the Federal Armed Forces University Munich at Neubiberg. All geometric and hydraulic values were calculated according to FROUDE's similarity law, which means that processes dominated by gravitation and inertia are represented similarly in nature and model. Frictional forces may be neglected in this case due to the coarse grain material of the river bed.

The choice of the model sand was guided by the idea to keep the part of the suspended material as low as possible, because the geometrical reduction would have contradicted natural conditions. Therefore an almost uniform sand was selected, the mean diameter ($d_m \approx d_{65} \approx 0.85$ mm) of which corresponds well with the geometric reduction of the d_m-value of the natural bed material ($d_m = 14.3$ mm). Fig. 1 contains the grain distribution curves of the natural and the model materials as well as the theoretical curve corresponding to the model scale 1 : 20.

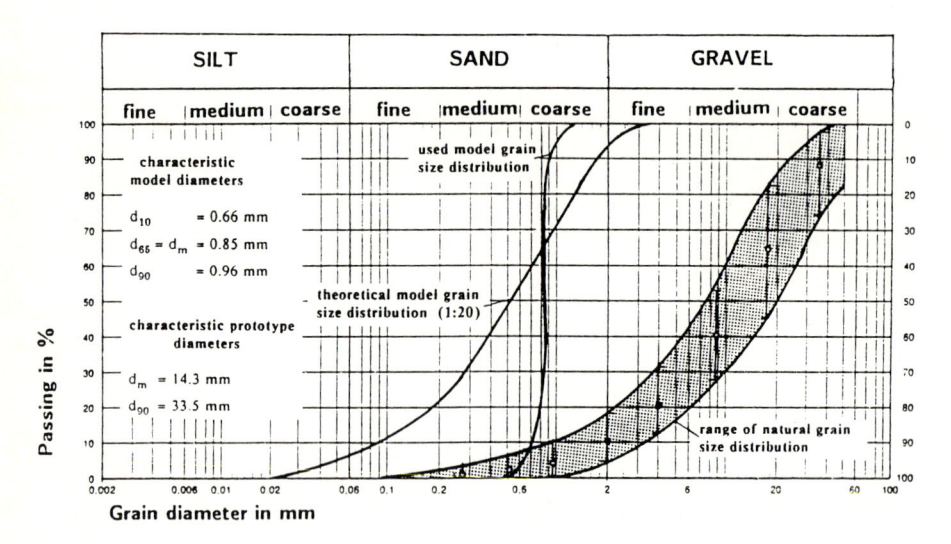

Fig. 1: Grain size distribution curves: natural grain, used model grain, theoretical model grain

3. Model Set-up
3.1 Model Description

The model consists of two equally long sections that are connected by a hinge. It is supported by two I-shaped steel girders, the elevation of which can be adjusted to any slope desired for different investigations. The simplified trapezoidal cross-section of the river has been placed into the frame of a rectangular cross-section. The lateral embankments are made of sheet iron. Roughness is simulated by coarse sands glued onto the surface. A bed material ground layer of 10 cm thickness prevents the ground plate from being washed free in erosion ranges. Two circular metal rods, one on each side of the flume, serve as rails for the measuring carriage. Fig. 2 shows a model cross-section and Fig. 3 presents a plan view of the model.

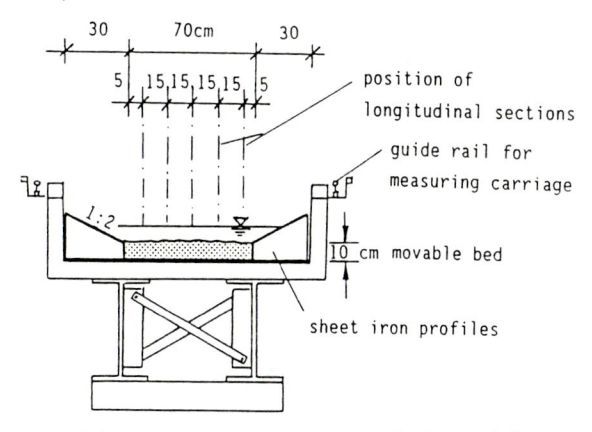

Fig. 2: Schematic cross-section of the model

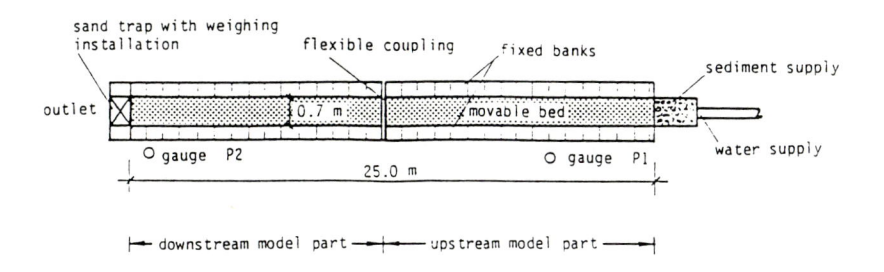

Fig. 3: Plan view of the model

3.2 Operating and Measuring Installations

The water supply for the model comes out of an elevated reservoir. Discharge regulation is controlled by an inductive flowmeter. Both are connected to the computer. At the model outlet the water pours down into a basement reservoir where it is pumped back into the elevated supply reservoir.

The water level at the model outlet is controlled by a nearby gauge and may be adjusted by an overflow plate. A second gauge is installed at the beginning of the measuring range, about 5 m away from the inflow and the sand supply. Measuring instruments are so-called vibrating water level followers (Delft) with an accuracy of ± 0.5 to 1 mm.

Fig. 4:   Profile follower

For the assessment of river bottom changes longitudinal profiles were registered both before and after a test (positions of these profiles are given in Fig. 3). Registration was performed using a Delft profile follower (Fig. 4) with a vertical adjustment velocity of up to 50 cm/s and an accuracy of ± 1 to 1.5 mm.

The input comes as a sand-water-mixture out of a plexiglass tube with a vertical slot, which is opened at a constant predefined speed and thus provides a constant inflow of material per unit of time (Fig. 5).

The collection device (sand trap) at the model outlet is a balance pan suspended from a pressure/strain transducer (Fig. 6). At certain variable time intervals signals are transmitted to the computer.

Fig. 5:   Input device for wet sand

Fig. 6:   Model outflow with bed load measuring device

4. Model Tests and Results
4.1 Tests without Constructional Elements

Tests without constructional elements served as the basis of comparison with the others that were to be performed later on and to contain different variations. At the beginning tests with constant discharges provided a transport-discharge relation (Fig. 7). They have been executed for three different slopes (S_I = 0.002, S_{II} = 0.004 and S_{III} = 0.006). In order to allow systematic comparisons regarding the sediment transport, the transport rates obtained in the calibration tests were used as input rates for the tests with constructional elements.

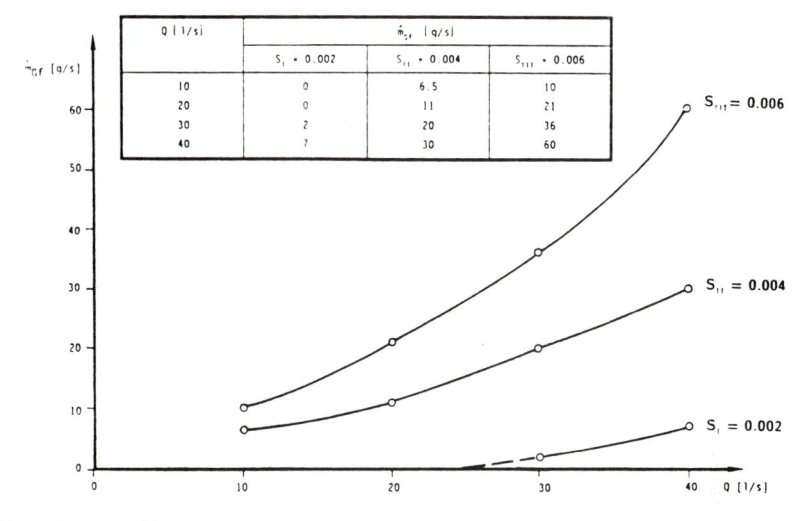

Fig. 7: Transport-discharge diagram for all three slopes

4.2 Tests with Constructional Elements
4.2.1 Test Programme and Performance

The test programme comprised investigations of ten series of different arrangements of constructional elements, each of which was to be tested for three discharges (Q = 10 l/s, 20 l/s, 40 l/s), three slopes (S = 0.002, 0.004, 0.006) and one characteristic grain diameter (d_{ch} = 0.85 mm).

Altogether this led to 90 different tests. Fig. 8 presents all variations of arrangements of constructional elements schematically in plan view and cross-sectional elevation (series I through X).

As actual construction works use boulders of about 1 m side length, all constructional elements were modelled using stones of 5 to 6 cm diameter, which corresponds to the 1 : 20 scale. Boulders and groynes are put directly upon the movable bed, whereas levelled and elevated ground sills are placed on the fixed solid ground plate, which lies 10 cm lower. Thus it can be checked whether some of the elements change their positions owing to erosion and scouring.

Ground Plan   **Cross Section**

I. Stones, alternate, distance 1 m

II. Groyns, on both sides, distance 1 m

III. Groyns, alternate, distance 1 m

IV. III. + Stones centric

V. Groyns, alternate, distance 3 m

VI. V. + Stones centric

VII. Levelled ground sills, distance 1 m

VIII. Ground sills, distance 1 m

IX. Levelled ground sills, distance 2.8 m

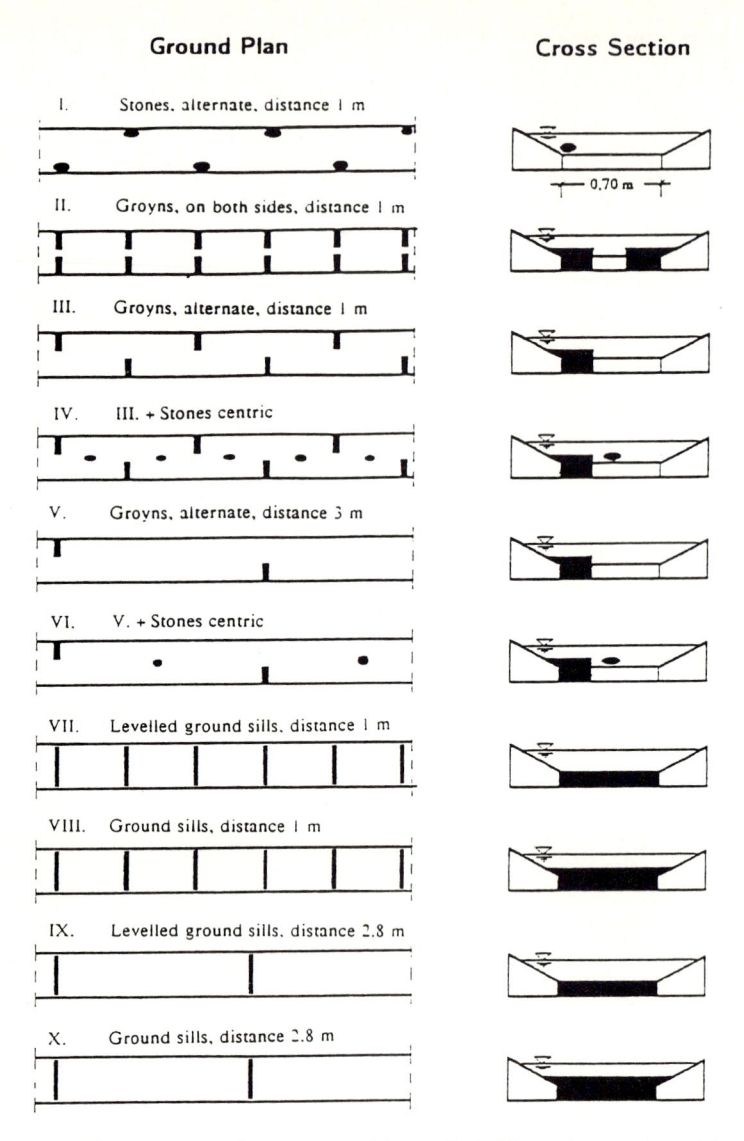

X. Ground sills, distance 2.8 m

Fig. 8: Plan view and cross-section of different constructional measures (schematic)

The execution of a test is similar with and without arranging constructional elements. Bed profiles were registered before the test after levelling the flume bed and after test execution. Comparing both registrations allows the assessment of bottom changes. Their differences show where erosion or sedimentation, occurred in the 16 m measuring range.

Every minute the discharge, the two water level gauges and the sediment transport are recorded and printed in a data sheet. Tab. 2 presents an example of such a data sheet.

Filename: WEI089.DAT					Start of measurement: 8.1.1992 13:10:06		
Time hh:mm:ss	[sec]	Q [l/s]	H1 [cm]	H2 [cm]	mass [kg]	m_{Gf} [g/s]	average [g/s]
13:10:14	0	40.1	10.7	10.8	3.14	11.1	6.5
13:11:14	60	40.1	10.7	10.8	3.48	5.8	6.5
13:12:14	120	40.0	10.7	10.8	4.00	8.4	6.7
13:13:14	180	40.2	10.7	10.7	4.99	16.6	7.6
13:14:14	240	40.0	10.7	10.7	5.27	4.6	7.3
13:15:14	300	40.1	10.7	10.7	5.69	7.2	7.3
13:16:14	360	40.1	10.7	10.7	6.12	7.1	7.2
13:17:14	420	40.1	10.7	10.7	6.46	5.6	7.1
13:18:14	480	40.1	10.7	10.7	6.80	5.6	7.0
13:19:14	540	40.1	10.7	10.7	7.10	5.0	7.1
13:20:14	600	40.1	10.7	10.7	7.67	9.5	6.8
13:21:14	660	39.9	10.7	10.7	7.78	1.9	6.8
13:22:14	720	40.1	10.8	10.7	8.18	6.6	6.9
13:23:14	780	40.2	10.8	10.7	8.75	9.3	7.0
13:24:14	840	40.1	10.8	10.7	9.28	8.9	6.9
13:25:14	900	40.1	10.8	10.7	9.47	3.2	6.8
13:26:14	960	40.1	10.8	10.7	9.85	6.4	6.6
13:27:14	1020	40.0	10.8	10.7	9.94	1.6	6.6
13:28:14	1080	40.2	10.8	10.6	10.31	6.1	6.6
13:29:14	1140	40.2	10.8	10.6	10.47	2.7	6.5
End of measurement: 8.1.1992 13:29:27							

Tab. 2: Data sheet of test WEI089.DAT. Q = 40 l/s, S_I = 0.002, Variant II

Column "m_{Gf}" shows the transport per minute and column "average" gives the average transport value since the start of the test. The values per minute scatter rather strongly around the quickly stabilized mean value. This is because generally bed load transport is not a uniform process, but rather takes place by thrusts, which is even increased by bed forms moving obliquely to the flume axis.

4.2.2 Test Evaluation

The comparison of test series results was made possible by introducing a number characterizing the degree of obstruction caused by the constructional elements. As these elements have different distances the degree of obstruction was calculated to be the ratio of "obstructed volume" (Fig. 9 "B") to "unobstructed volume" (Fig. 9 "A"). Reference length is one metre. Thus levelled ground sills have an obstructional degree of 0 %, as their crests are at an even level with the bottom elevation, and therefore do not obstruct the cross-sectional area at all.

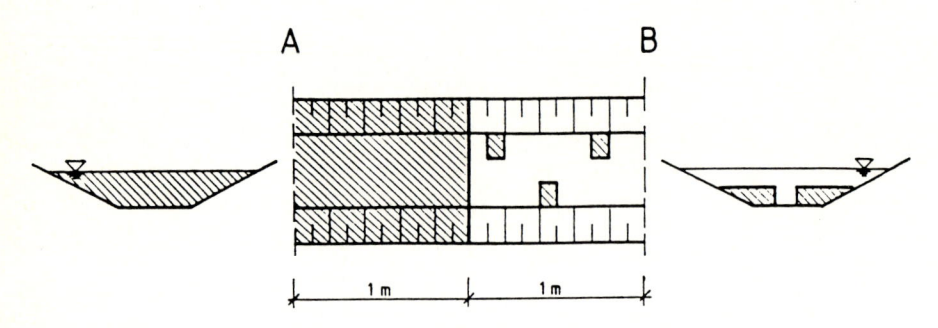

Fig. 9: A: unobstructed volume, B: obstructed volume (schematically)

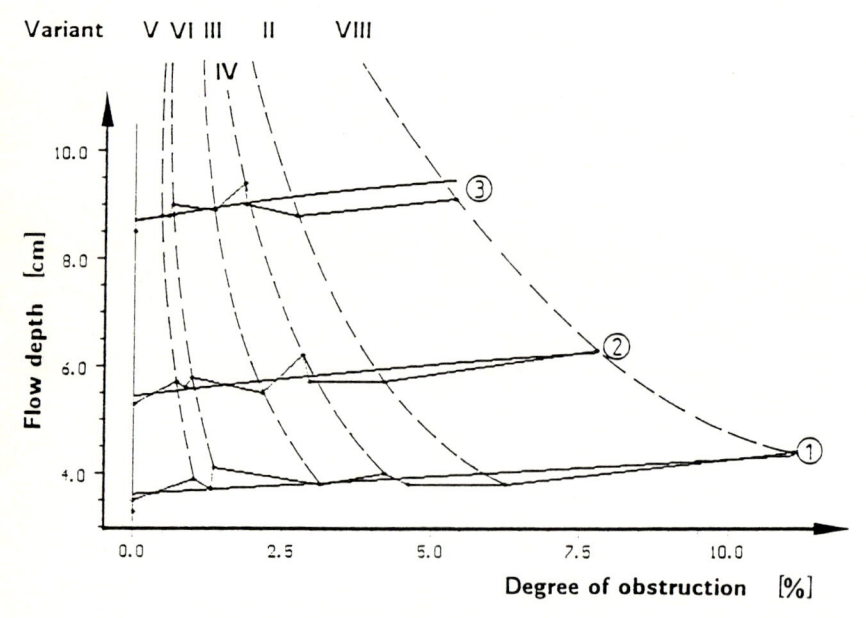

Fig. 10: Changes of flow depths versus degree of obstruction and corresponding fitting curves for slope S_{II} = 0.004 and discharges Q = 10 l/s (1), Q = 20 l/s (2) and Q = 40 l/s (3)

The same systematic tendency could be found in all tests performed so far. Water levels increase with an increasing degree of obstruction. Fig. 10 show the results from test with a slope S_{II} = 0.004. Measurement points in the diagram are connected by a polygonal line. The second degree fitting curves demonstrate the systematic tendency of the measurements. They are not meant to serve as calibration curves for the relation between degree of obstruction and water depth.

Flow depths served to compute velocities and FROUDE numbers as well as MANNING-STRICKLER coefficients.

For increasing degrees of obstruction there is a constant decrease of roughness coefficients, velocities, and FROUDE numbers. This tendency can be recognized for more or less all values calculated. As an example Figs. 11 and 12 show for slope S_{II} = 0.004 the dependence of velocities and MANNING-STRICKLER coefficients on the degree of obstruction.

Contrary to tests with fixed bottom and constant roughness, k_{St}-values do not increase but rather decrease with rising discharge and water level. Flume roughness grows for greater discharges due to bed form shaping. As flow depths are altogether rather low, i.e. between 3 and 11 cm, bed forms have a great influence on roughness. This influence was reduced by increasing water depths.

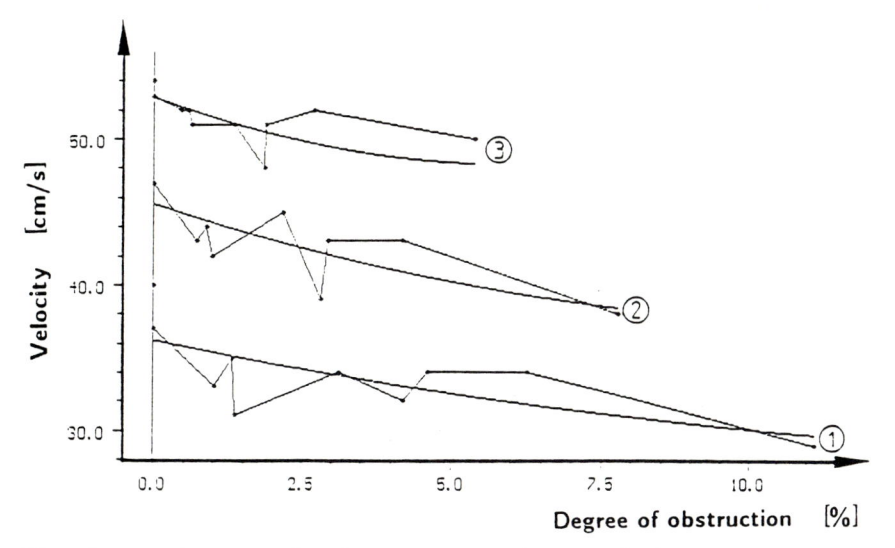

Fig. 11: Changes of velocity versus degree of obstruction and corresponding fitting curves for slope S_{II} = 0.004 and discharges Q = 10 l/s (1), Q = 20 l/s (2) and Q = 40 l/s (3)

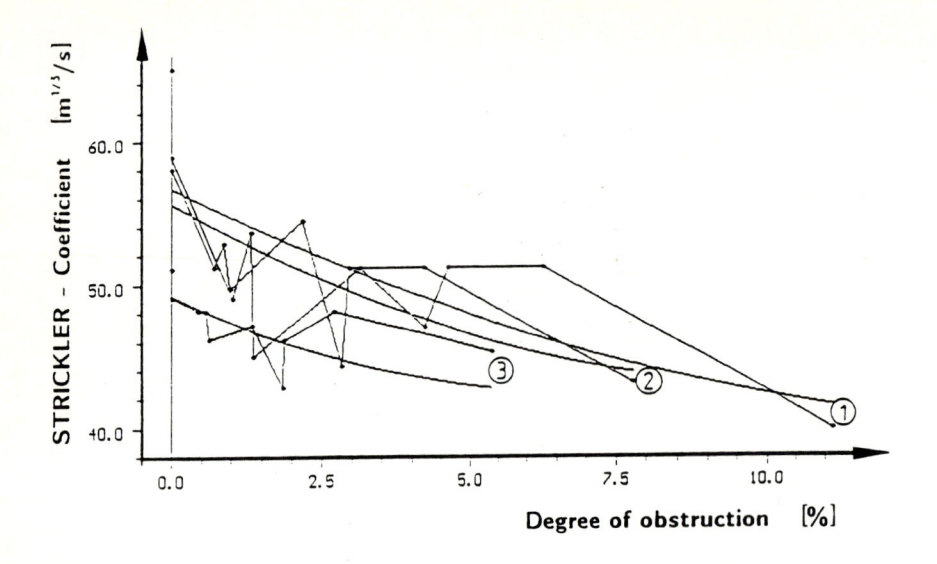

Fig. 12: Changes of roughness coefficients versus degree of obstruction and corresponding fitting curves for slope $S_{II} = 0.004$ and discharges Q=10 l/s (1), Q=20 l/s (2) and Q=40 l/s (3)

Changes of transport behaviour are strongly dependent on slope. A slope of 0.002 does not show any transport changes. For Q = 10 l/s and 20 l/s no transport at all takes place, as is the case without constructional elements. For Q = 40 l/s transport corresponds, within measuring tolerances, to an input rate of 7.5 g/s.

At a slope of 0.004 changes of transport capacities can be observed for small discharges, i.e. small flow depths. The supply rate was 6.5 g/s, the actual transport, however, averaged 3.5 g/s only. A direct relation to the degree of obstruction cannot be established. In the cases of Q = 20 l/s and Q = 40 l/s the influence of constructional measures was too small to produce any changes of transport behaviour. For these discharges the transport oscillated, within measuring tolerances, around the supply values, just as in the tests of $S_I = 0.002$.

The exception from the rule is given by series VIII, the one with the greatest obstructional degree. All tests show transport values clearly below the ones found at the beginning in the initial test series (Fig. 11).

5. Conclusion

The effects of renaturalisation works on the hydraulic and morphologic equilibrium of a river, can hardly be foreseen. It is therefore necessary to rely on simulations in physical or mathematical models to predict possible effects of constructional steps. The choice of the model type depends on which model is better suited to simulate nature properly.

Calculation procedures already exist to deal with overgrown cross-sections (DVWK Regel Nr. 127). The influence of constructional elements, however, still cannot be described exactly. First investigations on changes of transport capacity have been performed (EILERS, 1990), but have yet to be generalized.

In all test series the same systematic development could be observed. With an increase in the degree of obstruction the water depths also increased, and velocities, FROUDE numbers and roughness coefficients decreased. The results of the investigations on the transport capacity differ considerably. For the slope of 0.002 it is hardly possible to make any satisfying statement on the influence of the constructional elements on transport capacity owing to the late start of transport at discharges of $Q > 20$ l/s. In the case of $S_{II} = 0.004$ the influence of constructional elements was visible for very high degrees of obstruction only. The tests with a slope of $S_{III} = 0.006$ have not yet been completed.

The limitation to one characteristic grain size so far makes a global assessment of constructional measures for natural rehabilitation difficult. This is primarily because grain diameter and slope are the parameters which influence transport behaviour overproportionally. This has also been shown by a DFG research project. Therefore additional investigations with two more grain sizes are planned.

Nowadays theoretical knowledge on sediment transport is in progress. Nevertheless it is still impossible to generally describe the transport behaviour of a river. Therefore fundamental investigations have yet to be performed in combination with field observations and measurements.

6. References

BECHTELER W.
VOGEL G.
VOLLMERS H.-J.
Model Investigations on the Sediment Transport of a Lower Alpine River
Proceedings, International Workshop on Fluvial Hydraulics of Mountain Regions, Trento, Italy, 1989

DIN
Deutsche Norm, Hydrologie, Quantitative Begriffe, DIN 4049, Teil 103

DVWK
Hydraulische Berechnungen von Fließgewässern
DVWK-Merkblatt 220/1991, Verlag Paul Parey, 1991

DVWK
Geschiebemessungen
DVWK-Regeln 127/1992, Verlag Paul Parey, 1992

DVWK
Hydraulic Modelling
DVWK-Bulletin No. 7, Editor H. Kobus, Verlag Paul Parey, 1980

EILERS J.
Zur Berechnung offener Gerinne mit beweglicher Sohle und Uferbewuchs
Mitteilungen aus dem Leichtweiß-Institut für Wasserbau, Nr. 106, TU Braunschweig, 1990

VOLLMERS H.-J.
The State of the Art of Physical Modelling of Sediment Transport
Sediment Transport Modelling, Proc. International Symposium, New Orleans, 1989

VOLLMERS, H.-J.
Sediment Transport Equations and Annual Total Load, 5th Int. Symp. on River Sedimentation, Karlsruhe, 1992

ZANKE U.
Grundlagen der Sedimentbewegung
Springer-Verlag, Berlin Heidelberg, 1982

BEDLOAD TRANSPORT AND DISCHARGE IN THE ERLENBACH STREAM

Dieter Rickenmann
Swiss Federal Institute for Forest, Snow and Landscape
Research, 8903 Birmensdorf, Switzerland

ABSTRACT

As indicators of bedload transport, hydrophone measurements have been performed in the Erlenbach stream since 1986. Analysis of the data shows that the total number of hydrophone impulses for several flood events is proportional to the sediment volume accumulated in the retention basin below the measuring site. A relationship also exists between the hydrophone activity and the runoff volume of the flood events. Looking at a shorter time scale, the number of hydrophone impulses per minute varies considerably for a given flow rate, which seems to be characteristic for the pulsing nature of bedload transport. A seasonal variation is observed for both the threshold discharge for the beginning of transport and for the transported sediment volume.

INTRODUCTION

The Institute for Forest, Snow and Landscape Research operates three experimental catchments in a prealpine region of Central Switzerland. Discharge, water chemistry, precipitation and other meteorological and climatic parameters have been measured for about 20 years. In 1982, a sediment retention basin was constructed at the outlet of the Erlenbach catchment. This allows a measurement of the accumulated sediment transported by the stream, in regular intervals as well as after extreme flood events.

The Erlenbach torrent is situated in a flysch zone and has a catchment area of 0.70 km^2, extending from 1110 m a.s.l. to 1655 m a.s.l. The average channel gradient is about 20% and the average annual precipitation amounts to 2300 mm. A maximum peak discharge of about 12 m^3/s was observed in July 1984; it has an estimated recurrence interval of more than 100 years. During this flood event about 2000 m^3 of sediment were transported to the retention basin.

With regard to bedload transport, it is important to obtain information not only on sediment volumes per event but also on transport rates and their relation to stream discharge. Therefore a new measuring technique was developed. In 1986, a number of so-called hydrophones was installed in the bottom of the inlet channel to the sediment retention basin. The hydrophones are sensors which measure the acoustic signals due to the impact of bedload grains transported over the measuring cross section. The signal is recorded in the form of an electrical potential. It is determined how many times per minute the signal exceeds a given threshold value. The number of impulses per minute is expected to be proportional to the bedload transport intensity.

There are nine sensors distributed over the measuring cross-section, which is of parabolic type shape. Due to the geometric arrangement and the local flow conditions it is the sensor no. 3 (H3) which usually records the most impulses. Most of the other sensors are placed at a higher level on the sloping side of the cross section and only record signals during very intense bedload transport events. A comparison of events of different magnitude shows that the signals recorded with sensor no. 3 are representative also for the other sensors of the cross section. The installation of the instrument and a first analysis of the observations is described in more detail by Bänziger and Burch (1990).

INITIAL ANALYSIS OF THE HYDROPHONE DATA

The main interest of the hydrophone data first focussed on three questions: (1) What is the critical discharge at the beginning of bedload transport ? (2) Is there any relation between flow rate and hydrophone intensity (as an indicator of bedload transport rates) ? (3) What is the relation between the number of hydrophone impulses and transported sediment volumes ? A first analysis showed basically the following results:

* There is no clear trend between the hydrophone intensity (given as number of impulses per minute) and the corresponding flow rate. For a given flow rate, the hydrophone intensity can vary over more than one order of magnitude. A similar statement applies when measured transport rates are compared with the corresponding flow rate (e.g. Beschta, 1981; Sawada et al., 1985; Tacconi and Billi, 1987).

* The threshold flow rate for the beginning of bedload transport, Q_a, shows a variation by about a factor of 4. There is a trend for Q_a to be smaller in winter than in summer. Further, the variability of the threshold discharge appears to be larger in summer than in winter. In general, winter conditions last from December until about end of May.

* The survey of the accumulated sediment volumes in the retention basin allows the hydrophone measurements to be roughly "calibrated". The cumulative number of hydrophone

impulses is linearly proportional to the cumulative sediment volume (Fig. 1). The proportionality constant possibly depends on the magnitude of the events. The first period (with a sediment volume of about 550 m^3) includes the two largest floods, with a return period of around two years. For these two events, the total number of hydrophone impulses corresponds to a proportionally smaller sediment volume than for the rest of the observation period, representing events with a return period of about one year and less.

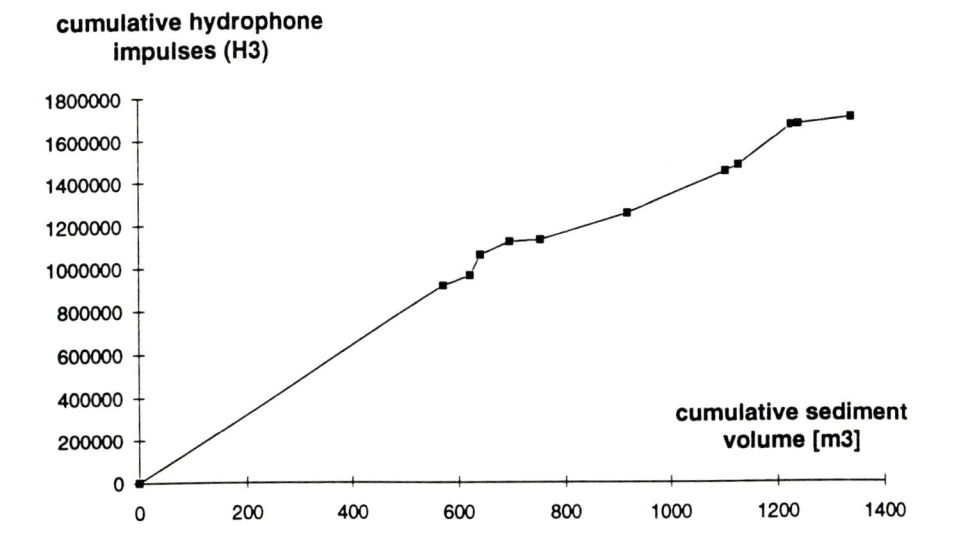

FIG. 1 Relation between cumulative hydrophone impulses and cumulative sediment volume, representing a rough calibration of the hydrophone data. The observation period represented in the figure extends from May 26, 1987 to July 4, 1991.

SOME CHARACTERISTICS OF EVENTS

Transport activity, duration and distribution of events

In general, bedload transport begins at a certain threshold discharge Q_a and continues until another threshold discharge Q_e is reached during the recession part of the hydrograph. However, there are also events during which bedload transport at the measuring site ceases for a certain time even though the flow rate is still increasing, and commences again on the rising or falling limb of the hydrograph. This is a characteristic sign of the pulsing nature of bedload transport also observed in other studies (Hayward and Sutherland, 1974; Beschta, 1981; Reid et al., 1985; Tacconi and Billi, 1987).

In order to distinguish between consecutive events, a single event is generally defined by a hydrograph composed of one rising and one falling limb. Using this definition, the analysed bedload transport events are usually more than one hour apart; in most cases events are separated by more than one day.

A total number of 142 events were analysed for the period from October 1986 to December 1991. Looking at all the observations a distinction could be made between events occuring during the summer and those occuring under winter conditions. The latter category includes snowmelt runoff events and rain on snow events when at least about half of the catchment area is covered by snow.

Characteristic mean values and idealised hydrographs are shown in Fig. 2. The summer events are about twice as frequent as the winter events. For the average summer event there is bedload transport during about 100 minutes. The average peak discharge, Q_p, is almost 1 m^3/s, the threshold discharge at beginning of transport, Q_a, is about 480 l/s and the one at the end of transport, Q_e, is about 640 l/s. The average event occuring during winter conditions shows (intermittent) hydrophone activity during about 160 minutes, and the respective discharge values are: Q_p = 520 l/s, Q_a = 360 l/s and Q_e = 430 l/s (Fig. 2). For these events, there are often intermittent periods with no bedload transport at all: on the average these breaks amount to about 50 minutes of the total event duration of 160 minutes. If the sum of the hydrophone impulses (sensor no. 3) is assumed to be

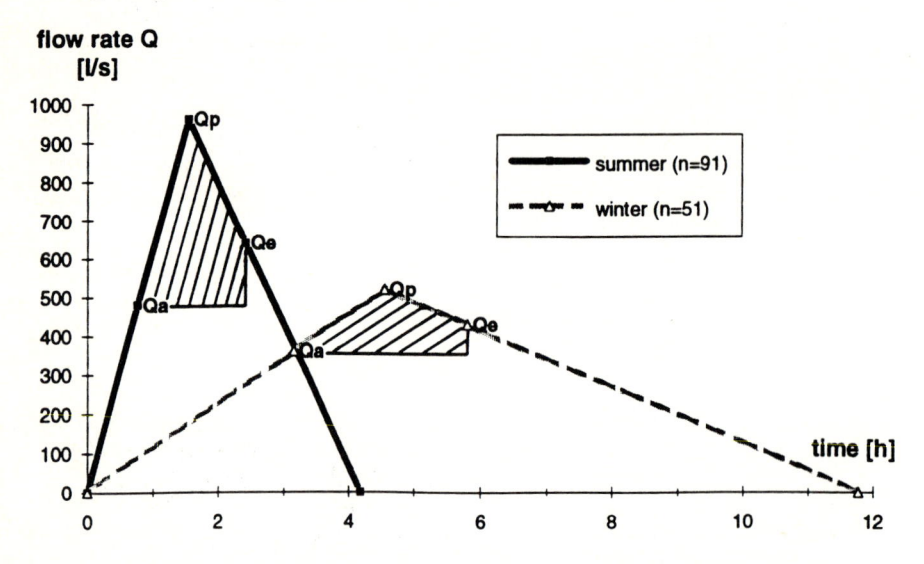

FIG. 2 Idealised mean hydrographs of bedload transport events, for summer and winter conditions, respectively. (See text for explanation of characteristic values.)

indicative of the total sediment volume passing the measuring site, then only about 10 % of the total sediment is transported during winter conditions.

Looking at the critical discharge Q_a at beginning of bedload transport, a clear seasonal trend can be observed (Bänziger and Burch, 1991). During winter conditions both the average value and the variation of Q_a is much smaller than in summer. A possible explanation could be the frozen soil so that it is mainly finer grains not protruding much into the bed which are available for transport. A therefore more limited grain size distribution available for transport during winter conditions could also account for the observed intermittent breaks with no transport activity during the same event.

Characteristics of a few selected events

The hydrograph and the sedimentgraph as represented by the hydrophone measurements (of sensor no. 3) of two events is shown in Fig. 3. In one event (Fig. 3a) the maximum bedload transport intensity occured a few minutes before the peak discharge, whereas the situation was reversed in the other event (Fig. 3b). In the latter case there are two peaks in water runoff followed by two peaks in transport intensity. It can be seen that in both events the discharge at negligible or ceasing bedload transport is higher than the one at beginning of transport. This is generally the case with most bedload transport events. It should be noted that both events occurred in July and represent summer conditions.

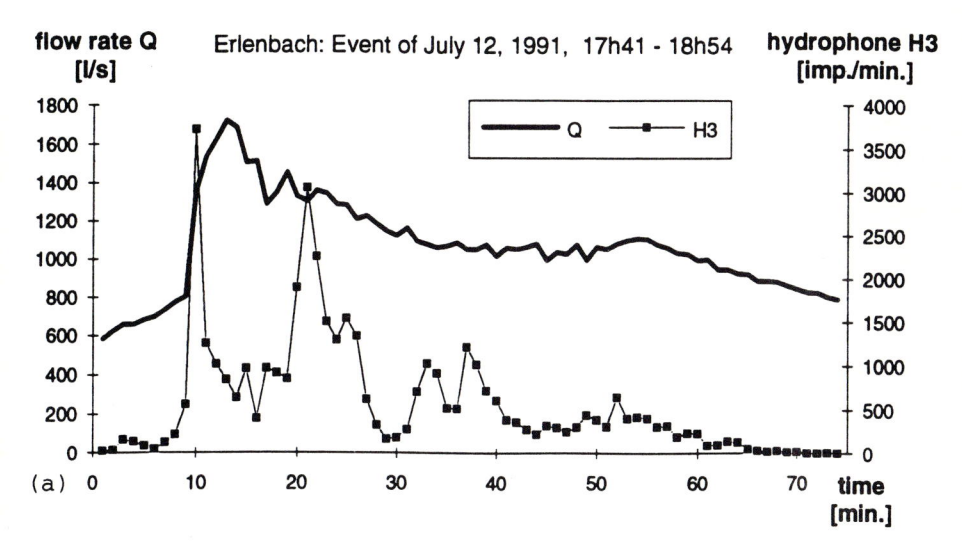

FIG. 3 Two examples of flood events with bedload transport in the Erlenbach stream. The hydrophone intensity (impulses/minute) represents the bedload transport intensity.

(Fig. 3b)

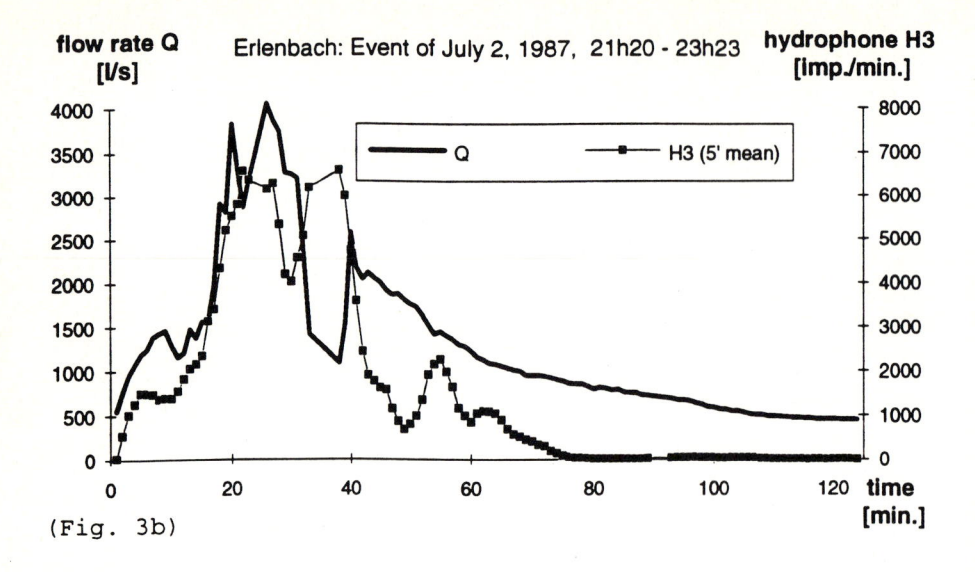

The bedload transport intensities generally show larger fluctuations than the flow rate. To smooth this effect, moving average values over 5 minutes of the hydrophone impulses were determined in some cases, as for example in Fig. 3b. Using the same averaging procedure, hydrophone intensities are plotted against flow rate in Fig. 4, for the same events as before. There is a trend for higher transport intensities at the same discharge on the rising limb than on the falling limb of the hydrograph. Such a hysteresis effect has also been noted for suspended load transport (Beschta, 1987). The trend for a hysteretic behaviour is also confirmed by other events; but the picture can be quite complex (also if an averaging procedure is used), particularly at discharges close to the peak flow rate. The somewhat erratic behaviour seems to be indicative of the pulsing nature of bedload transport. The fact that the maximum bedload transport intensities are generally not in phase with the peak discharge, may also be due to the wave-like transport and further be the result of average grain velocities being smaller than mean fluid velocities.

For comparison, an event which occurred during winter conditions is shown in Fig. 5. The first bedload activity was recorded at 5 a.m. Rainfall intensities were about 5 to 10 mm/h for 10 hours until the end of bedload activity at 3 p.m. It can be seen that there is only very infrequent bed load transport during the first 8 hours. There was only continuous bedload transport when the flow rate had reached about 1 m^3/s or about 2.5 times the threshold discharge at 5 a.m. It should be noted, however, that this event represents an extreme case of all the analysed winter events, for which the average event duration with bedload transport activity is about two and a half hours (s. also Fig. 2).

hydrophone H3 Erlenbach, bedload transport and discharge
[Imp./min.]

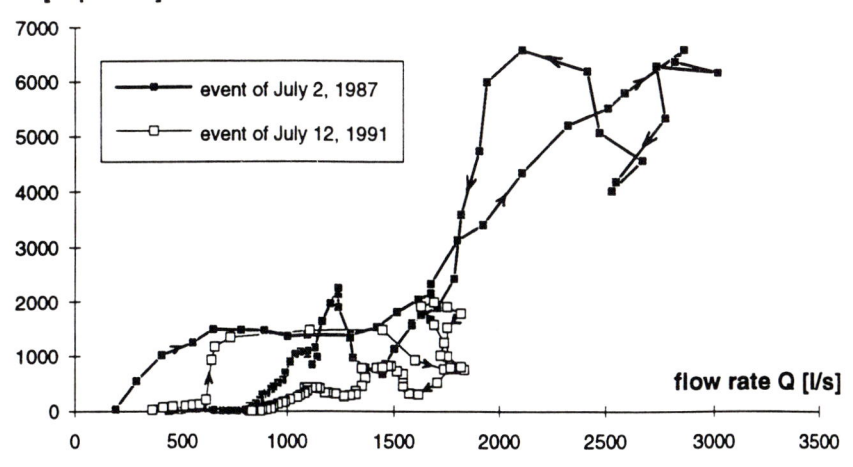

FIG. 4 Hydrophone impulses, as an indicator of bedload transport intensities, in relation to flow rate, as recorded for the two flood events shown in Fig. 3. Arrows indicate sequence of measurements.

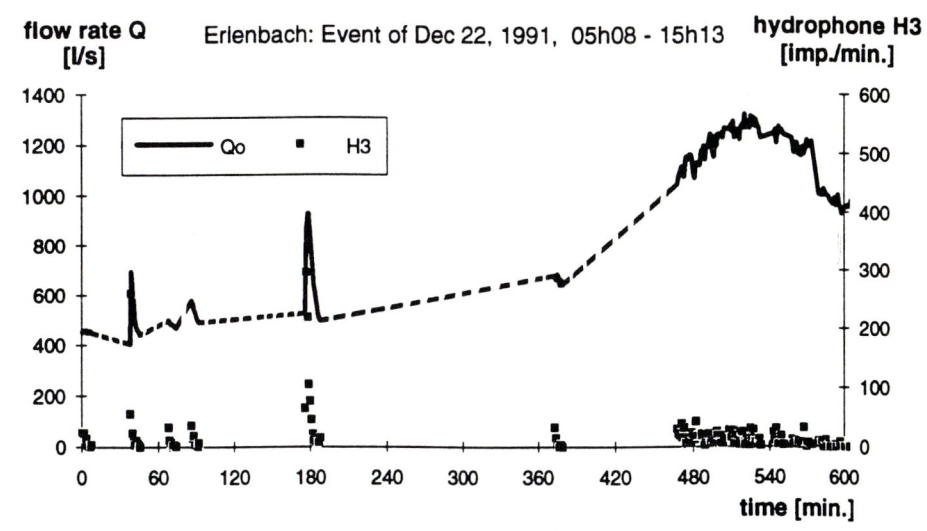

FIG. 5 Example of flood event that occurred during winter conditions. Note infrequent periods of bedload transport activity.

RELATION BETWEEN STREAMFLOW, HYDROPHONE INTENSITY AND BEDLOAD TRANSPORT

Integrated discharge and sum of hydrophone impulses

Though there seems to be no simple relation between flow rate and bedload transport intensity (as expressed by the number of hydrophone impulses per minute), one may expect a clearer trend if the data is averaged over a longer time period. The sum of the hydrophone impulses per event is plotted against the integrated water discharge per event in Fig. 6. It is evident that there is a (non-linear) relation between the two parameters although the scatter is large, the variation extending over more than one order of magnitude. Not surprisingly the winter events show somewhat lower sums of hydrophone impulses for the same runoff volume even though they have a smaller mean threshold discharge. It is quite probable that less sediment is available for transport during winter conditions which may be associated with temperatures generally below the freezing point. It may be expected that predominantly finer bed material is transported during such events.

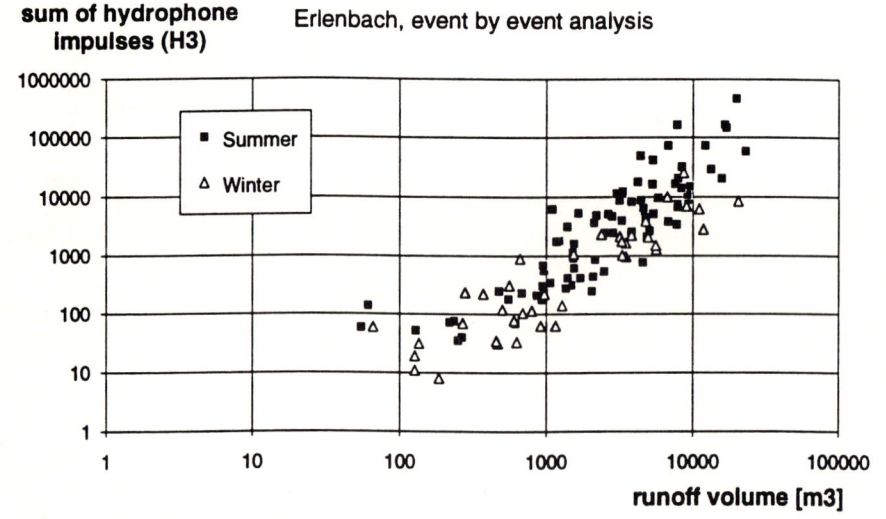

FIG. 6 Sum of hydrophone impulses over flood duration with bedload transport activity in relation to corresponding runoff volume.

If only the effective discharge responsible for bedload transport is considered, the scatter is considerably reduced (Fig. 7). Here, the runoff volume was calculated with the flow rate difference above the threshold discharge $(Q-Q_a)$, for the time period with bedload activity for each event (shaded area in Fig. 2). Considering summer conditions only, the data could be described best by two separate relationships. For larger effective runoff volumes (with generally

$Q_p \gg Q_a$) the correlation is better than for flow rates close to the threshold discharge (smaller effective runoff volumes). This finding is in agreement with the results of flume studies which show the difficulty of predicting bed-load transport rates at discharges close to critical conditions at the beginning of grain movement.

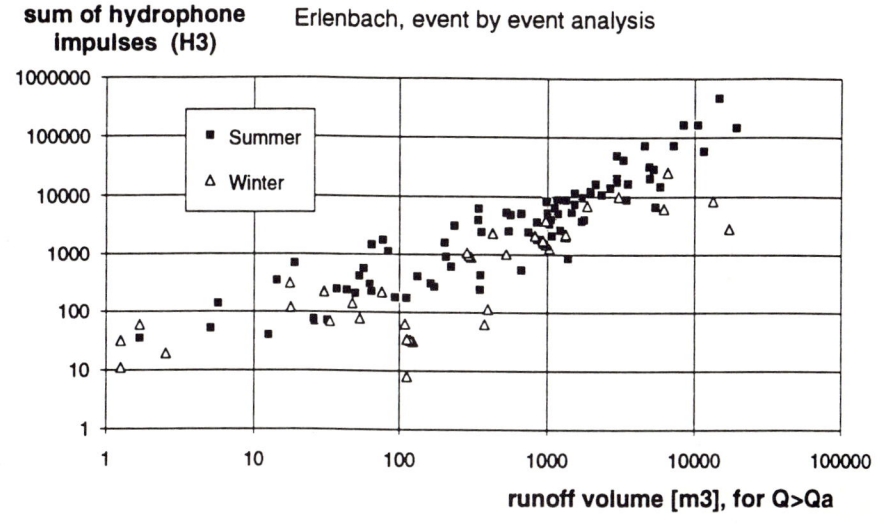

FIG. 7 Sum of hydrophone impulses over flood duration with bedload transport activity, shown in relation to the effective runoff volume above the threshold discharge for beginning of transport, Q_a.

For predictive purposes, no exact information on the flood hydrograph would generally be available. It is therefore interesting to look at the effect of introducing the average values of the threshold discharge Q_a for both the summer and winter conditions and of approximating the hydrograph by a triangle shape. As can be expected the scatter is larger than on Fig. 7, but similar to the case of using the total runoff volume (as on Fig. 6). This finding is important with regard to a predictive situation where the shape of the hydrograph is not well known and the threshold discharge can only be estimated.

Relation between hydrophone impulses and sediment volume

The total sediment discharge was determined 11 times during the period from May 1987 to April 1991 by surveying the debris volume accumulated in the sediment basin at the outlet of the Erlenbach catchment. Consequently the sum of the hydrophone impulses was determined for the time spans corresponding to the observation periods of the volume survey. The relation between the hydrophone activity and the sediment volume for these periods is shown in Fig. 8 (see also Fig. 1). Again the scatter appears to be somewhat larger for events associated with smaller bedload volumes.

sum of hydrophone impulses (H3) Erlenbach, observ. periods of volume survey

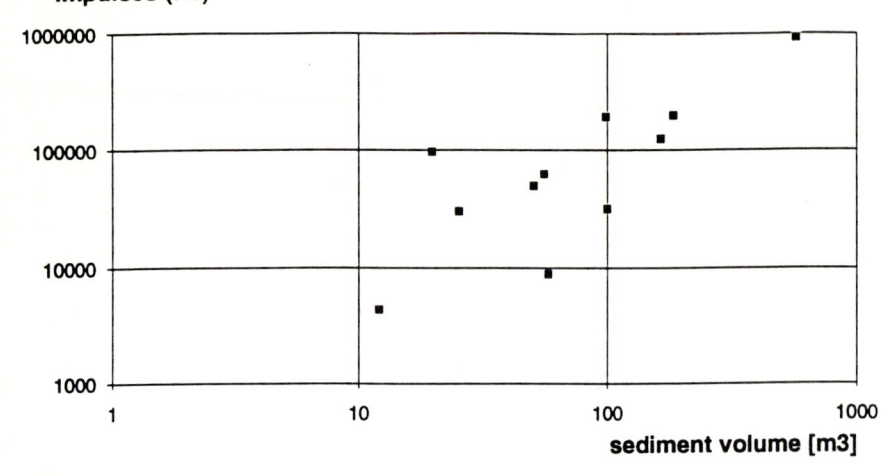

[scatter plot: sum of hydrophone impulses (H3) vs sediment volume [m3], log-log axes]

FIG. 8 Sum of hydrophone impulses, determined for the observation periods between two consecutive sediment volume surveys, shown in relation to the sediment volume accumulated in the retention basin within the same period.

As has been shown above (Fig. 6 and 7), the hydrophone activity is related to the runoff volume. For larger floods there seems to be trend of producing more hydrophone impulses than expected from extrapolation of the data for smaller floods. Plotting the two parameters in a similar way for the periods according to the volume survey (Fig. 9) implies a further averaging of the data. Fig. 9 confirms the correlation between hydrophone impulses and runoff volume and shows a reduction of the scatter by using larger summation units.

Relation between discharge and sediment volume

As can be expected from the previous analysis there should also be a relationship between sediment volume and runoff volume. The diagram in Fig. 10 shows a clear correlation between these two parameters. A further observation included in Fig. 10 is the largest flood event (of July 25, 1984) recorded so far in the Erlenbach torrent. This event had a peak discharge of about 12 m^3/s and transported 2000 m^3 of sediment into the retention basin. With an estimated reccurence interval of more than 100 years it was much larger than any event during the hydrophone observation period. The data point of the 1984 event shows a larger sediment volume than would be expected by extrapolation of

sum of hydrophone impulses (H3) Erlenbach, observ. periods of volume survey

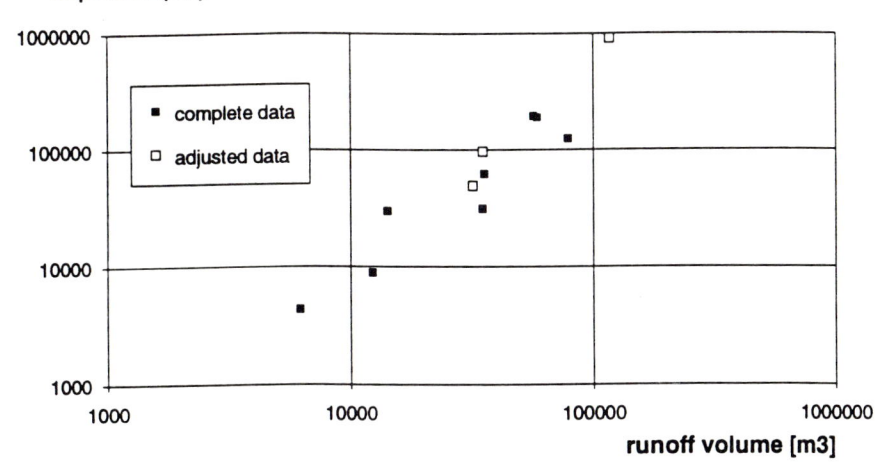

FIG. 9 Sum of hydrophone impulses, determined for the observation periods between two consecutive sediment volume surveys, shown in relation to the corresponding total runoff volume during periods of bedload transport activity. For the points labelled "adjusted data" a minor part of the discharge data was reconstructed.

sediment volume [m3] Erlenbach, observ. periods of volume survey

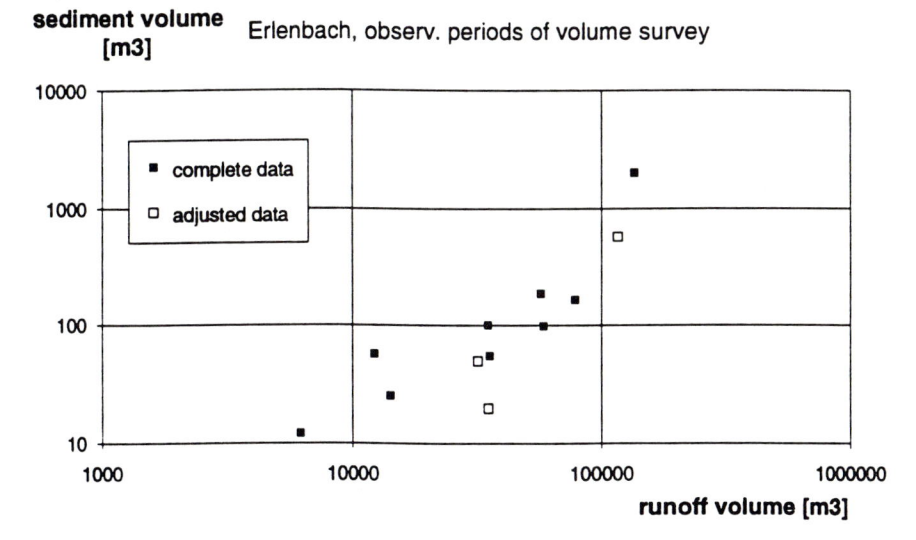

FIG 10 Sediment volume accumulated in the retention basin, shown in relation to the corresponding total runoff volume during periods of bedload transport activity.

the other data points referring to smaller floods. This finding is in line with the tendency of larger floods to produce proportionally more hydrophone impulses than smaller floods (Fig. 6 and 7).

It can be expected that the scatter in the plots with total runoff volume as a parameter (Fig. 9 and 10) will be somewhat reduced if it is substituted by the effective stormflow volume above the threshold discharge for beginning of bedload transport ($Q > Q_a$). Due to incomplete discharge data sets this analysis has not yet been performed.

Discussion of main results

A number of bedload transport formulae have been proposed which relate the (maximum) transport rate to the flow rate and the slope (Rickenmann, 1990). Some of them are based on hydraulic laboratory tests performed in steep flumes with slopes up to 20%. They generally indicate a linear relationship between bedload transport rate and flow rate or effective flow rate above the threshold discharge Q_a.

If these equations are applicable also for the natural conditions in a torrent channel, one might also expect a linear relation between observed sediment volume (G) and runoff volume (V_r) for one or several flood events. The data given in Fig. 10 support a proportionality of the type $G = C \cdot V_r^b$, where C is a constant and the exponent b may roughly vary between 1 and 2, with a tendency of b to be closer to 2 for larger floods. Considering the event by event analysis (Fig. 6 and 7) similar relationships can be determined between the sum of hydrophone impulses and the runoff volume.

Worldwide, there are only very few field data available concerning continuous observations of bedload transport rates. Data sets exist for Oak Creek in Oregon, USA, and for Virginio Creek in Italy. Milhous (1987) found a relationship $Q_{BL} = C (Q-Q_a)_m^2$ to apply for the Oak Creek data and suggested that a similar relationship also holds for the Virginio Creek. Here Q_{BL} is the average hourly bedload transport rate and $(Q-Q_a)_m$ the effective flow rate averaged over the flood event. He also noted that the exponent of the discharge factor depends on the value of the threshold discharge Q_a.

Bedload transport formulae based on laboratory experiments indicate a dependence of the bedload transport rate on $S^{1.5}$ or S^2 (Rickenmann, 1990), where S is the bed slope. Oak Creek and Virginio Creek both have a channel slope at the measuring site of about 1%. In the Erlenbach stream, with a bed slope of 19% at the measuring site, higher bedload transport rates can be expected, the slope factor implying a ratio between about 80 and 360. For a few flood events the hydrophone data of the Erlenbach stream has been converted into transport rates by using Fig. 1 as a rough calibration relation. If the slope factor is taken into account, these bedload transport rates are found to be of the same order of magnitude as those measured in Oak Creek and Virginio Creek.

The above findings indicate that bedload transport data obtained in natural streams may be at least roughly described by semi-empirical formulae based on flume experiments, if the fluctuations observed in the field measurements are smoothed over a sufficiently long time period as for example the flood duration.

CONCLUSIONS

In the Erlenbach stream the signal of bedload grains transported through the measuring cross section is recorded by hydrophones. It has been shown that the transported sediment volume is related to total number of hydrophone impulses for a given observation period. The hydrophone data are a continuous record of grain transport activity and they may be expected to represent bedload transport rates in the stream. Specific experiments will be performed in order to better define a calibration relation which allows to determine transport rates from the hydrophone data.

Considering short time intervals and comparing hydrophone intensities with flow rates, great variations and fluctuations are observed. There seems to be generally a higher transport activity during the rising limb than during the falling limb of the hydrograph. By averaging the data on an event by event basis, a clear trend between the number of hydrophone impulses and the runoff volume is revealed.

Both the threshold discharge at the beginning of transport and the hydrophone acitivity (and thus the transported sediment) appear to be different in summer and in winter conditions.

It is possible that a part of the overall variation may be explained by introducing further parameters in the analysis such as peak discharge, flood free time before the event and steepness of the rising part of the hydrograph. Such an analysis will be performed in the near future.

Semi-empirical formulae based on laboratory data may be used to describe field observation if the measured bedload transport and flow rates are averaged over the duration of the flood period.

Acknowledgements: My sincere thanks go to to my predecessors and coworkers of the Swiss Institute for Forest, Snow and Landscape Research involved in setting up the hydrophone measurements and in charge of the data collection. Without their initiative and help such a valuable set of measurements would not be available today. The presented analysis is part of a research project within the EROSLOPE-Program of the EEC.

REFERENCES:

Bänziger, R., Burch, H. (1990): Acoustic sensors (hydrophones) as indicators for bed load transport in a mountain torrent. Proc. Int. Symp. on Hydrology in Mountainous Regions, I Hydrological Measurements; the Water Cycle. IAHS Publ. no. 193, pp. 207-214.

Bänziger, R., Burch, H. (1991): Geschiebetransport in Wildbächen. Schweiz. Ingenieur u. Architekt, Heft 24, pp. 576-579.

Beschta, R.L. (1981): Patterns of sediment and organic-matter transport in Oregon Coast Range Streams. Proc. Int. Symp. on Erosion and Sediment Transport in Pacific Rim Steeplands, IAHS Publ. no. 132, pp. 179-188.

Beschta, R.L. (1987): Conceptual models of sediment transport in streams. In: C.R. Thorne, J.C. Bathurst and R.D. Hey (editors), Sediment transport in gravel-bed rivers, John Wiley and Sons, Chichester, pp. 387-420.

Hayward, J.A., Sutherland, A.J. (1974): The Torlesse Stream vortex-tube sediment trap. J. Hydrol., 13: 1, pp. 41-53.

Milhous, R.T. (1987): Discussion of Tacconi,P., Billi, P. (1987) 'Bed load transport measurements by the vortex-tube trap on Virginio Creek, Italy', In: C.R. Thorne, J.C. Bathurst and R.D. Hey (editors), Sediment transport in gravel-bed rivers, John Wiley and Sons, Chichester, pp. 611-614.

Reid, I., Frostick, L.E., Layman, J.T. (1985): The incidence and nature of bedload transport during flood flows in coarse-grained alluvial channels. Earth Surf. Proc. and Landf., Vol. 10, pp. 33-44.

Rickenmann, D. (1990): Bedload transport capacity of slurry flows at steep slopes. Mitt. der Versuchsanstalt für Wasserbau, Hydrologie und Glaziologie, ETH Zürich, Nr. 103, 249 pp.

Sawada, T., Ashida, K., Takahashi, T. (1985): Sediment transport in mountain basins. Proc. Int. Symp. on Erosion, Debris Flow and Disaster Prevention, Tsukuba, Japan, pp. 139-144.

Tacconi, P., Billi,P. (1987): Bed load transport measurements by the vortex-tube trap on Virginio Creek, Italy. in: C.R. Thorne, J.C. Bathurst and R.D. Hey (editors), Sediment transport in gravel-bed rivers, John Wiley and Sons, Chichester, pp. 583-616.

SEDIMENT TRANSPORT AND WATER DISCHARGE DURING HIGH FLOWS IN AN INSTRUMENTED WATERSHED

Vincenzo D'Agostino, Mario A. Lenzi

University of Padova, Department of Land and Agro-forest Environments, Water Resources and Soil Conservation Division, Via Gradenigo 6, 35131 Padova, Italy

Lorenzo Marchi

CNR, Institute for the Prevention of Geological and Hydrological Hazards, Corso Stati Uniti 4, 35020 Padova, Italy

Abstract

Some significant flood events which occurred from 1987 to 1992 in a small instrumented watershed in the Eastern Italian Alps (Rio Cordon, watershed area = 5 km^2) are considered. An experimental station for recording water and sediment discharge has been operating in this catchment since 1986. The relationship between water discharge and bedload transport can be analysed from the continuous recording of water discharge and the hourly measurement of coarse sediment (minimum size exceeding 20 mm) which is deposited at the recording station in an open storage area. It is possible to recognise the discharge threshold for starting coarse sediment transport during the rising limb of the hydrograph. In addition, a preliminary analytical relation between water discharge and volumes of coarse bedload transport is inferred. With respect to suspended sediment, a description is made of the procedure for turbidimeter calibration, and data recorded in 1990 and 1991 are given and discussed.

1. Introduction

Continuous measurement of coarse sediment transport meets one of the main objectives of the researches started in the Rio Cordon instrumented watershed (Figure 1). The experimental station for the recording of water and sediment discharge in the Rio Cordon, described in a previous paper (Fattorelli et al., 1988), operates by separating coarse bedload from water and fine sediment. The measuring device consists of an inlet channel, a grille installed at the downstream end of the inlet channel and allowing coarse sediment to be separated from fine sediment, a storage area for coarse sediment deposition, an outlet channel to convey water and fine sediment passing through the grille. Coarse sediment with a minimum size exceeding 20 mm slides over the grille and develops a small debris fan in the storage area (Figure 2). The height of accumulated debris is evaluated, during a flood event, by means of ultrasonic gauges fitted on an overhead travelling crane that moves over the storage area. The ultrasonic sensors record the distance from the travelling crane to accumulated debris: based on these measurements, height and volume of deposited material can easily be assessed (Lenzi et al., 1990). Water and fine sediment are directed to the outlet channel; gauges for recording water stages and turbidity are installed in the outlet channel. Data on suspended sediment concentration are also provided by water samples collected in the outlet channel. Sand and fine gravel, discharged as bedload in the outlet channel, are not measured by the installed instrumentation; an upgrading of the measuring station has been planned in order to record fine bedload.

This paper presents experimental data, recorded in the Rio Cordon from 1987 to 1992, on water discharge, coarse sediment transport and suspended sediment during flood events and during snowmelt periods. The possibility of developing a preliminary relationship, for hourly observations, between water discharge and coarse bedload transport is also discussed.

2. Measurement of bedload transport

The paper presents data from four flood events with coarse bed transport (Table 1). A complete recording of coarse bedload transport for the flood of June 1991 was prevented by the malfunctioning of the data acquisition

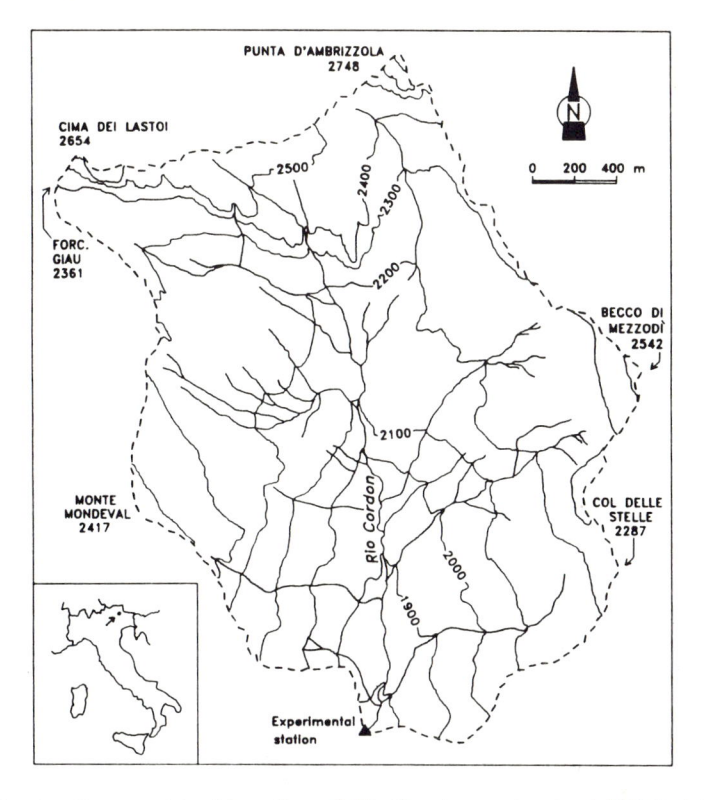

Fig. 1 - Geographical location of Rio Cordon watershed (NE Italy)

Fig. 2 - Coarse sediment accumulation after the event of June 17, 1991

system. For this event, however, it was possible to determine the total volume of the material accumulated in the storage area, by means of a topographic survey. The hydrographs of the four floods and the deposition pattern of coarse bedload transport in the storage area for the events of October 1987, July 1989 and October 1992 are presented in Figures 3, 4, 5 and 6.

Table 1 - Main characteristics of studied events

Date	Peak discharge [m^3 s^{-1}]	Duration of coarse bedload transport [hrs]	Deposited volume [m^3]
11 Oct. 1987	5.15	8	50
03 July 1989	4.42	27	85
17 June 1991	4.00	20	39
05 Oct. 1992	2.91	10	9.3

The particle size distributions of the material which reached the storage area during the floods of July 1989, June 1991 and October 1992 were calculated using a bulk by weight procedure (Church et al., 1987) and are shown in Figure 7. While the pattern of the two curves is similar, it should be observed that material transported in the floods of July 1991 and of October 1992 has a finer particle size distribution than the event of July 1989. This fact, and also the smaller volume of coarse material transported in the latter events, can be partly explained both by the lower value of the peak discharge and by the shorter duration of the maximum discharges.

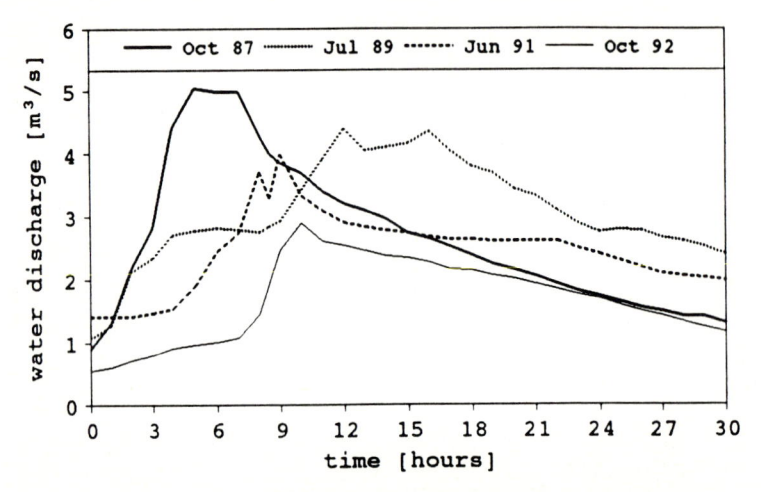

Fig. 3 - Comparison of hydrograph for the events of October 11, 1987; July 3, 1989; June 17, 1991 and October 5, 1992.

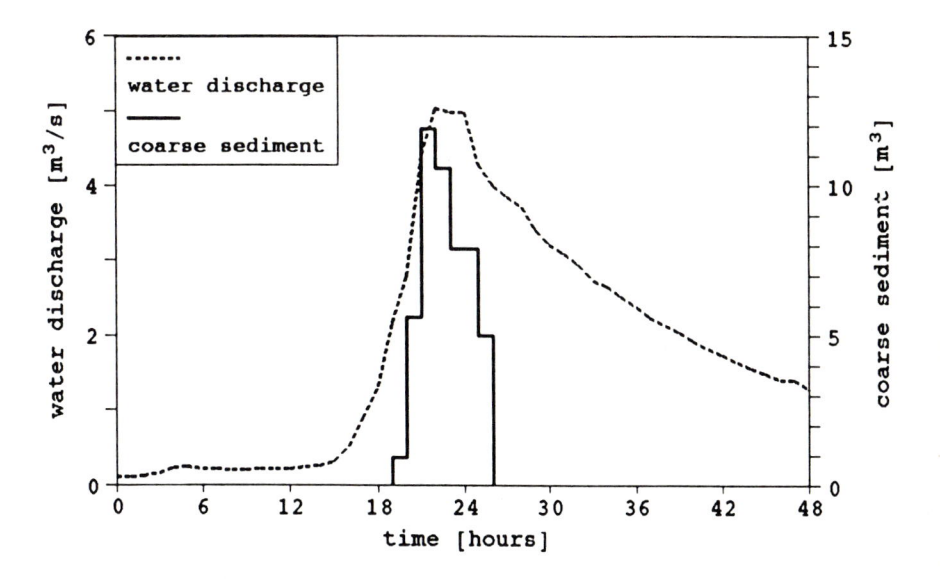

Fig. 4 - Water discharge and coarse sediment accumulation during the event of October 11, 1987

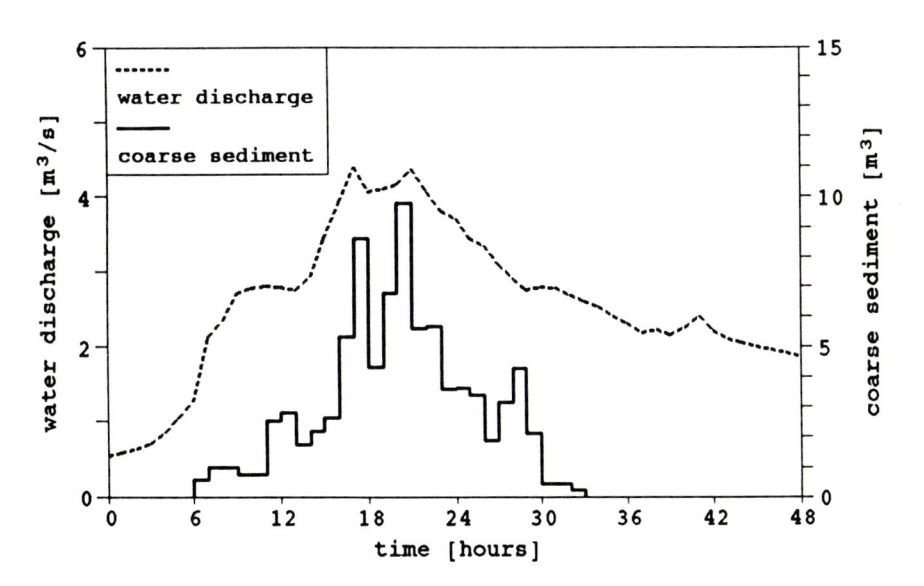

Fig. 5 - Water discharge and coarse sediment accumulation during the event of July 3, 1989

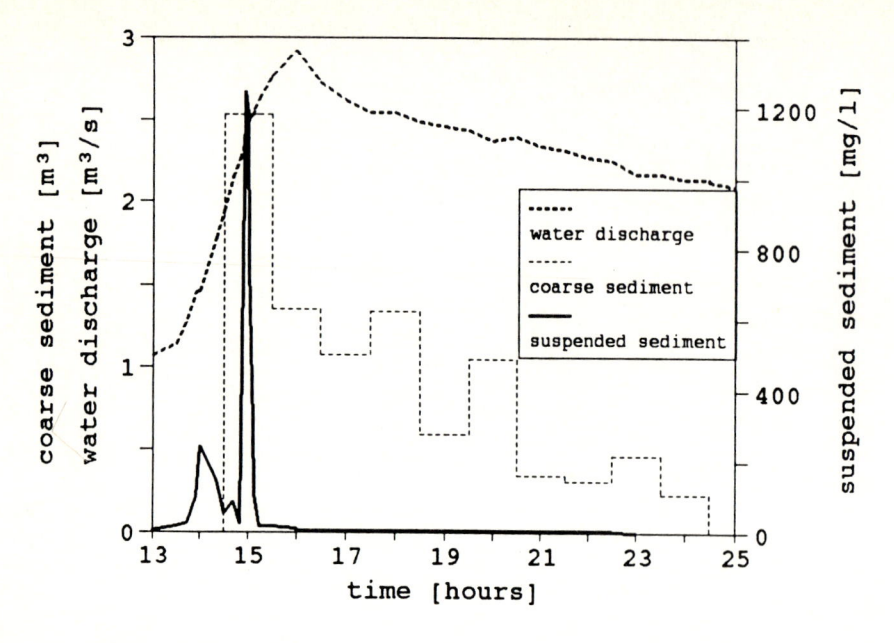

Fig. 6 - Water discharge, coarse sediment accumulation and suspended sediment concentration during the event of October 5, 1992.

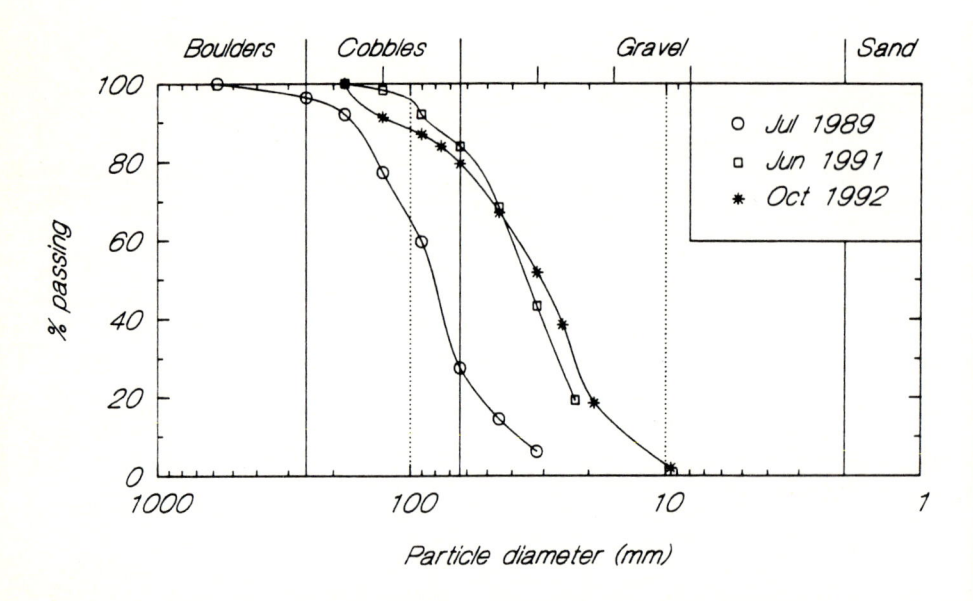

Fig. 7 - Particle size distribution of the sediment accumulated during flood events.

3. Correlation between discharge and coarse bedload transport

Though the number of flood events transporting considerable volumes of coarse sediment was small, an analytical relation between the recorded flood hydrographs and the respective progressive sediment deposition could still be derived.

The recording interval of what could be defined as a "sediment-graph" of coarse material is about 1 hour, since it depends on the time necessary for the overhead travelling crane to cover a complete passage on the storage area. Thus, flood hydrographs (obtained from water levels recorded at 5 minute intervals) were first transformed into hourly observations.

These data were used to search for an expression which could quantify:

A) the correlation between volumes of coarse bedload and water discharges for single hourly observations;

B) the total quantity of coarse bedload transported by flood events.

The analysis was made first for the two events of October 1987 and July 1989 and the formula obtained was then applied to test condition B for the June 1991 event and conditions A and B for the October 1992 event.

The relation found is of the type:

$$Q_s = [(3600\, Q)^x / K_1] - K_2 \qquad (1)$$

where:
Q_s = coarse bedload transport; [m^3 hr^{-1}]
Q = average hourly water discharge; [m^3 s^{-1}]
x, K_1, K_2 = the event calibration constants.

From (1), threshold discharge Q_{cr} is deduced for the beginning of coarse bedload transport ($Q_s = 0$), which is given by the expression:

$$Q_{cr} = [(K_1 K_2)^{1/x}] / 3600 \qquad (2)$$

The calibration performed on the first two events supplied the values:

$$K_1 = 4000;\ x = 1.1;$$

$K_2 = 4.0;\ Q_{cr} = 1.8\ m^3\ s^{-1}$ for the flood of October 1987;
$K_2 = 5.0;\ Q_{cr} = 2.2\ m^3\ s^{-1}$ for the flood of July 1989.

The relation achieved was then applied to the June 1991 and October 1992 floods, adopting a value of 4.5 for K_2, equal to the average of the coefficient calibrated on the previous events. A total volume of 37 m^3 against the 39 m^3 recorded was estimated for the event of June 1991.

From the observation of the graph (Figure 8) which compares cumulated volumes measured and those computed through (1), a satisfactory estimate for total volumes (percentage error amounts to 16% for the 1987 event and is almost zero for those of 1989 and 1992) can be inferred for the considered events.

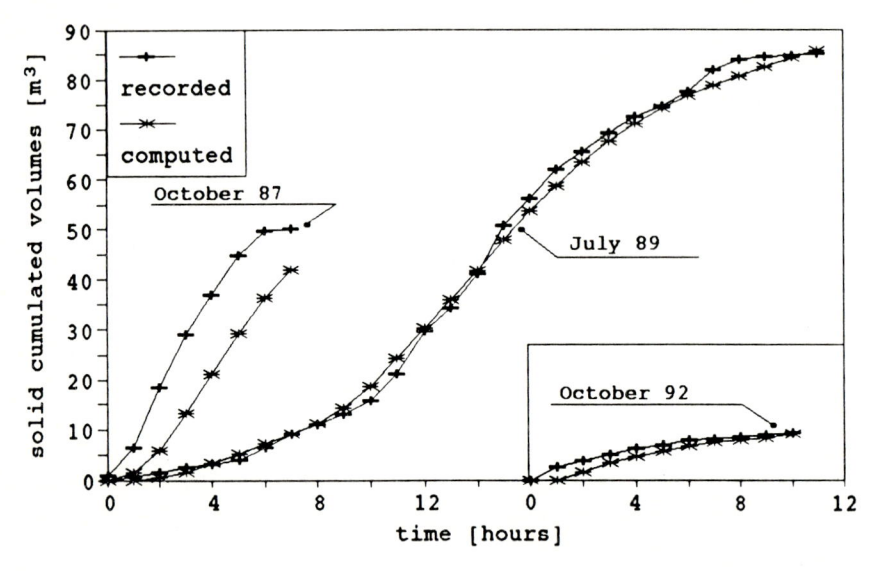

Fig. 8 - Coarse sediment deposited during the event of October 11, 1987; July 3, 1989 and October 5, 1992: comparison of recorded and computed cumulated volumes

It should be pointed out that (1) was used for the rising limb of the flood hydrographs as soon as the respective Q_{cr} value was exceeded. For the recession limb of the hydrographs, the computation of the volume was restricted only to the moment when the overhead travelling crane did not record sediment storage any longer. In fact, field observations showed that the critical threshold for transport cessation is higher than for the starting phase. Recorded flood events did not affect channel stability: transported sediment was supplied by minor amounts of loose material available in the stream bed. After the removal of such sediment, transport stopped in spite of rather high values of water discharge. That is significantly illustrated by the very short event of October 1987, for which coarse transport stopped when discharge still amounted to 3.8 m^3 s^{-1}. Coarse sediment transport for the 1989 event stopped for a water discharge of about 2.7 m^3 s^{-1}, which, however, was superior to the discharge value recorded at the beginning of transport. The same condition occurred for the minor flood of October 1992 (Q_{cr} = 1.9 m^3 s^{-1}; water discharge at the cessation of bedload transport = 2.14 m^3 s^{-1}).

The formula (1) obtained is very sensitive to variations of the index x, which is kept constant in the calibration of the two events. Instead, once K_1 is fixed, the K_2 factor can be used slightly to vary the value of the hourly average critical discharge. This depends on available solid material and on riverbed conditions before the flood occurs.

In order to complete and integrate the findings that can be inferred from the graph, Table 2 shows some statistical indexes obtained by comparing the series of hourly values of recorded volumes (V_R) with the respective computed volumes (V_C) for the events of October 1987, July 1989 and October 1992. The correlation coefficient, mean absolute error E_A and mean error E_R were determined as follows:

$$E_A = \Sigma \, |V_C - V_R| \, / \, (N\overline{V_R})$$
$$E_R = \Sigma \, (V_C - V_R) \, / \, (N\overline{V_R})$$

Based on the results achieved, it can be inferred that a satisfactory mean error, equal to -0.05, was obtained, though the correlation coefficient was not very high. Absolute differences between V_R and V_C are found to be more marked when transport begins and ends.

Table 2 - Statistical indexes for the evaluation of the formula (1)

Number of observations	N = 45
Mean recorded volume	$\overline{V_R}$ = 3.21 m^3 hr^{-1}
Mean computed volume	$\overline{V_C}$ = 3.04 m^3 hr^{-1}
Correlation coefficient	R = 0.77
Mean absolute error	E_A = 0.39
Mean error	E_R = - 0.05

It should be pointed out that the relation between bedload transport and water discharges recorded in the Cordon catchment for hourly data might be missing for short time intervals (5-10 minutes), since, in this watershed as in other cases (Ergenzinger, 1988), accentuated fluctuations in bedload transport were observed when water discharge variations are absent. More numerous records will allow further checking of the proposed relation both in terms of reliability of computed volumes and of knowledge of the parameters required for the calibration of the same.

4. Measurement of suspended sediment

Suspended sediment is continuously measured by means of a light absorption turbidimeter. Sufficiently continuous and reliable recorded data are available only for the latest period, when the instrument was placed in the outlet channel of the station, well-protected by a suitable metal grille. The turbidimeter measures the percentage values of intensity loss of the light emitted by a light source due to the turbidity induced by suspended solid particles in the flowing water.

It was necessary to develop a calibration curve of the instrument to meet the particular conditions which characterise the Cordon experimental station, with particular regard to the characteristics of suspended material. Sediment was supplied to the channel reach immediately upstream of the recording station in May 1989, during the snowmelt period, in the presence of comparatively high discharges (close to 1 m^3 s^{-1}). The material employed had been collected from beneath the bed and from the banks of the torrent. Its characteristics were similar to those of the sediment measured in normal station operation.

The passage through the grille afforded a complete mixing up of water and fine sediment. Some samples were then collected in the outlet channel and percentage adsorption values measured by the turbidimeter were read; a calibration curve was obtained by means of a suitable automatic interpolation procedure.

Data on transport of suspended sediment recorded in May and June 1990 and during the June-November period of 1991 are analysed here. In May-June 1990 water discharges were not particularly high, nevertheless suspended sediment transport reached quite remarkable values, amounting to 1968 t month^{-1} and 471 t month^{-1} respectively. Figures 9 and 10 show hourly values of water discharges and of suspended sediment concentration. In the first twenty days of May, water discharges illustrate the daily oscillations that characterise snowmelt discharges: the considerable

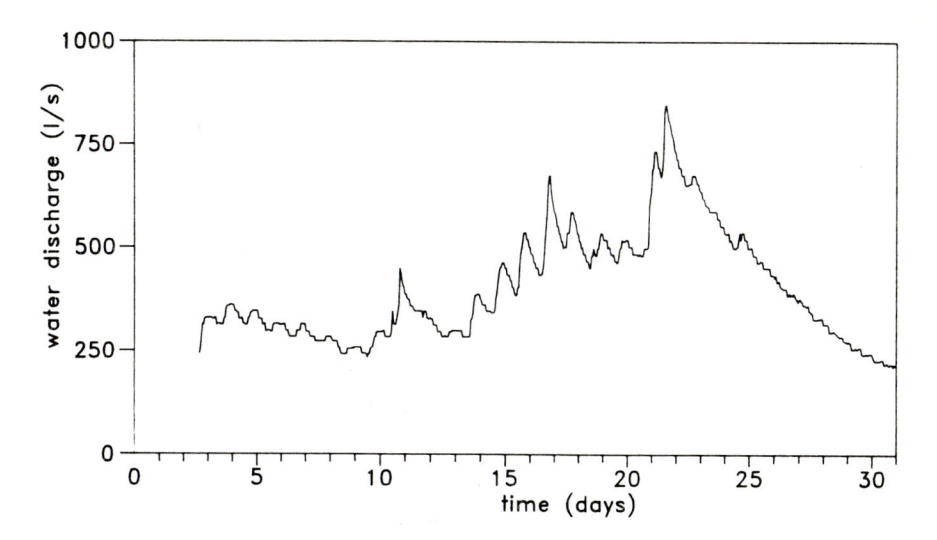

Fig. 9 - Water discharge in May 1990

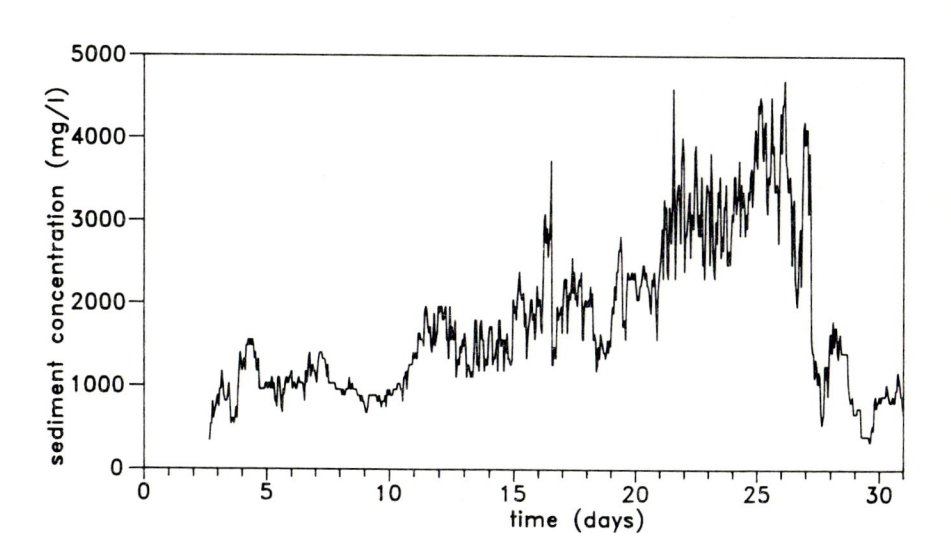

Fig. 10 - Suspended sediment concentration in May 1990

contribution to watershed sediment yield from snowmelt discharges in many catchments is well known (Johnson et al., 1985; Bathurst et al., 1986). In the course of the two months considered, it was possible to observe a fair agreement between the patterns of the two variables, though marked oscillations were present in sediment concentration values. In addition, in most cases it was possible to observe that the values of suspended sediment concentration reached their peak at the same time as the maximum values of water discharges. Though, in literature, the most frequent situation shows that maximum values of sediment concentration occur before peak floods, as in the flood of October 5, 1992 in the Rio Cordon (Figure 6), there are some studies which report conditions where the two peaks occur at the same time (Tropeano, 1991).

Again, Figure 9 shows that a stage of water discharge exhaustion is matched by persisting high values of suspended sediment concentration. This situation was not found in the subsequent month of June nor in data recorded during 1991. That might be attributed to an increased availability of loose sediment, coming from source sediment areas which contributed to catchment runoff or to fine material found in the torrent bed as a result of erosion or bank erosion phenomena during the period considered.

Figure 11 shows the scatter diagram of discharges and of sediment concentration for the months of May and June 1990. 1254 data values were available, corresponding to 86% of the hourly surveys that could be recorded in the two months under consideration. There is a rather marked dispersion of sediment concentration values; the linear relationship between the two variables is characterised by a coefficient of determination equal to 0.615.

In 1991, available data for suspended sediment transport covered the June to November period, though with some gaps (Table 3). The available data show that suspended transport in the experimental Rio Cordon watershed has values which, on the whole, are in agreement with the physical characteristics of the catchment and that it provides a significant contribution to sediment yield. The value recorded in June 1991 proved inferior to the one of the corresponding month in 1990, though average discharges were rather high and a rather remarkable flood event occurred.

A better interpretation of the relationships between water discharges and sediment transport in the watershed under investigation can be provided by future data collection at the recording station, and by field studies in the watershed areas where contribution from various sediment sources will be evaluated.

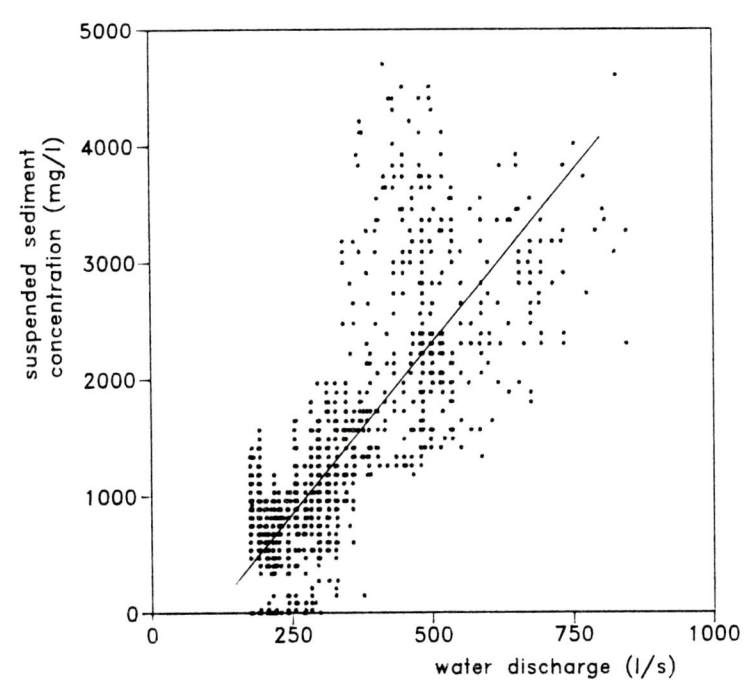

Fig. 11 - Suspended sediment concentration relationships for the period May - June 1990

Table 3 - Suspended transport in 1991

Month	Average water discharge [m^3 s^{-1}]	Suspended sediment transport [t month^{-1}]
June	1.12	64.2
July	0.43	5.6
August	0.20	3.0
September	0.11	0.8
October	0.15	[0.0] (*)
November (**)	0.12	0.2

(*) Gaps in records
(**) Records up to November 20

5. Conclusions

The early years of station operation have confirmed the soundness of the approach used for measuring bedload transport. Although the number of significant flood events recorded in the Rio Cordon watershed is restricted, some suitable data are available thanks to the continuous recording of coarse transport. These can be used for analysing the relationships between water discharges and coarse bedload transport. The refinement of procedures in station management and the improvement of data acquisition techniques will lead to improved data collection in the course of future research. A particular necessity regards the improved collection of data on suspended sediment transport, which have so far been available for a very restricted period. Reliable measurements of the finest bedload, allowing the estimation of total sediment yield, will be possible after a sedimentation basin is built at the end of the outlet channel of the measuring station.

6. Acknowledgements

Our thanks go to the Experimental Centre for Avalanche Control and Hydrological Defence of Veneto Region for the support given. This study was partly funded by the Italian Ministry of University and Research (60 % funds, Prof. G. Benini) and was carried out as a preliminary contribution in the preparation of an EEC project on erosion and sediment transport on steep mountain watersheds (EROSLOPE).

7. References

Bathurst J.C., Leeks G.J.L., Newson M.D., 1986. *Relationship between sediment supply and sediment transport for the Roaring River, Colorado, USA*. IAHS Publ. no. 159, 105-117.

Church M.A., McLean D.G., Wolcott J.F., 1987. *River bed gravel:sampling and analysis.* In: Sediment Transport in Gravel Bed Rivers. C.R. Thorne, J.C. Bathurst, R.D. Hey, eds, J. Wiley and Sons, New York, 43-88.

Ergenzinger P., 1988: *The nature of coarse bed load transport.* IAHS publ. no. 174, 207-216.

Fattorelli S., Keller H.M., Lenzi M., Marchi L., 1988. *An experimental station for the automatic recording of water and sediment discharge in a small alpine watershed.* Hydrol. Sciences Journal, 33, 6, pp. 607-617.

Johnson C.W., Gordon N.D., Hanson C.L., 1985: *Northwest rangeland sediment yield analysis by the MUSLE.* Trans of the ASAE, Vol. 28 (6), 1889-1895.

Lenzi M.A., Marchi L., Scussel G.R., 1990: *Measurement of coarse sediment transport in a small alpine stream.* IAHS Publ. No. 193, 283-290.

Tropeano D., 1991: *High flow events and sediment transport in small streams in the 'Tertiary basin' area in Piedmont (Northwest Italy).* Earth Surface Processes and Landforms, Vol. 16 (4) 323-339.

LUMINOPHOR EXPERIMENTS IN THE SAALACH AND SALZACH RIVERS

F.H. Weiss
Bayer. Landesamt für Wasserwirtschaft
Lazarettstr. 67, D-80236 München

Abstract

Both the Salzach and the Saalach Rivers show a serious lack of bed load and downcutting effects after river training in the last century and dam construction. As a countermeasure bed material feeding downstream of a reservoir has been practised. Its effectiveness was controlled using a special tracer method. Since 1986 research has been carried out in cooperation with the Federal Bundesanstalt für Wasserbau in Karlsruhe. Because of some difficulties in the investigated diversion reach, first results were only obtained two years after insertion of the tracer material. The applied luminophoric method turned out to be feasible for detecting the transport of the added material into the Salzach River over a distance of about 60 km. On condition that there adequate effective discharges occur, the transport process takes approximately one flood period. The disadvantages of this method - that field detection is not practicable and that a great many samples have to be taken, dried and examined under ultraviolet light- have to be accepted.

Introduction

The Saalach River is a tributary of the Salzach River and rises in the Austrian Alps south of Kitzbühel at an altitude of about 2000 m. It runs through the Pinzgau Valley with the towns of Saalfelden and Lofer and reaches the Bavarian border at Steinpass. This bed load bearing river enters the Kibling reservoir (completed in 1913) near Reichenhall. Here the whole bed load is deposited. The result is a steady degradation process in the downstream reach (Weiss 1989), continuing to the Salzach River. The mouth of the Saalach River is about 7 km downstream of Salzburg. Since 1985 bed material has been added to the Saalach River in order to re-establish bed load transport downstream of the dam and especially in the Salzach River, which also displays a serious lack of bed load and downcutting effects (Weiss & Mangelsdorf 1982). Although conditions for bed material feeding such as grain size distribution, transport length and possibilities for dispensing, are favourable (Weiss 1986) at the chosen adding point in the scour of the dam, some questions have required clarification.

Because of the changes in morphology since the dam closure in 1913 we did not know exactly how far the added material would be transported by the river, which part of it would pass the Saalach mouth and arrive at the target reach in the Salzach River and how long sediment transport would take.

Methodology

Because the usual cross section surveys over a marked distance of 200 m are too rough a method for monitoring transport behaviour, a tracer method was employed. There are three different tracer methods for detecting sediment transport (Mangelsdorf, Scheurmann & Weiss 1970): marking by radioactive material (Fahse 1987), coating by luminophors (Ruck 1979), and incorporating magnets into the pebbles (Bunte & Ergenzinger 1989). Because of the labour-intensive nature of the marking procedure, the extremely sensitive environment (near the health resort of Bad Reichenhall), and the short half-life of the tracer material, the radioactive method had to be ruled out. As there is no natural magnetic bed load in the Saalach River and otherwise the introduction of magntic cores into the pebbles is very laborious, the magnetic tracer method also seemed unsuitable.

The remaining and environmentally more compatible possibility is the luminophoric method. It consists of spot-coating bed load or sediment particles with a luminescent pigment with the aid of a suitable cementing material and adding the material to the river before bed load transport begins. If a bed load sample is then investigated under ultraviolet light, the marked grain particles will show up clearly and can be recorded. Tests with cementing materials such as agar-agar, bone glue, starch, gelatine, artificial resins, nitro-lacquers, etc. have shown that these lead to agglomeration of individual grains. Water-glass resulted in the thinnest coating on the grains and the longest adherence. Rolling tests showed only minor attrition, so that the full luminescence was still preserved after about 160 km of simulated transport. This is caused by the tendency of the pigment particles to enter the surface pores of the carrier particles after repeated drying and washing procedure during the application in the mixer. The colours yellow, red, green and orange were found to be suitable, so that it is possible to differentiate between materials added at different intervals. Blue is not sufficiently distinct because of the similar glow of organic particles in the sediment samples. In tests along the Rhine, marked gravels could be traced for up to 8 km below the point of insertion.

Results

Research was carried out by the Federal Bundesanstalt für Wasserbau in Karlsruhe with the local aid of the Wasserwirtschaftsamt Traunstein and accompanied by the Bavarian Landesamt für Wasserwirtschaft. In April 1986 the check-material was coated with yellow luminophoric colour. The coating procedure for 7 t took about two weeks. The coated control material was reinstalled in May 1986. In the first two years of investigation some difficulties occurred. Although the point of addition for sediment feeding in the deep scour of the dam (15 m) is favourable for picking up the material by floods in relation to the transport ability of the river, there is also a disadvantage. The dam diverts a discharge of 58 m^3s^{-1} to a power plant situated about 2 km downstream. This means that the flood frequency or the occurrence of effective discharge is reduced in this diversion reach. So we had to wait for 2 years until the required effective discharge occurred. The subsequent sampling activities were unsuccessful, because there is a widening of the channel some metres downstream of the scour. This caused renewed deposition of the material just entrained and covered the control material with a one-metre-thick layer. So this cover layer had first to be scraped away by a bulldozer and the next point of addition had to be located downstream of this reach.

In order to speed up the research procedure and to achieve quicker results, additional red-coloured control material was introduced 10 km downstream in September 1987. First results in detecting the yellow particles were achieved by extensive sampling in September 1988. The results of the next sampling in November 1988 were extended to the Saalach mouth. They showed that the luminophor tracers were distributed over the whole reach with different grain sizes. The precise counting of the last samples, extended as far as the Salzach River in December 1988, resulted mainly from fragmentary material and some marked pebbles. The maximum transport length was 32 km.

Further sampling proceeded step by step, i.e. if the result of the sample located farthest downstream was positive, the next sampling reach could be extended some kilometres. Sampling was only possible within the gravel bars.

Near Tittmoning (river km 28.7-24.8) 16 red and 3 yellow pebbles (grain size 20-100 mm) and 10 red and 20 yellow fine particles (3-20 mm) were discovered in October 1989. The next sampling near Burghausen (river km 17.3-9.2) resulted in

10 coarse red particles (20-100 mm); yellow particles were missing. Most of the detected grains were located at km 9.2. Thus the maximum of the proven transport distance, measured from the point of addition (Saalach river km 10), is about 60 km. In the last samples from the headwater of the Simbach-Braunau power plant no marked particles were discovered.

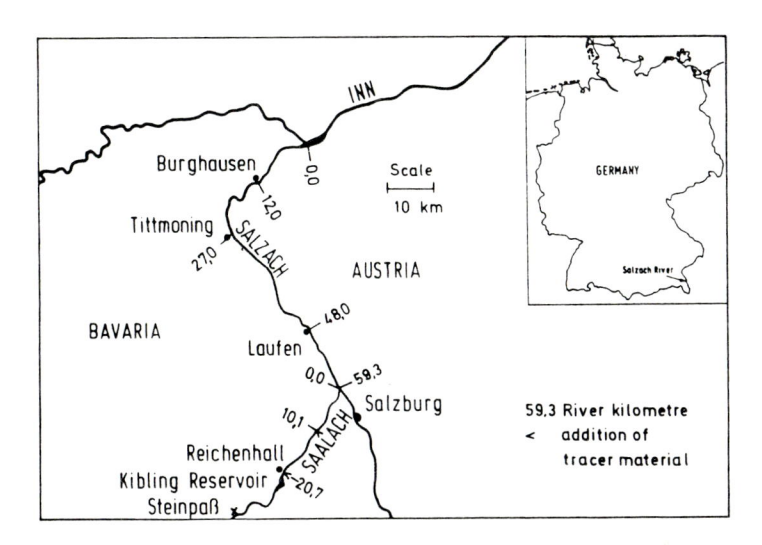

Fig. 1: Location map

In the previous investigations we gained qualitative results, i.e. the proof that grain feeding works, and in addition we obtained the maximum transport length. In the following expansion of the investigation we aimed to cover the quantitative side and to determine the transport time more precisely. For this reason additional control material (about 8 t, grain size 20-100 mm) was coated with green colour and, in June 1990, it was put in at the same location as the red material 10 km downstream of the Kibling dam.

Full mobilization of the green material was already detected at the end of July, but because of the high water level the samples could not be taken until the end of September. The sampling stretch was from the Saalach mouth (km 59.3) to km 9.3 downstream from Burghausen, at the same places as for the yellow and red material. Additional samples were taken from the right bank of the Salzach opposite Laufen. The results and a comparison with the samples before the input of the green material are included in Table 1.

colour		(yellow) red					(yellow, red) green			
date of insertion		(6.5.1986) 8.9.1987					6.6.1990			
date of sampling	15.6., 12.10.89, 1.2.90						25.9., 8.11.90			
sampling reach km	mass of samples kg	number of detected grains ⌀ 20 - 100 mm		valency number (3):(2) (4):(2)		mass of samples kg	number of detected grains ⌀ 20 - 100 mm		valency number (8):(7) (9):(7)	
		yellow	red	yellow	red		red	green	red	green
(1)	(2)	(3)	(4)	(5)	(6)	(7)	(8)	(9)	(10)	(11)
Saalach Mouth 59,4 - 59,2	576	2	4	4	7	648	5	10	5	9
Laufen 48,0 - 47,7	1 746	2	11	2	7	414	-	-	-	-
Oberndorf 47,7	-	-	-	-	-	324	1	2	3	7
Tittmoning 30,0 - 27,2	2 250 0 3 - 20 mm: 1 890	3 20	16 10	2 11	8 6	378 -	- -	2 -	- -	6 -
Raitenhaslach 17,5	738	-	2	-	3	252	-	2	-	8
Burghausen 12,2	720	-	1	-	2	342	-	2	-	6
9,3	720	-	7	-	10	720	3	4	5	6
Burghausen 17,5 - 9,3 total	2 178	-	10	-	5	1 314	3	8	3	6

Table 1: Evaluation of the detected grains

description	location	sampling reach Salzach km	degree of dilution in percent
input (coated: added material) total 1986/88	downstream of Kibling Reservoir		0,015
only 1987			0,037
output (weight of detected grains: sample weight)	Saalach Mouth	59,4 - 59,2	0,092
	Laufen	48,0 - 47,7	0,066
	Tittmoning	30,0 - 27,2	0,075
total samples	Burghausen	17,3 - 9,2	0,041
single sample		17,3	0,024
single sample		12,15	0,012
single sample		9,2	0,086

Table 2: Results of calculating the degree of dilution.

In order to eliminate the influence of the different masses of the samples, in the columns (5, 6, 10 and 11) the number of the recovered marked grains was divided by the mass of the sample. The result was rounded up to the next higher number. The "valency numbers" thus obtained for the green-coated grains show that this material is well represented in the different sampling points of the stretch and that the transport along the reach of 60 km took about one longer lasting flood period. The transport length was already detected with the yellow and red material, but these tracers had been in the river for a much longer period of time. The fact that less red material than green was found perhaps may have been influenced by the duration of the experiment. Because the marked material was added about three years ago, the glow intensity of the red colour and also the adhesive strength of the pigment particles at the grain could have decreased considerably.

A first attempt to quantify the detected marked material was done by calculating the dilution. This means that we related the mass or the weight of the detected pebbles to the total sample weight (Table 2).

The dilution percentage ranges from 0.09 to 0.02. This output was then compared with the input. The input dilution is the relation of the coated material to the added material in the scour of the dam for sediment feeding. This relation also exhibits a similar order of magnitude. The result is only a rough balance. By achieving the same dilution percentage downstream as far as Burghausen we provided first evidence of the transport of the fed material from the scour of the Kibling dam as far as Burghausen.

Conclusions

With the luminophoric method it has been possible to prove the transport of the added material through the various weirs into the Salzach River over a distance of about 60 km. To mobilize the added material, adequate effective discharges are necessary. Sediment transport lasted about one flood period. This result could have been influenced by the fact that bed erosion in both rivers has concentrated in a deep channel within the fine-grained sublayer, where bed load transport primarily takes place.

It has been of prime importance for the application of the luminophors that, in contrast to radioactive material, there is no discernible risk either for the environment or for the staff during the coating, adding, sampling and counting procedure. The luminophoric method has its disadvantages - field detection is not practicable, a great many samples have to be taken, dried and controlled separately, grain by grain, under ultraviolet light - but these have to be accepted.

References

Bunte, K. & Ergenzinger, P. 1989. New tracer techniques for particles in gravel bed rivers. Bulletin de la Sociétè Geographique de Liège 25: 85-90.

Fahse, H. 1987. Traceruntersuchungen in der Natur. Mitt.-Blatt der BAW, Nr. 60

Mangelsdorf, J., Scheurmann, K. & Weiss, F.H. 1990. River Morphology. A Guide for Geroscientists and Engineers. Springer Verlag, Heidelberg

Ruck, K.W. 1979. Ingenieurgeologische Naturversuche mit markiertem Geschiebe. Ber. 2, Nat. Tag., Ing. Geol., 283-296, Fellbach

Weiss, F.H. & Mangelsdorf, J. 1982. Morphological investigations on the lower Salzach River downstream of Salzburg.
Proceedings of the Exeter Symp., IAHS Publ. no. 137

Weiss, F.H. 1986. Riverbed degradations downstream hydraulic structures, determination and possibilities for management.
Proc. of the 3rd Int. Symp. on River Sedimentation, Jackson/Miss.

Weiss, F.H., 1989. Störungen des morphologischen Gleichgewichts der unteren Saalach durch anthropogene Einflüsse, in: Informationsbericht des Bayer. landesamtes für Wasserwirtschaft Nr. 2/89, München

Weiss, F.H., 1990. Investigations on sediment transport by luminophors in the lower Saalach River, Bavaria. Sand Transport in Rivers, Estuaries and the Sea. A.A. Balkema, Rotterdam/Brookfield

THE DOWNSTREAM FINING OF GRAVEL-BED SEDIMENTS IN THE ALPINE RHINE RIVER

Matjaž Mikoš
Swiss Federal Institute of Technology
Laboratory of Hydraulics, Hydrology and Glaciology
CH - 8092 Zurich, Switzerland
(formerly Institute for Water Management, Hajdrihova 28
SLO - 61000 Ljubljana, Slovenia)

ABSTRACT

The field study of the downstream changes of size, shape and petrographic composition of the Alpine Rhine sediment concerned the reach between the confluences with the Landquart River and the Ill River. The results show the relative importance of sorting effects at the beginning and the end of the study reach. In the middle part of the study reach, where sorting effects seem to be neglibile, the weight reduction coefficient a_w reaches values definitely lower than $0.010 / km$. Surprisingly, only one lithology is dominant, namely limestone with approximately 5 to 50 % quartz or silica. Differences between carbonate and non-carbonate lithology groups is only of lesser significance.

The laboratory abrasion experiments were performed in a specially adjusted tumbling mill of 690 liter volume. Sediment mixtures with grain sizes ranging from 2 to 128 mm were used and the initial mass was usually about 400 kg. The test mixtures were sieved from the bulk samples taken from the alternate bars of the Alpine Rhine. The abrasion distance travelled in the laboratory abrasion set-up corresponded to a range of 40 to 80 kilometers. The mean weight reduction coefficient a_w of each mixture was then determined from the observed weight changes. Their values $a_w = 0.005 - 0.018 / km$ confirm the weight reduction coefficients resulting from the field study.

1. INTRODUCTION

Downstream fining in gravel-bed rivers is a well known fact, at least since Darcel (1857) published the results of the field study in the River Seine. Since first attempt of Daubrée (1879) with a simple abrasion mill experiment, to explain downstream fining as an effect of fluvial abrasion only, a lot of theoretical research work, as well as practical laboratory and field investigations have been done. Nowadays it is commonly accepted, that not only *fluvial abrasion* of sediments causes downstream fining. At least equally important factors are also *selective transport*, *physical* and *chemical weathering* as well as *sediment supply* by tributaries, bank erosion or bed degradation. The main problem is to determine their relative importance in each specific field situation. The most important contributing authors for the theory of abrasion of coarse river sediments are among others : Sternberg (1875), Schoklitsch (1933), Wentworth (1919), Krumbein (1941), Kuenen (1956), Stelczer (1968), Schumm and Stevens (1973), Shaw and Kellerhals (1982) and Parker (1991a,b). Despite all the new ideas, arisen after Sternberg, the usage of so called "classical Sternberg's law" is still very popular due to its easiness to be applied in an everyday engineering practice.

The main objective of the presented combined field and laboratory study of the fluvial abrasion of coarse sediments was to re-evaluate the abrasion coefficient as a part of the calibration of the numerical simulation model for sediment budget of the Alpine Rhine River between Reichenau and the mouth into the Lake of Constance. The Alpine Rhine is a typical regulated alpine gravel-bed river with alternate bars. For the simulation a 1-D numerical model for degrading and aggrading natural rivers, developed at ETH Zurich (Hunziker (1992)), was used. Until nowadays the abrasion coefficient for the Alpine Rhine in the study reach was taken as high as $a_{w1} = 0.046 / km$, based solely on the old field measurements. Lately, as a part of the numerical simulation, much lower values for the abrasion coefficient seemed to be more acceptable, as low as $a_{w1} = 0.016 / km$. Because of the direct influence of fluvial abrasion on the transported sediment quantity, the main question at the beginning of the study was : Can suggested lower values for the abrasion coefficient in the Alpine Rhine be confirmed and if possible also specified for all important lithologies, using new extensive field data, laboratory abrasion experiments, and last but not least a better definition of which characteristic diameter of the grain-size distribution of gravel-bed sediments should be used for the determination of the abrasion coefficient ?

2. FIELD STUDY

After the confluence of the Hinterrhein River and the Vorderrhein River at Reichenau, the Alpine Rhine flows 91.2 km long to the Lake of Constance. The important tributaries on this way are the Plessur River, the Landquart River, the Tamina River, the Ill River and the Frutz River. The mean bed slope decreases from 0.004 to 0.001. Accordingly to the 6123 km^2 big catchment area, the Alpine Rhine transports every year on average more than 3 millions m^3 sediment, mainly sand and mud into the Lake of Constance. The gravel mining data, published by the "Internationale Rheinregulierung" (1990), show that from that amount only approximately 40.000 m^3 are transported as bed load.

As a study area for the field investigation on the abrasion characteristics of the Alpine Rhine sediments, a 85 to 115 m wide study reach between the confluence of the Landquart River (Rhine km 23.2) and the Ill River (Rhine km 65.2) was chosen. This 42 km long reach has only one important tributary, the Tamina River. The reach shows repeated sections of degradation and aggradation downstream and upstream of the Ellhorn sill control structure at Rhine km 33.9 and the Buchs sill control structure at Rhine km 49.2. Since there are no important tributaries in the study reach, presenting a prismatic channel, it has been assumed to be good and long enough for a field study on fluvial abrasion. In agreement with the main objective of the study, that is the re-evaluation of the abrasion coefficient for the Alpine Rhine, two major aims of the field study have been defined :
- to attempt to find a stretch in the study reach, where fluvial abrasion may be taken as the predominantly cause of the downstream fining of bed load. The stretch should not be chosen immediately downstream of the confluence with the Landquart River, and ought to be long enough, so that expected low abrasion rates can be measured precisely enough, and since a variety of lithologies in the Alpine Rhine sediment was expected,
- to try to determine the abrasion coefficient for the whole sediment and separately for each important lithology, so that the effect of each lithology on the abrasion rate can be evaluated.

To gain a proper image of poorly sorted gravel sediments along the study reach, 11 practically equally spaced sites along it were chosen. They were situated on the *alternate gravel bars,* whose typical length was up to 100 m (photo of a typical one is given in Mikoš (1993)). While taking the samples site # 3 was not found suitable for the study, since the maximal clast size was much lower than at other sites. The grain-size analysis of

gravel sediments was carried out using sampling procedures proposed by Kellerhals and Bray (1971) and thus (see Table 1)
- at 10 sites a *sample* of the largest clasts, found on each alternate bar (over 20 grains),
- at 10 sites three differently coarse *transect samples* of the coarse surface layer (at least 100 grains greater than 8 mm, that lie under the straight line in the flow direction) and
- at 7 sites a *bulk sample* of the fine subsurface layer (approximately 0.5 m^3) were taken.

The main differentiating factor between in the field found lithologies was the percentage of contented quartz resp. calcite in each lithology. They were then grouped into 8 classes (in brackets are given their number fractions) :
1. metamorphic (3 %) - mainly quartz-rich gneisses and slates, subordinate also igneous rocks : diorite and peridotite
2. quartzites (16 %) - practically pure quartz or rocks dominated by vein quartz, partly with some calcite
3. calcite (13 %) - calcite from veins, frequently with some quartz
4. sandstone 1 (7 %) - practically pure sandstone or siliclastic rocks, with max. 5 % calcite
5. sandstone 2 (7 %) - sandstone or siliclastic rocks, with approximately 5 to 50 % calcite
6. limestone 1 (48 %) - limestone with approximately 5 to 50 % quartz or silica
7. limestone 2 (3 %) - without quartz, mainly very fine-grained limestone
8. dolomite (3 %) - mainly without quartz, very fine-grained

The first 5 lithologies are defined as *non carbonate* and the rest as *carbonate lithologies*.

site #	Rhine km	local bed slope* bed stability		largest clasts	transect samples	bulk sample
1	23.8	0.0031	E	yes	yes	yes
2	27.0	0.0031	E	yes	(yes)**	yes
3	30.2	0.0025	E	no	no	no
4	32.5	0.0024	A	yes	yes	yes
5	37.2	0.0047	E	yes	yes	no
6	42.4	0.0009	A	yes	yes	yes
7	47.0	0.0016	A	yes	yes	no
8	51.4	0.0024	A	yes	yes	yes
9	56.6	0.0025	A	yes	yes	no
10	60.8	0.0023	A	yes	yes	yes
11	64.9	0.0026	A	yes	yes	yes

legend : * ... measured in the 400 m long section   E ... erosion in the last 15 years
        ** ... later discarded   A ... aggradation in the last 15 years

TABLE 1 : Summary of the field work.

The largest clasts

As the clasts were not removed to the laboratory, 3 perpendicular axes of each clast (largest *a*, intermediate *b* and smallest *c*) were measured with callipers, and their lithology determined in the field. The intermediate axis of the maximal clast and the arithmetic mean of the intermediate axes of 20 largest clasts, respectively, were taken as the sample's mean diameters and are given in Figure 1. The sample in site # 2 exhibits a very low maximum clasts size and low variance. So it was assumed that site # 2 represents a locally sorted finer sediment and the transect samples from that site could not be used in further study.

The transect and bulk samples

The same measurements (3 axes, lithological determination) as for the samples of the largest clasts in the field, were done for the transect samples in the laboratory. Additionally the weight of each of 3356 sediment grains was measured. The mean density of the transect samples was 2670 kg/m^3, determined by weighing them under water and in air.

One of the main aims of the field study was to determine the abrasion coefficient for the whole Rhine sediment and separately for each important lithology. For that reason the transect samples, taken at each site, were put together and then split into eight lithologies, described before. Since a transect sample should have at least 100 grains to be representative, only for the most abundant lithology, namely limestone 1, succeded to build 9 new and representative transect samples (site # 2 was excluded). So the total number of the transect samples was 39.

Since the transect samples were taken for the study of fluvial abrasion of moving sediment, it was not necessary to apply modern methods (e.g. photo-sieving from Ibbeken and Schleyer (1987)) to get their grain-size distributions. That is why only a simple *frequency-by-number size distribution* for each of 39 transect samples was determined and the arithmetic mean of the intermediate axes of all clasts from one sample was taken as the sample's mean diameter. A frequency-by-number distribution can be transformed into a frequency-by-weight distribution of a fictious bulk sample of the parent bed material through an appropriate conversion method, described by Fehr (1986,1987), which has been sucessfully used in recent Swiss investigations. The method can be summarised as follows (Fehr (1986)) : determination of the coarser grains (from 8 mm up to 1 m) by a transect-by-number analysis in the field, its conversion to a volumetric-by-weight analysis of the subsurface and prediction of the finer grains by a fixed overlapping of the converted curve of the coarse grains and a Fuller curve.

DOWNSTREAM FINING - ALPINE RHINE RIVER

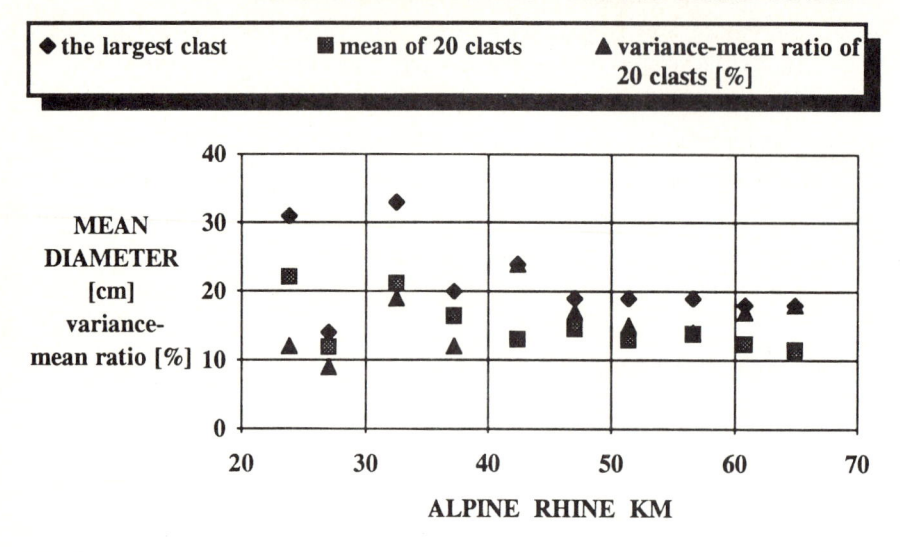

◆ the largest clast ■ mean of 20 clasts ▲ variance-mean ratio of 20 clasts [%]

Figure 1 : The intermediate axes of the samples of the largest clasts in the study reach.

DOWNSTREAM FINING - ALPINE RHINE RIVER

◆ frequency-by-weight bulk sample ■ frequency-by-number transect samples ▲ frequency-by-weight converted bulk samples

Figure 2 : The mean diameters of the transect samples, the converted bulk samples and the bulk samples in the study reach.

The result of such a conversion of the surface transect samples were the *frequency-by-weight size distributions* of so called *converted bulk samples*. The arithmetic mean was taken as the mean diameter of the size distribution.

The *frequency-by-weight size distributions* of the *bulk samples* were determined directly by a sieve analysis. The arithmetic mean was taken as the mean diameter of the size distribution. The mean diameters of different grain-size distributions of the samples are given in Figure 2.

3. DOWNSTREAM FINING

In general, the coarse sediment carried by streams appears to decrease downstream. The changes could be summarized as changes in :
- *the form of sediment grains* (their shape, roundness and surface texture descriptors)
- *the grain-size distribution* (its statistical parameters - moments : mean, variance, skewness, kurtosis etc.)
- *the petrographic composition*

Changes in the form of sediment grains

The analysis of 3 perpendicular axes of the 3356 grains from the transect samples shows, that on average, the ratio between perpendicular axes $a : b : c$ is $1.42 : 1 : 0.54$ and that the shape of the sediment grains is practical ellipsoidical. No roundness or surface texture descriptors were analysed, only two shape descriptors for all but the bulk samples were determined :

maximal projection sphericity : $Sph = \sqrt[3]{\dfrac{bc}{a^2}}$ ... [-] (3.1)

Corey shape factor : $SF = \dfrac{c}{\sqrt{ab}}$ ... [-] (3.2)

The analysis shows, that the grain shape :
- remains approximately constant along the study reach,

- is strongly different between different grain-size classes within the transect samples (grains, coarser than 64 mm, were moderately spherical, and grains, smaller than 64 mm, were well or very poor spherical), and
- is less strongly different between the non-carbonate (SF = 0.51 and Sph = 0.68) and carbonate lithologies (SF = 0.43 and Sph = 0.63).

Changes in the grain - size distributions

The most often used model for describing the downstream fining of river sediments is that according to the *Sternberg's exponential law*, which could be in his original form written as :

$$W = W_0 e^{-a_w x} \quad ... \text{[kg]} \tag{3.3}$$

where W_0 is the initial weight of the characteristic sediment grain at distance zero, and W is the weight of that grain at a distance x downstream, a_w being the weight reduction coefficient. The main disadvantage of such a model is the fact, that only the changes in the characteristic size of the sediment (largest or mean diameter) could be modelled and not the changes of the whole grain-size distribution.

The downstream fining itself is an effect of three important factors (Pettijohn (1975)) :

1. ***fluvial abrasion*** (general term meaning wearing away or attrition - applicable to any mechanical process of size reduction), where different researchers distinguished between processes like :
   - abrasion (restricted), impact and grinding - Marshall (1927)
   - solution, attrition, chipping and splitting - Wadell (1932)
   - splitting, crushing, chipping, cracking, grinding, solution and sand blasting - Kuenen (1956)
   - abrasion during transport and in place - Stelczer (1968), Schumm and Stevens (1973)

2. ***selective transport*** (selective entertainment, hiding effects, equal mobility) - Deigaard and Fredsoe (1978), Troutman (1980), Parker (1991a,b) etc.

3. ***weathering*** (physical, chemical) - Bradley (1970)

not mentioning a ***sediment supply*** by tributaries, bank erosion or bed degradation.

Following Church and Kellerhals (1978) and dependent on whether grain weight or size is involved in the particular study, exponential law in eq.(3.3) may be rewritten as a sum of relative contributions of the above mentioned factors of the downstream fining :

$$D = D_0 e^{-a_d x} \quad \text{... [mm]} \qquad a_d = a_{d1} + a_{d2} + a_{d3} \quad \text{... [-/km]} \qquad (3.4)$$
$$W = W_0 e^{-a_w x} \quad \text{... [kg]} \qquad a_w = a_{w1} + a_{w2} + a_{w3} \quad \text{... [-/km]} \qquad (3.5)$$

$$a_w = 3 a_d \quad \text{... [-/km]} \qquad \text{assuming constant grain shape and density} \qquad (3.6)$$
$$a_w \approx a_{w1} \quad \text{... [-/km]} \qquad \text{taking into account only fluvial abrasion} \qquad (3.7)$$

where $a_{d1}(a_{w1})$, $a_{d2}(a_{w2})$ and $a_{d3}(a_{w3})$ represent the size (weight) reduction coefficient due to fluvial abrasion, selective transport and weathering, respectively, D and W represent size and weight of the characteristic sediment grain, respectively, and x represents distance along the river. It has been shown (Mikoš (1993,in prep.)) that the eq.(3.6) is strictly valid only for single sediment grains and their characteristic diameters. If a sieve analysis has been used for the determination of characteristic diameters of the grain-size distribution, then the following conversion equation should be used instead of the eq.(3.6) :

$$a_w \cong 1.53 a_d \quad \text{... [-/km]} \qquad (3.8)$$

Therefore eqs.(3.4) and (3.6) were used to obtain reduction coefficients from the mean diameters of the samples of the largest clasts and the transect samples, respectively. Eqs.(3.4) and (3.8) were used with the mean diameters of the converted bulk samples and the bulk samples. All reduction coefficients are given in Table 2.

Changes in the petrographic composition

An new approach to the determination of the abrasion coefficients is represented by the study of changes in the petrographic composition along a river. Gölz and Tippner (1985) have determined the abrasion as a mass (volume) loss ΔV. The abrasion coefficient a_{w1} can be than defined by the loss of the original volume V_0 along distance x as follows :

$$a_{w1} = \frac{\ln V_0 - \ln(V_0 - \Delta V)}{x} \quad \text{... [-/km]} \qquad (3.9)$$

They also defined a ratio b between the volume loss of non-carbonate and carbonate sediment. Experimentally was found, that a value $b = 0.33 - 0.50$ would be appropriate for the Upper Rhine River sediment.

Because only the middle part of the study reach, between site 4 (km 32.5) and site 9 (km 56.6), can be used efficiently for a field abrasion study, the changes in the petrographic composition along the study reach were restricted to this part only. So first the carbonate content at the beginning $[C_0]$ and at the end $[C_s]$ of this 24.1 km long middle part of the study reach was determined. For that purpose the changes of the petrographic composition of different particle size classes of the transect samples were used. These changes, together with possible ratios $b = 0.33 - 0.50$ were used to determine corresponding total volume losses and from eq.(3.9) an interval of possible abrasion coefficients a_{w1-t} (see Table 2).

basis for the determination of reduction coefficients	R - squared [-]	size reduction coef. a_d [-/km]	weight reduction coef. a_w [-/km]
OLD FIELD DATA $a_{w1} \approx a_w$	-	0.015	$a_{w1-o} = 0.046$
NEW FIELD DATA			
largest clasts	0.831	0.0157	$a_{w-l} = 0.047$
mean of 3 transect samples	0.083	0.0030	$a_{w-t} = 0.009$
mean of 3 converted bulk samples - all lithologies :			
km 23.8 - km 64.9	0.558	0.0099	$a_{w-cb} = 0.015$
km 32.5 - km 51.4	0.745	0.0116	$a_{w-cb2} = 0.018$
mean of 3 converted bulk samples - limestone 1 :			
km 32.5 - km 51.4	0.726	0.0109	$a_{w-cb2-C} = 0.017$
bulk samples :			
km 23.8 - km 64.9	0.816	0.0154	$a_{w-b} = 0.024$
km 23.8 - km 32.5	2 points	0.0453	$a_{w-b1} = 0.069$
km 32.5 - km 56.6	0.637	0.0024	$a_{w-b2} = 0.004$
km 56.6 - km 64.9	2 points	0.0678	$a_{w-b3} = 0.104$
changes in petrographic composition of the transect samples	-	-	$a_{w1-t} =$ 0.014 - 0.026
LABORATORY DATA			
mass loss of the model mixtures	> 0.99	-	$a_{w1-m} =$ 0.005/0.018

Table 2 : A comparison of the reduction coefficients in the study reach.

In addition, an average carbonate and a non-carbonate content, respectively, of the transect samples in the study reach, but only for the movable part (fraction 8 to 128 mm) :
- carbonate content (limestone 1 & 2, dolomite) : $0.5 \, ([C_O] + [C_S]) = 55\%$
- non-carbonate content (other lithologies) : $0.5 \, ([NC_O] + [NC_S]) = 45\%$

can be used to express the total weight reduction coefficient as a sum of two partial coefficients, for carbonate and non-carbonate lithologies, respectively :

$$a_w = 0.55 a_{w-C} + 0.45 a_{w-NC} \quad \text{... [-/km]} \qquad \text{... (3.10)}$$

It should be first assumed that the petrographic compositions of the transect samples and the converted bulk samples, respectively, are similar. Then the results from field measurements for the mean of 3 converted bulk samples (see Table 2) may be used : the total weight reduction coefficient (all lithologies) a_{w-cb} and the partial coefficient for the carbonate part (represented by lithology limestone 1) a_{w-cb-C}. The weight reduction coefficient for the non-carbonate part can be determined for the subreach between sites 4 and 8 (km 32.5 - km 51.4) as :

$$a_{w-cb2-NC} = (a_{w-cb2} - 0.55 a_{w-cb2-C})/0.45 = 0.0191 = 1.14 \, a_{w-cb2-C}$$

$$\text{... (3.11)}$$

4. LABORATORY STUDY

Two very different research devices have been used so far for the abrasion experiments of river sediments in a laboratory : **tumbling mill** (Daubrée (1879), Schoklitsch (1933), Krumbein (1941)) and **circular flume** (Kuenen (1956)). The greatest practical problem associated with a tumbling mill abrasion experiment is the precise determination of the real abrasion distance sediment grains travel inside the mill. Nevertheless, when abrasion experiments with sediment mixtures are planned, provided the abrasion distance can be simply evaluated, tumbling mill should be given priority. Following these reasons, a common concrete mixer with a horizontally rotating cylindric drum was chosen as a research device. The diameter of the drum was 1.05 m, the axial length 0.80 m and the volume 690 l. The mixer was then adjusted to abrasion mill experiments. Full description of the laboratory study is to be found in Mikoš (1993). The problem of the determination of the abrasion distance was solved experimentally (Mikoš and Jaeggi (in prep.)). The aim of the abrasion experiment was to determine the abrasion coefficients for the Alpine Rhine sediment. For that purpose 4 different **model mixtures** from 2 to 128 mm and approx. 400

kgs in weight were used. They were sieved from the bulk samples from the sites # 1, 2, 4 and 6, and then rotated in the abrasion mill for 40 to 80 km. The abrasion coefficient a_{wl-m} according to eqs.(3.5) and (3.7) was obtained from the mass losses of the model mixtures (see Table 2).

5. DISCUSSION

The first important conclusion of the field study is the division of the study reach into three distinct parts :
- first part of the reach, approximately 10 km long (site 1 - site 4), with very high reduction coefficients, having mean bed slope 0.003 and exhibiting strong erosion,
- middle part of the reach, approximately 24 km long (site 4 - site 9), with low reduction coefficients, having mean bed slope in the upper part about 0.0023 resp. 0.0024, decreasing then downstream to 0.0012, and exhibiting moderate aggradation, and
- last part of the reach, approximately 10 km long (site 9 - site 11), with again very high reduction coefficients, having mean bed slope 0.0014 and exhibiting strong aggradation.

The abrasion coefficient $a_{wl-o} = 0.046 / km$, used until now for the study reach, corresponds to the newly determined weight reduction coefficient for the largest clasts $a_{w-l} = 0.047 / km$ (at R-sq = 0.831). This would also be true for the bulk samples in the whole reach $a_{w-b} = 0.024 / km$ (at R-sq = 0.816), if eq.(3.6) would apply instead of eq.(3.8). Anyway, the old value for the abrasion coefficient should be rejected. The differences among reduction coefficients for each part of the study reach suggest sorting effects. It is obvious that strong sorting effects are present at the beginning of the study reach, that is downstream of the confluence with the River Landquart, which is the Rhine's most important sediment supplier. That is not the case at the end of the study reach, that is upstream the confluence with the River Ill, another very important sediment supplier. But current investigations on sorting done by Hunziker (in prep.) show that strong aggradation also comprehends sorting effects. Due to strong sorting effects reduction coefficients cannot agree with abrasion coefficients. This is even more true, as the weight reduction coefficient of the samples of largest clasts takes practically the same value $a_{w-l} = 0.047 / km$ (at R-sq = 0.831). Which new values should be used then ?

In any case our search should be restricted to the middle part of the study reach, where fluvial abrasion should play a more important role. The surface transect samples provide a lower value $a_{w-t} = \boldsymbol{0.009 \ / \ km}$, but they may be predominantly defined through selective transport (downstream decreasing of the maximum clast size and above all vertical sorting) and therefore they cannot be used directly for a fluvial abrasion study (R-sq = 0.083 for the mean of 3 samples at the site). If it should be so, that would be purely chance. Transect samples should be first converted to converted bulk samples ($a_{w-cb} = \boldsymbol{0.015 \ / \ km}$ at R-sq = 0.558). But it should be noticed, that converted bulk samples are very sensitive to the "reasonable" choice of conversion parameters and therefore must be treated with care. This may be the reason why the reduction coefficient for converted bulk samples is much higher than that for bulk samples ($a_{w-b2} = \boldsymbol{0.004 \ / \ km}$ at R-sq = 0.637). It seems at that moment that for the purposes of a study of fluvial abrasion only densely arranged bulk samples can adequately represent the lithological and granulometrical variety in the sediment population. Thus the second conclusion of the field study is that the abrasion coefficient for the middle part of the study reach seems to be limited to be lower than $\boldsymbol{0.010 \ / \ km}$.

The difference between computed weight reduction coefficients for the non-carbonate and carbonate parts of the Rhine sediment is much smaller, as could be proposed from the changes of its petrographic composition. Remembering that sufficiently big samples must be taken, the downstream changes in a petrographic composition can be caused in general not only by abrasion, but by selective transport or additional sediment supply as well. Selective transport does not seem to be the right answer, since the grain shape in the study reach remains practically constant and is not a function of site or lithology. In fact, the grain-size distribution of the carbonate lithologies is a little bit coarser than that of the non-carbonate lithologies, and carbonate grains are less rounded, whatever can be neglected. So it can be assumed that the selective transport is lithologically independent. Then is the ratio b between weight reduction coefficients for carbonate and non-carbonate lithologies the same as the ratio for the abrasion coefficients for these two groups of rocks. Using eq.(3.11) is the ratio b as follows :

$$\frac{a_{w-NC}}{a_{w-C}} = \frac{a_{w1-NC}}{a_{w1-C}} = b \approx 1 \qquad \ldots (3.12)$$

In Table 2 used ratios b between volume losses of non-carbonate and carbonate parts of the sediment are then taken too low. The ratio b for the gravel sediments of the Alpine Rhine River lies outside the interval experimentally found for the Upper Rhine : b = 0.33 - 0.50 (Gölz and Tippner (1985)). That means, that the Alpine Rhine sediment is already selected, while less durable lithologies are already abraded away. The third important conclusion of the field study is that only one lithology is dominant, namely limestone with approximately 5 to 50 % quartz or silica, and that the differences between carbonate and non-carbonate lithology groups is only of lesser significance.

The observed variability in the petrographic composition may be, on the one hand, simply explained by the fact that the field transect samples were possibly big enough for the determination of the grain size distribution (Gale and Hoare (1992)) but too small for the more precise determination of the petrographic composition, and that rather bulk samples should be used instead. On the other hand, the strong variability seems to be caused by the strong degrading, which reenters old sediment deposits and mixes them with new sediment coming from upstream. So it can be concluded, that the changes of the sediment petrographic composition which took place gives an unreal high abrasion coefficient $a_{w1-t} \geq 0.025 \ / \ km$.

In the laboratory abrasion experiments for model mixtures determined weight reduction coefficients $a_{w1-m} = 0.005 \div 0.018 \ / \ km$ (see Table 2) correspond well with new field data and are even a little bit higher than the weight reduction coefficients determined for the bulk samples. The main conclusion may be that tumbling mill abrasion experiments remain useful tool for fluvial abrasion investigations.

ACKNOWLEDGMENTS

This project was undertaken with the financial support from the Canton St.Gallen and Scholarship Nr. 88-901-269-B from Swiss Federal Institute of Technology in Zürich.

REFERENCES

Bradley, W.C. (1970) : "Efect of weathering on abrasion of granitic gravel, Colorado River (Texas)", Bulletin of Geological Society of America, V.81, pp.61-80
Church, M. and Kellerhals, R. (1978) : "On the statistics of grain size variation along a gravel river", Canadian Journal of Earth Sciences, V.15, pp.1151-1160
Darcel, M. (1857) : Annales des ponts et chaussées, 3e série, V.14, p.108 cf.

Daubrée, A. (1879) : "Etudes synthétiques de géologie expérimentale", Dunod, Paris (German translation : Gurtl, A. (1880) : "Synthetische Studien zur Experimental-Geologie von A,Daubrée", Verlag von Friedrich Vieweg und Sohn)

Deigaard, R. and Fredsoe, J. (1978) : "Longitudinal grain sorting by current in alluvial streams", Nordic Hydrology, V.9, pp.7-16

Fehr, R. (1986) : "A method for sampling very coarse sediments in order to reduce scale effects in movable bed models", Proc. of the Symposium on Scale Effects in Modelling Sediment Transport Phenomena, Int. Assoc. of Hydr. Research, 25-28 August 1986, Toronto, pp.383-397

Fehr, R. (1987) : "Geschiebeanalysen in Gebirgsflüssen", Mitteilungen VAW ETH Zurich, Nr.92

Gale, S.J. and Hoare, P.G. (1992) : "Bulk sampling of coarse clastic sediments for particle-size analysis", Earth Surface Processes and Landforms, V.17, pp.729-733

Gölz, E. and Tippner, M. (1985) : "Korngrössen, Abrieb und Erosion am Oberrhein", DGM, V.29/4, pp.115-122

Hunziker, R. (1992) : "Modellierung des Schwebstoff- und Geschiebehaushalts", Schriftenenreihe EAWAG - Dübendorf, Schweiz, Nr.4, pp.163-181

Hunziker, R. (in prep.) : "Morphologisches Verhalten kiesführender Flüssen, untersucht mit Hilfe numerischer Modellierung", Diss. ETH Zurich

Ibbeken, H. and Schleyer, R. (1986) : "Photo-Sieving : A method for grain-size analysis of coarse-grained, unconsolidated bedding surfaces", Earth Surface Processes and Landforms, V.11, pp.59-77

Internationale Rheinregulierung, Rheinbauleitung Lustenau (Austria) (1990) : "Geschiebe-bilanzen 1966-1989"

Kellerhals, R. and Bray, D.I. (1971) : "Sampling procedures for coarse sediments", Journal of the Hydraulics Division, Proc. of the ASCE, V.97, pp.1165-1180

Kuenen, Ph.H. (1956) : "Experimental abrasion of pebbles. 2.Rolling by current", Journal of Geology, V.64, pp.336-368

Krumbein, W.C. (1941) : "The effects of abrasion on the size, shape and roundness of rock fragments", Journal of Geology, V.49, pp.482-520

Marshall, P. (1927) : "The wearing of beach gravels", Tran. and Proc. of the New Zealand Institute, Vol.58/1-2, pp.507-532

Mikoš, M. (1993) : "Fluvial abrasion of gravel sediments", Mitteilungen VAW ETH Zurich, Nr.123, 322 p.

Mikoš, M. (in prep) : "An approach to convert size reduction coefficients due to fluvial abrasion into weight reduction rates", submitted to the Journal of Sedimentary Research

Mikoš, M. and Jaeggi, M.N.R (in prep.) : "Experimental study on motion of sediment mixtures in a tumbling mill as typical laboratory abrasion set-up", to be submitted to the Journal of Hydraulic Research

Parker, G. (1991a) : "Selective sorting and abrasion of river gravel. I : Theory", Journal of Hydraulic Engineering, ASCE, V.117/2, pp.131-149

Parker, G. (1991b) : "Selective sorting and abrasion of river gravel. II : Application", Journal of Hydraulic Engineering, ASCE, V.117/2, pp.150-171

Pettijohn, E.J. (1975) : "Sedimentary Rocks", 3rd. ed., Harper & Row, 628 p.

Schoklitsch, A. (1933) : "Ueber die Verkleinerung der Geschiebe in Flussläufen", Sitzungsberichte der Akademie der Wissenschaften in Wien, Math.-naturw. Klasse, Abt. IIa, Band 142, Heft 8, pp.343-366

Schumm, S.A. and Stevens, M.A. (1973) : "Abrasion in place : a mechanism for rounding and size reduction of coarse sediments in rivers", Geology, pp.37-40

Shaw, J. and Kellerhals, R. (1982) : "The Composition of Recent Alluvial Gravels in Alberta River Beds", Bulletin 41, Alberta Research Council, Edmonton, Alberta, 151 p.

Stelczer, K. (1968) : "Der Geschiebeabschliff", Die Wasserwirtschaft, V.38/9, pp.260-269

Sternberg, H. (1875) : "Untersuchungen über Längen- und Querprofil geschiebeführender Flüsse", Zeitschrift für Bauwesen, V.25, pp.484-506

Troutman, B.M. (1980) : "A stochastic model for particle sorting and related phenomena", Water Resources Research, V.16, pp.65-76

Wadell, H. (1932) : "Volume, shape and roundness of rock particles", J. of Geology, V.40, pp.443-451

Wenthwort, C.K. (1919) : "A laboratory and field study of cobble abrasion", Journal of Geology, V.27, pp.507-521

RIVER CHANNEL ADJUSTMENT AND SEDIMENT BUDGET IN RESPONSE TO A CATASTROPHIC FLOOD EVENT (LAINBACH CATCHMENT, SOUTHERN BAVARIA)

Karl-Heinz Schmidt
Fachrichtung Physische Geographie, Freie Universität Berlin
Altensteinstr. 19, 14195 Berlin

Abstract: In summer 1990 the Lainbach catchment experienced a thunderstorm-induced flood with a recurrence interval far in excess of 100 years of a magnitude never registered before. The Lainbach has been an experimental field site for many years, so there was precise information on pre-flood conditions and the effects of the flood could be observed directly in the field. For the 120 m long measuring reach a new topographic map was drawn after the flood on the same level of resolution (0.1 metre contours) as the basemap produced in 1988. A comparison gave detailed information on river bed changes. Using pre-flood and post-flood longitudinal and cross profile surveys, sediment budgets were calculated for the mountain reaches of the Lainbach and Kotlaine. The main sources of eroded channel sediments lay upstream of the broken check dams. River segments with low slopes were generally most affected by high accumulation rates. In the geodetically surveyed reaches considerably more material was accumulated than eroded, resulting in an overall mass balance surplus of about 35000 m^3. A large part of this material was mobilized in the tributary channels and slope rills. Slope failures were only of minor importance, characterizing this catastrophic event as a channel flood. The flood caused a fundamental transformation of the channel system.

INTRODUCTION

Geomorphological, hydrological and sedimentological aspects of large floods can be viewed in the light of magnitude and frequency concepts. Events of very high magnitude and low frequency may change a fluvial system in equilibrium to a new system state, in which the processes of lower magnitude and higher frequency adapt to the conditions created by the rare flood, until a new equilibrium is reached. In their terminology CHORLEY & KENNEDY (1971) used the term "metastable dynamic equilibrium" for this sequence of system states. A formative event of extreme effectiveness that passes a given system threshold may result in a substantial river metamorphosis. The growth of geoscientific interest in high magnitude events has been highlighted in two recent compilations of papers on geomorphological, hydrological and sedimentological aspects of floods (BAKER, KOCHEL & PATTON 1988; BEVEN & CARLING 1989).

The present paper on a catastrophic flood in a small catchment describes and analyses the hydrological setting, the geomorphological adjustment processes in the long and cross profiles, aspects of sediment movement and deposition, the role of torrent control structures and the effects of their destruction. In small catchments like the Lainbach basin (Figure 1) flood results are brought about almost instantaneously in the course of a few minutes or hours. This makes direct observations and on-site measurements of flood related processes extremely difficult, and reliable field data on rare events are scarce. Two of the prime essentials of flood investigations are (1) to have precise information on pre-flood conditions, which is necessary for the evaluation of net changes caused by the rare event, and (2) to be at the flood site as early as possible, preferably already during the flood event. Both of these requirements were met during the Lainbach flood of 1990.

ENVIRONMENTAL SETTING

The Lainbach catchment is located in the Bavarian forealps near Benediktbeuern about 60 km south of Munich, Germany. It is tributary to the Loisach river and has a total catchment area of 18.8 km^2 (Figure 1). Altitudes range between 1800 and 675 metres. The catchment is underlain by varying lithologies from resistant limestones to soft mudstones; sediment is mainly supplied by thick glacial valley fills located in the central parts of the catchment (FELIX et al. 1988). Mass movements and debris flows from the over-steepened valley sides contribute material of very different size and angularity to the river system.

The Lainbach is a steep mountain torrent, a "Wildbach" in the German terminology, its gradient frequently exceeding 2%. The long profiles of the Lainbach and the Kotlaine, a major tributary, were geodetically surveyed by the regional water management authorities (Wasserwirtschaftsamt Weilheim) in 1984. The Lainbach and the Kotlaine have been regulated by torrent control structures. River bed slope has been reduced by constructing erosion control check dams with overfall heights of up to four metres. In the lower part of the Lainbach barriers were especially numerous with up to 20 units per kilometre.

For the detailed bedload movement and river adjustment studies, an experimental reach, about 120 m in length, was established and instrumented downstream of the confluence of the two principal tributaries, the Kotlaine (6.2 km^2) and the Schmiedlaine (9.4 km^2) (Figure 1) at an altitude between 750 and 740 m. The reach was geodetically and photogrammetrically surveyed in great detail before measuring activities started in 1988 and a topographic map showing height differences as small as 0.1 metres was drawn, which later served as background information for the evaluation of the flood results in the experimental reach (cf. Figure 5a). Before the flood the river profile had the characteristics of a step-pool-system (cf. Figure 3a), a type of river recently described by CHIN (1989) and GRANT et al. (1990). It consisted of a series of reaches with steeper gradients (up to 10 %) (steps) and intervening reaches with gentler slopes (< 2 %) (pools) (SCHMIDT & ERGENZINGER 1992). Field observations and measurements before the 1990 rare flood showed that the steps generally remained stable during small and moderate events.

Floods usually occur after snowmelt or long-lasting frontal precipitation and after severe local summer thunderstorms. Mean discharge at the measuring site is about 1 $m^3 s^{-1}$. During the twenty years of record the highest flood discharges measured at the Kotlaine and Schmiedlaine gages were 43 and 26 $m^3 s^{-1}$ respectively (BECHT 1991). Flood peaks in the Schmiedlaine are attenuated by karst aquifers in the upstream areas. At the measuring site no flood exceeded 60 $m^3 s^{-1}$.

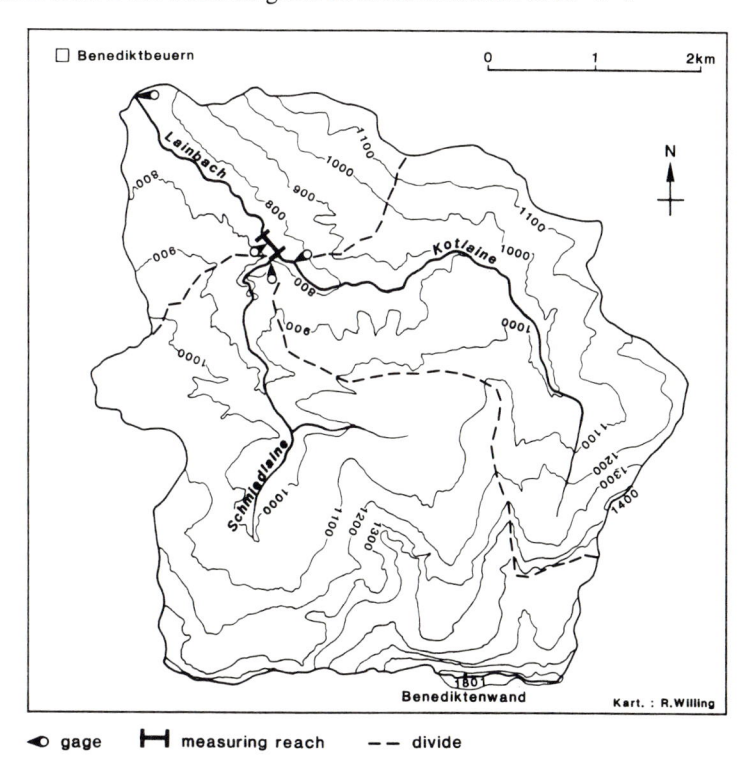

Figure 1: The Lainbach catchment with the location of the measuring reach

RECORD OF THE CATASTROPHIC FLOOD IN SUMMER 1990

During the night of June 30 to July 1 the Lainbach catchment experienced a flood of a magnitude never registered or observed before. The flood was caused by an extremely intensive local thunderstorm downpour mixed with hail. The thunderstorm covered an area of about 50 km^2. Within one hour the precipitation amount reached 90 mm with a maximum 15-minute intensity of 52 mm (BECHT 1991). Precipitation started at 8.15 p.m. with its intensity strongly increasing until 9 p.m. At that time several members of our research team started for the measuring site. We reached the main gage at river kilometre 4.4 at 9.15 pm, where the water level did not exceed 40 cm, which corresponds to a relatively low flood stage. Ten minutes later we reached the measuring site (2.2 km upstream of the main gage), where at that time the stage had already risen to 160 cm (low water level: 15 cm; highest stage previously observed: 80 cm). Automatic stage recording had been interrupted. All other downstream and upstream gages were also destroyed, so no chart record of the flood hydrograph is available. Beginning with our arrival (9.25 p.m.), changes in stage were recorded by direct observation resulting in the graph shown in Figure 2a. Surface velocities of more than 5 m s^{-1} were measured. Applying the MANNING-equation and our on-site observations a roughly approximated flood hydrograph was determined (Figure 2b). For a better comparison with the long-term record the estimated discharge of the 20-year flood is indicated. After our arrival stage continued to rise and at 9.45 flood water crossed the retainment wall and covered the forest road, peak discharge was reached at 10.05 p.m. with a stage of 235 cm and a cross-sectional area of 55 m^2. Estimated peak discharge amounted to 165 m^3 s^{-1}. This is almost three times the highest discharge measured before, the flood had a recurrence interval far in excess of 100 years. Expressed as specific yield, the extremely high value of about 10600 l s^{-1} km^{-2} was attained at the measuring site.

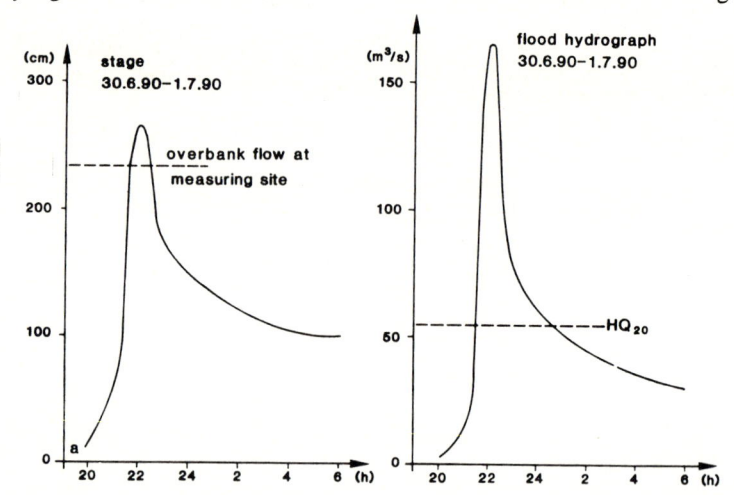

Figure 2: (a) Stage graph of the 1990-flood and (b) calculated flood hydrograph. The discharge of the 20-year flood (HQ$_{20}$) is indicated.

The high rainfall intensities resulted in quick runoff concentration with a very rapid rise of the hydrograph (Figure 2b). Peak discharge at the measuring site occurred

Figure 3: (a) Upstream view of the measuring reach before the flood. Two bridges had been constructed on steel girders for velocity measurements, sounding of river bed roughness changes and suspended sediment sampling. The bridge on the photo was built over a pool. Note the large boulders on the river bed in the step reach upstream of the bridge. The photo was taken a few days before the flood from the second bridge, which was located downstream over a step. (b) Measuring reach two days after the flood. A complete transformation of river bed morphology had occurred. The retaining wall and the forest road were destroyed along the lower part of the reach. Near the former location of the bridges the wall was still intact and could be used for stage observations during the flood and the determination of the cross-sectional flow area. Note the newly formed gravel bar on the left bank (right side of the photo) in the upper part of the reach.

less than two hours after the onset of precipitation and approximately one hour after the period of highest intensities in the headwater areas. The proportion of direct runoff surpassed 90% of total rainfall on measuring plots with lag times of only ten minutes (BECHT 1991). In river bends super-elevation could be observed, this was especially pronounced at the mouth of the Schmiedlaine, where the high water mark on the right bank was 90 cm higher than on the left bank.

Figure 4: Destroyed check dam at river kilometre 4.4 (main gage).

The flood caused serious damage to private and public property. The flood was so vigorous that most of our measuring equipment was destroyed and swept away including two specially constructed bridges resting on steel girders (Figure 3a) and the stationary antenna system for the radio tracer system (cf. SCHMIDT & ERGENZINGER 1990). Overbank flow occurred along the entire courses of the Kotlaine and Lainbach and most parts of the forest road, which had been stabilized by bank protection structures, were destroyed, undermined or covered by gravel. At the end of the flood we found ourselves on a peninsula formed by remnants of the forest road projecting into the expanded river bed (Figure 3b). Bridges were blocked by sediments and drifting logs causing overflow and bypass of the flood water. Undercuts were most frequent on the concave banks of bends. Most of the torrent control structures in the Lainbach and Kotlaine were destroyed (Figure 4), some of them were virtually shattered by the huge boulders (a-axes of more than 2 metres length, and weights of up to 5 tons) that were smashed against them. Flood water flowed around a number of check dams on one or both sides, which thus lost their slope control function. In the Kotlaine 40 % of the check dams were affected, in the mountain part of the Lainbach upstream of the main gage six of ten check dams were totally demolished, two check dams were by-passed by the flood waters. In the foreland on the alluvial fan, bed protection structures were not so much affected, twenty-three of thirty-four remained intact, many of them were covered by sediments. Major damage in the foreland was caused by blockage of a railway bridge, the discharge was dammed, the nearby village of Ried was flooded. Water flowed over the bridge, which was undercut and destroyed.

FLOOD EFFECTS IN THE MEASURING REACH

On July 5, only five days after the flood, the post-flood geodetic survey began. For the measuring reach a new topographic map was drawn on the same level of resolution as the base map produced in 1988 (Figures 5a, b). A comparison between the two maps showed which river bed changes had occurred as a consequence of the catastrophic flood and how the segments of predominant erosion or aggradation were distributed (Figure 5c).

On the 1988 map (Figure 5a) the thalweg of the Lainbach is clearly visible and there is a clear distinction between step and pool segments (e.g. step segments near the 746 m-contour and near the 747.5 m-contour and a pool segment downstream of the 746.5 m-contour and another one near the 745.5 m-contour). The steps are characterized by strong cross profile irregularities. The step-pool sequence reflects a system's state adjusted to the "normal" flood conditions of the past decades. River bed changes during the small and moderate floods were mainly confined to the thalweg and adjacent bank and bar positions. The measuring reach was a system in equilibrium, which received as much material from upstream as it lost to the downstream reaches when balancing the work done by several floods.

The extremely formative event of June 30, 1990, effected not only a modification but a complete transformation of river bed morphology. After the flood the step-pool sequence had disappeared and no thalweg is discernible on the reach (Figure 5b, see also Figure 3b). The supply-limited step-pool system changed to a braided system with abundant sediment supply. The shape of the cross profiles became much smoother and in parts of the reach a gently sloping surface was formed (e.g. between the 747 and 748 m contours). The delineation of the mass balance of the measuring reach (Figure 5c) shows that there are great contrasts in amounts of erosion or accumulation in the different sections of the reach. The 0 m contour separates areas of net erosion from areas of net aggradation. The spot of the highest sedimentation thickness (2 m) is located a few metres downstream of the mouth of the Schmiedlaine on the left bank (cf. Figures 3b and 10)). In the lower end of the experimental reach erosion dominates, because a few tens of metres downstream a check dam about three metres in height was destroyed initiating headward erosion.

In a comparison of the long profiles of the measuring reach before and after the flood the distribution of zones of erosion or accumulation is also clearly visible

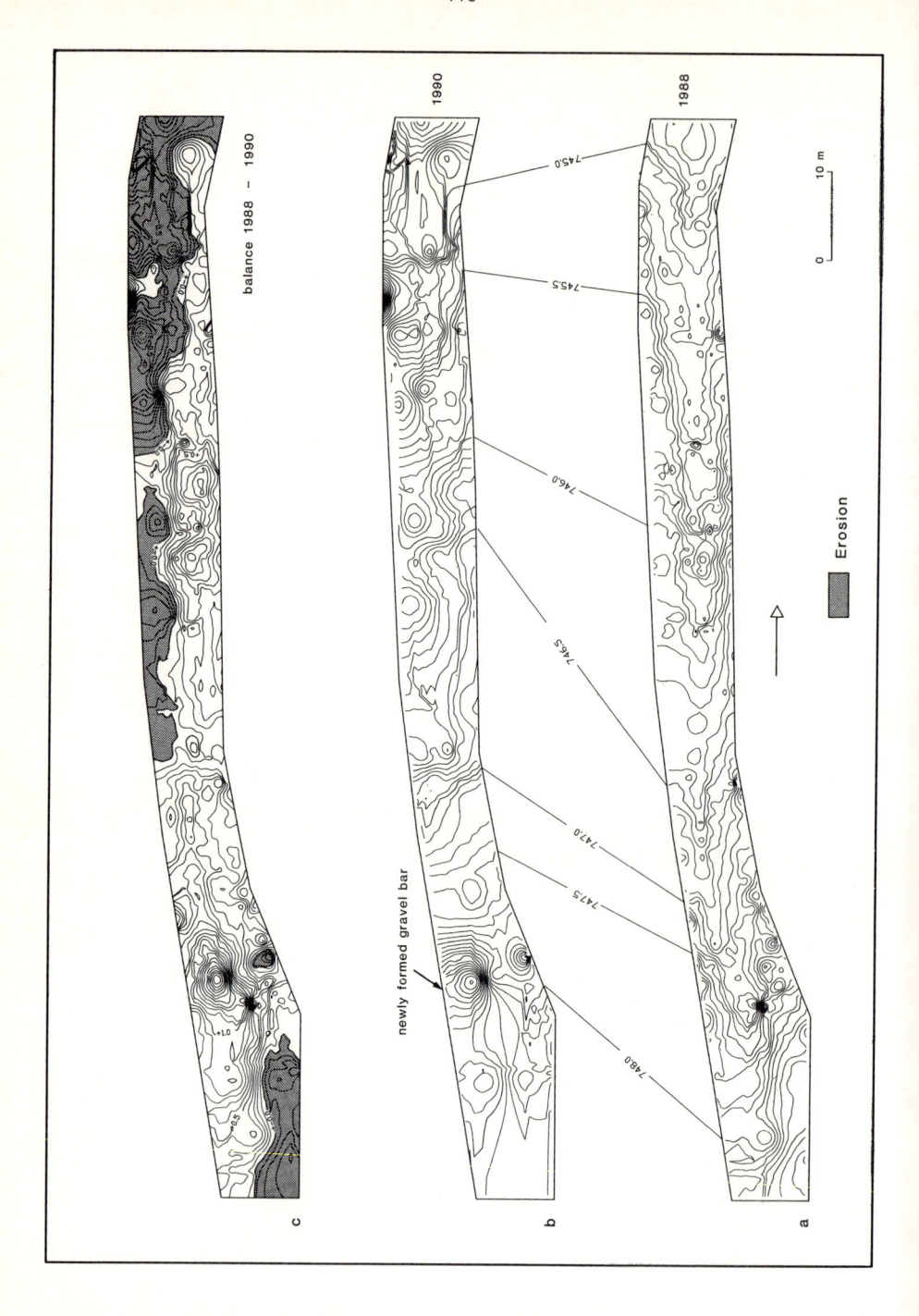

Figure 5: (a) Micromorphological map of the measuring reach before the flood and (b) after the flood and (c) sediment budget. Areas of erosion are shaded.

(Figure 6). Taken overall, aggradation processes have been dominant (Figure 5c) and the slope of the river bed has increased substantially (Figure 6). The volume of eroded material amounts to 156 m^3 and the volume of accumulated material to 670 m^3 resulting in an accumulation surplus of 514 m^3 (cf. Figure 9). The mean slope on the reach increased from 2 % before the flood to 3.5 % after the flood. The small and moderate events of the post-flood period have already caused some modifications of river bed morphology. In 1991 the knickpoint of headward erosion had moved as far upstream as the central part of the measuring reach (cf. BUSSKAMP, this volume).

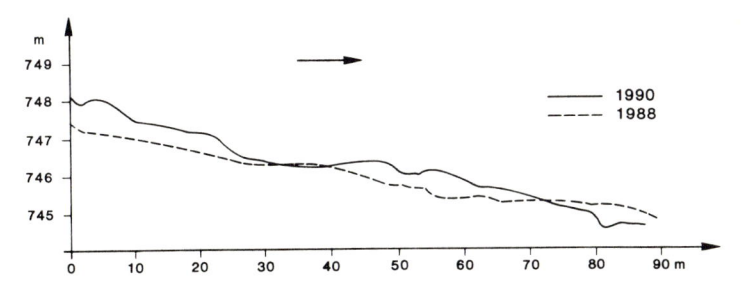

Figure 6: Slope of the measuring reach before and after the flood. In the lower part of the reach erosion occurred, because a few tens of metres downstream a check dam was destroyed causing headward knickpoint migration.

RIVER BED CHANGES AND SEDIMENT REDISTRIBUTION IN THE LAINBACH AND KOTLAINE

River bed changes were much more effective in the mountain parts of the river system than in the foreland area. This presentation focuses on the torrential reaches of the Lainbach and Kotlaine. More than 4 kilometres of the long profiles of the rivers from the Lainbach main gage (kilometre 4.4) to river kilometre 8.44 on the Kotlaine were geodetically surveyed immediately after the flood (cf. Figure 7). The survey also included 100 cross profile measurements, 60 on the Lainbach and 40 on the Kotlaine. Especially in the reaches with predominant accumulation the cross profiles possess relatively even shapes with no marked thalwegs. The post-flood long profile was compared with the 1984-long profile. There is no information on river bed elevation changes in the intervening years, but considering the flood record it is reasonable to assume that no major changes occurred during that time. The comparison of the profiles shows where river bed slope was changed and where stretches of erosion or aggradation are located (Figure 7). The diagram also provides a first rough approximation of the mass balance along the river course by determining the areas of erosion and accumulation and multiplying these data by the mean channel width.

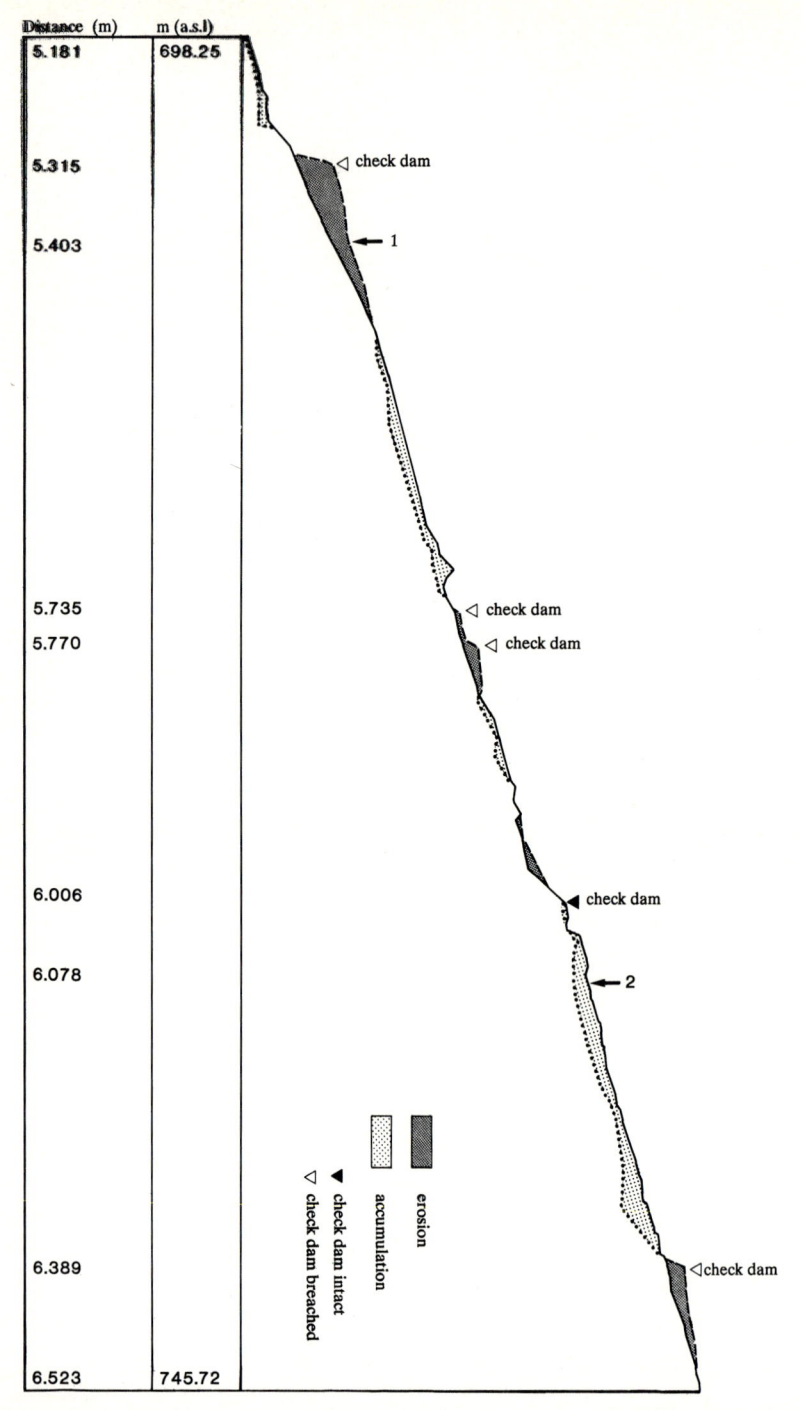

Figure 7: Longitudinal profile of the Lainbach from river kilometre 5.181 to river kilometre 6.523. There is a tenfold vertical exaggeration in the profile. Reaches of erosion or accumulation and the locations of the check dams are indicated.

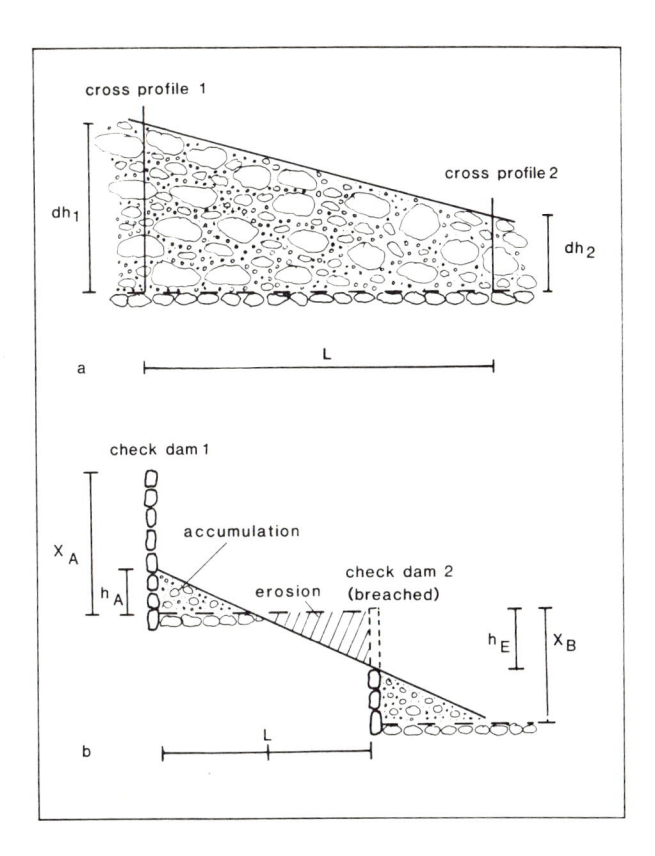

Figure 8: Illustration of volume calculations (a) between two surveyed cross profiles and (b) between check dams with stretches of accumulation and erosion.

Actually two other methods were applied to estimate the bedload budget of river sections (1) between two adjacent measured cross profiles or (2) between two check dams. The second, more reliable method was only applicable in river sections where check dams in close spatial proximity had remained intact. In most mountain segments, however, the "cross profile" approach had to be employed. For calculating the accumulated or eroded material, the volume was inferred to be a regular geometrical body between two cross profiles. The height of the accumulated (eroded) material at the deepest point of the profile is multiplied by the active channel width for both cross profiles. The mean area of the cross profiles is then multiplied by the length of the section (Figure 8a).

$V = 0.5 \cdot (w_1 dh_1 + w_2 dh_2) \cdot L$

V = volume of material; w = active channel width; L = length of section
dh = height difference of long profile surveys 1984 and 1990

The disadvantage of this method is that a rectangular shape of the profile is assumed. This source of error becomes much smaller when the lower and upper ends of the section are bounded by check dams. Then the volume of accumulated material is wedge-shaped and the second method of calculation can be applied.

$V_A = 0.5\, h_A w_m L$

V_A = volume of accumulated material;
h_A = height of accumulated material at the foot of the upper check dam;
w_m = mean width; L = length between check dams

When a check dam was destroyed during the flood, the calculation is composed of two elements (Figure 8b).

$V = (0.5 \cdot h_A L_1 w_{m1}) - (0.5 \cdot h_E L_2 w_{m2})$

V = net volume of accumulation/erosion
h_A = height of accumulation at upper check dam;
h_E = height of erosion at breached check dam;
L_1 = length of section with accumulation; L_2 = length of section with erosion;
w_{m1}, w_{m2} = respective mean widths

These calculations were made for almost 100 sections along the Lainbach and Kotlaine. The estimation of transported volumes resulted in sediment budgets for the surveyed river reaches. In the Kotlaine, between river kilometre 8.44 and 6.6 a total volume of 3584 m^3 was eroded, 30845 m^3 were accumulated, resulting in a net balance with a surplus of 27261 m^3, which is equivalent to a mean accumulation height of 0.9 m. In the mountain part of the Lainbach (river kilometre 6.66 to 4.4) 14967 m^3 were eroded and 23069 m^3 were accumulated, resulting in a sediment budget surplus of 8102 m^3 (Figure 9).

Along the river course the main sources of eroded sediments lay upstream of the broken check dams. Large quantities of stored material were mobilized. This applies especially to the mountain reach of the Lainbach, where two thirds of the torrent control structures were destroyed. Figure 9 shows the sediment budget and redistribution of the Lainbach from the confluence of the Kotlaine and Schmiedlaine downstream to the main gage. The stretch of greatest erosion lies upstream of the highest breached check dam. Here a volume of 6445 m^3 was removed and the river bed is 330 cm lower than before the flood. Erosion cut into the underlying bedrock. The stretch of greatest accumulation (12029 m^3) lies upstream of an intact barrier in a reach with a low river bed slope and a relatively wide valley bottom. Areas of low slope were generally most affected by high accumulation rates. More than 60 % of the entire accumulation in the Kotlaine (19219 m^3) are found in a reach about 400 m long with a below average slope. Here the mean height of accumulation amounts to 230 cm.

Figure 9: Sediment budget of the mountain part of the Lainbach from the measuring reach to the main gage. The vertical scale shows heights of accumulation or erosion per square metre. The absolute volumes are indicated for individual reaches. As the width of the individual segments varies, the volume represented by the area of the boxes is not exactly to scale.

121

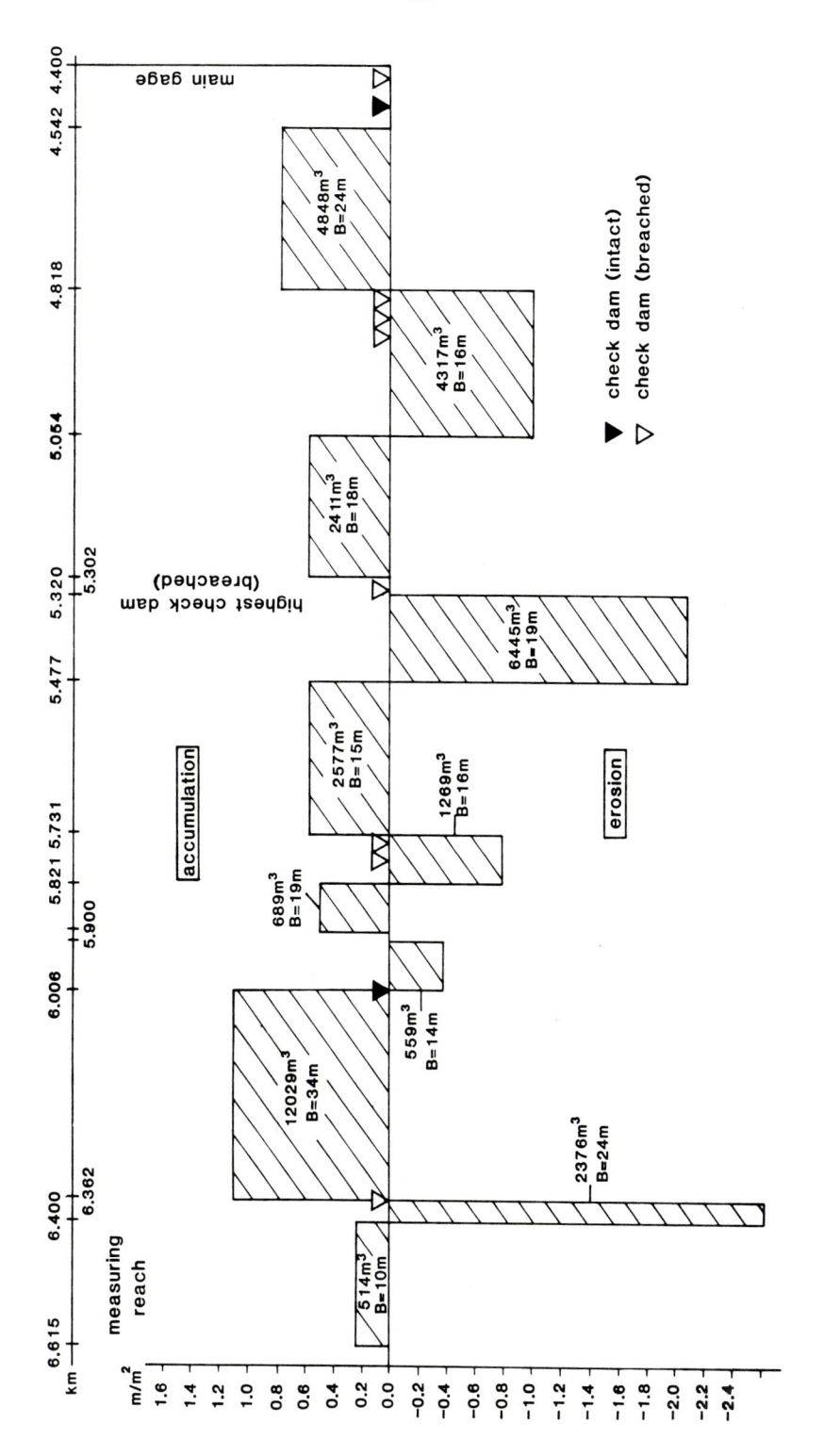

SEDIMENT SOURCES

In the geodetically surveyed reaches of the Kotlaine and Lainbach (kilometre 4.4 to 8.44) considerably more material was accumulated than eroded, resulting in an overall mass balance surplus of about 35000 m^3. Downstream of the main gage another 14000 m^3 were accumulated, and for the Schmiedlaine an additional surplus of bedload material of 5500 m^3 was estimated. Thus a total amount of about 55000 m^3 of flood sediments not originating from the main channel bed was deposited in the inspected reaches. The suspended sediment transported to the Loisach river is not included in the calculations.

It should be noted that the estimates of material eroded or accumulated in the different sub-reaches are subject to assumptions (e.g. regular geometrical bodies) which might cause error in the order of 10 to 20%. Another possible source of error can be viewed in the length of the time interval between the two surveys (1984-1990) of the longitudinal profile. During this time considerable amounts of bedload might have been temporarily stored in the channel bed. Especially short events of convective rainfall tend to have a relatively small sediment delivery ratio with high local storage (cf. BECHT & WETZEL 1989). But field observation in the pre-flood years showed that gravel bar morphology did not change fundamentally and the detailed measurements on the experimental reach gave no evidence of substantial aggradation when balancing the effects of several floods (ERGENZINGER 1992).

During the 1990 flood, the proportion of sediment supplied by the forest-covered parts of the Lainbach catchment was much more substantial than under conditions of less extreme rainfall. Measurements on experimental plots demonstrated that the unvegetated plots yielded only 2.5 times more material than the forest covered sites (20-25 t/ha compared to 7-10 t/ha), under "normal" conditions the ratio between unvegetated and vegetated plots lies close to 1000:1 (BECHT 1991). When the rates of sediment production of the test plots are extrapolated to the entire catchment a value of 21000 tons is obtained, which corresponds to a volume of only 8400 m^3 when bedrock density (2.5) is used for transformation or to a volume of 13000 m^3, when the loose packing is taken into consideration. In any case the total amount of accumulated material in the surveyed reaches is not accounted for.

Additional sources of sediment must be sought in alluvial fans on the sides of the main valley, in the few slope failures and in the channel beds of small tributaries. Indeed the field observations during the flood on the slopes along our measuring reach showed that all of the usually dormant slope rills had been activated and transported bedload material. Most of the small tributaries have vertically and laterally enlarged their channels releasing great amounts of material that had been stored for decades or centuries. The flood left scars in tributary rills and valleys that will be sources of sediment for floods in subsequent years.

Figure 10: Exposure in a ditch in the newly formed gravel bar. The ditch was excavated during the search for the magnetic tracers. The tracers lay at the base of the exposure. Above them there was an up to 25 cm thick layer of gravel and pebbles (below the water line) followed by a layer of pebbles and cobbles. Above the cobbles 50 cm of particles of mixed sizes with gravel, pebbles and few cobbles followed. The bar was covered by cobbles and boulders with a-axis lengths of more than 80 cm. The exposure was 110 cm high, the maximum thickness of the gravel bar was 2 metres.

RECOVERY OF THE MAGNETIC TRACERS AFTER THE CATASTROPHIC FLOOD

In the year preceding the flood, the river reach downstream of the confluence of the Kotlaine and Schmiedlaine had been a site of experiments with artificial magnetic tracers belonging to different weight and shape classes (GINTZ & SCHMIDT 1991, SCHMIDT & ERGENZINGER 1990, 1992). Before the catastrophic flood 960 concrete tracers of different weights (500 g, 1000 g) and shapes (rod, ellipsoid, ball, plate) had been placed in different positions in the river bed (pool, step channel, stoss sides of large boulders, gravel bar) to investigate the combined effects of these controlling factors on transport lengths and probabilities of erosion. After the catastrophic flood only few tracers were recovered. The gravel bar downstream of the mouth of the Schmiedlaine on the right bank (cf. Figure 5), where the 240 "gravel bar" tracers had been deposited, was completely eroded by the forces exerted by the Schmiedlaine flood waters flowing against the opposite bank. The tracers in the step channel were also affected by this erosional impact. Only 32 (13 %) of this sample were found below the newly formed gravel bar on the left bank of the Lainbach (cf. Figure 5). Here they were covered by 140 cm of flood sediments. A great number of the tracers (123 = 51%) placed in the pool position were also buried by these flood deposits. Only 19 (8%) of the tracers, which were placed on the stoss sides of large boulders on both sides of the river, were found below the large gravel bar. Here altogether 174 tracers (18% of the total population) were recovered.

Digging for the tracers below the huge pile of gravel and large boulders was very strenuous, and not all of the buried tracers were recovered. On top of the layer with the recovered tracers the gravel bar consists of layers of different grain size composition (Figure 10). Apparently the tracers were hit by a fast-moving thick bedload sheet and were moved downstream for not more than 20 metres and then covered by the upstream material.

The measuring reach and the Lainbach downstream to river kilometre 3.9 were systematically inspected with magnetic detectors. In the foreland reaches of the Lainbach only some sporadic search efforts were made in sites where accumulation had occurred. Here no tracers were found. In the 2.5 kilometres long reach, which was inspected in greater detail, only 62 (6.5%) tracers were found, many of them (27) downstream of the first broken check dam (river kilometre 6.389) in and on a gravel bar, which was formed during the flood. They had been transported 300 to 350 metres downstream from the positions of emplacement. The tracers were distributed over the entire gravel bar and were located in different depths from 5 to 90 centimetres below the bar surface. There was no relation between depth of burial and the point of deposition on the gravel bar. Tracers from all starting positions were found in the gravel bar deposits. The remaining 35 tracers were distributed over the inspected reach without major concentrations in specific sites. Burial depths varied from 10 to 60 centimetres. Taken overall 236 magnetic tracers (about 25%) were retrieved after the flood. A great number was probably transported out of the surveyed reach or was buried too deep for the magnetic detectors or for recovery.

During the pre-catastrophic flood measurements the platy tracers had the least chance of being eroded and had the shortest mean transport distances (SCHMIDT & ERGENZINGER 1990, 1992). During the catastrophic flood, when the entire river bed was in motion and most of the bedload particles travelled in suspension, the platy tracers had no transport disadvantage. The number (35) of tracers recovered downstream of the gravel bar was too small to make a statistically sound analysis of travel length differences between the shape categories. But it was noticeable that the plates travelled the greatest mean distance.

CATEGORIZATION OF THE FLOOD AND CONCLUSION

Diagnostic features of debris flows (WILLIAMS & COSTA 1988) are missing in the flood deposits. Only when large amounts of stored sediments were suddenly released by the destruction of check dams may the flood have acquired the character of a debris flow event for short distances. This was especially the case downstream of the highest breached check dam at river kilometre 5.315. On the whole the event was a water-dominated flood which carried large quantities of fines and coarse grained debris.

In a discussion on the effectiveness of floods NEWSON (1980) distinguished between "slope" and "channel" floods. He stated that high intensity rainfall with the occurrence of rapid runoff by surface routes is important for the generation of effective channel floods. The input of slope material is small leaving channel erosion and sediment redistribution by the floodwave as the major activity (NEWSON 1980, 10). Slope failures are of major importance in slope floods and hence the effects of this kind of event are mainly observed on the slopes. A prerequisite for slope floods is a high antecedent moisture content of the slope deposits with increased pore water pressures. When additional precipitation falls and soil saturation increases, slides and flows are triggered. In the 1990 Lainbach flood slope failures were mainly caused by undercuts in river bends, which largely occurred in later stages of the flood, major slope failures on the upper parts of the slopes in the unconsolidated glacial sediments were not observed. The highly formative effects of the flood in the tributary reaches and in the main channels make the event a characteristic example of a channel flood.

The effectiveness of the 1990 flood was not documented by just a modification of the channel bed but by a fundamental transformation of the channel system. In some reaches a supply-limited step pool system was transformed into a braided system with an abundant sediment supply. After the catastrophic flood the subsequent moderate and small events have begun to re-establish a stepped longitudinal profile not yet resembling the former more regularly spaced step-pool sequence. Two different scales of magnitudes of events have been observed in the system during the field investigations, on the one hand the catastrophic flood with a recurrence interval of more than 100 years and on the other the "normal" sequence of moderate and not extreme events. A great number of the latter type of events will be needed to re-install a well-operating step pool system, before an extreme event may again disrupt the newly established equilibrium and initiate a new cycle in the succession of changing system states. Whether the term "metastable dynamic equilibrium"

(CHORLEY & KENNEDY 1971) is applicable to this sequence is a matter of discussion, as the Lainbach is gradually returning to a system state similar to the level that existed prior to the event. One important change in the energy state of the Lainbach and Kotlaine has occurred as a consequence of the destruction of many of the torrent control structures. Higher slopes in the reaches between the destroyed check dams result in an increase in stream power and bedload transport effectiveness.

Acknowledgements: I should like to thank the Deutsche Forschungsgemeinschaft and the Freie Universität Berlin for the financial support of the post-flood measurements and the Wasserwirtschaftsamt Weilheim for copies of their field surveys. Special thanks go to the members of the Lainbach group who assisted during the field and computer work.

REFERENCES

BAKER, V.R., KOCHEL, R.C. & PATTON, P.C. (eds.) (1988): Flood Geomorphology. - New York (Wiley).

BECHT, M. (1991): Auswirkungen und Ursachen von Katastrophenhochwassern in kleinen, alpinen Einzugsgebieten. - Z. Geomorph. N.F Suppl. Bd. 89: 49-61, Berlin, Stuttgart.

BECHT, M. & WETZEL, K.F. (1989): Der Einfluß von Muren, Schneeschmelze und Regenniederschlägen auf die Sedimentbilanz eines randalpinen Wildbacheinzugsgebietes. - Die Erde 120: 189-202, Berlin.

BEVEN, K. & CARLING, P. (eds.) (1989): Floods - Hydrological, Sedimentological and Geomorphological Implications. - Chichester (Wiley).

BUSSKAMP; R. (1993): The influence of channel steps on coarse bed load transport in mountain torrents: Case study using the radio tracer technique PETSY. (this volume).

CHIN, A. (1989): Step pools in stream channels. - Progress in Physical Geography 13: 391-407, London.

CHORLEY, R.J. & KENNEDY, B. (1971): Physical Geography. - A Systems Approach. - London (Prentice Hall).

ERGENZINGER, P. (1992): River bed adjustment in a step-pool system. - In: BILLI, P., HEY, R.D., THORNE, C.R. & TACCONI, P. (eds): Dynamics of Gravel-Bed Rivers, 415-430, Chichester (Wiley).

FELIX, R., PRIESMEIER, K., WAGNER, O., VOGT, H. & WILHELM, F. (1988): Abschlußbericht des Teilprojektes A2, Sonderforschungsbereich 81 (TUM). - Münchener Geogr. Abhandlungen B 6, München.

GINTZ, D. & SCHMIDT, K.-H. (1991): Grobgeschiebetransport in einem Gebirgsbach als Funktion von Gerinnebettform und Geschiebemorphometrie. - Z. Geomorph. N.F. Suppl. Bd. 89: 63-72, Berlin, Stuttgart.

GRANT, G.E., SWANSON, F.J. & WOLMAN, M.G. (1990): Pattern and origin of stepped morphology in high-gradient streams, Western Cascades, Oregon. - Geol. Soc. Am. Bull. 102: 340-352, Boulder.

NEWSON, M. (1980): The geomorphological effectiveness of floods - a contribution stimulated by two recent events in Mid-Wales. - Earth Surface Processes and Landforms 5:1-16, Chichester.

SCHMIDT, K.-H., BLEY, D., BUSSKAMP, R., ERGENZINGER, P. & GINTZ, D. (1992): Feststofftransport und Flußbettdynamik in Wildbachsystemen.- Das Beispiel des Lainbachs in Oberbayern. - Die Erde 123: 17-28, Berlin.

SCHMIDT, K.-H. & ERGENZINGER, P. (1990): Magnettracer und Radiotracer -Die Leistungen neuer Meßsysteme für die fluviale Dynamik. - Die Geowissenschaften, 8: 96-102, Weinheim.

SCHMIDT, K.-H. & ERGENZINGER, P. (1992): Bedload entrainment, travel lengths, step lengths, rest periods studied with passive (iron, magnetic) and active (radio) tracer techniques. - Earth Surface Processes and Landforms 17: 147-165, Chichester.

WILLIAMS, G.P. & COSTA, J.E. (1988): Geomorphic measurements after floods. - In: BAKER, V.R., KOCHEL, R.C. & PATTON, P.C. (eds.): Flood Geomorphology: 63-77, New York..

THE INFLUENCE OF CHANNEL STEPS ON COARSE BED LOAD TRANSPORT IN MOUNTAIN TORRENTS: CASE STUDY USING THE RADIO TRACER TECHNIQUE 'PETSY'

Ralf Busskamp

B.E.R.G., Institut für Geographische Wissenschaften, Freie Universität Berlin, Grunewaldstr. 35, D- 12165 Berlin

Abstract

Direct measurements of the movements of bed load particles in a mountain torrent, using the radio tracer technique 'PETSY', show the influence of channel (bottom) steps on coarse bed load transport. This effect is demonstrated through an analysis of the step length of tagged cobbles. A significant change in the mean step length of particle is noted and the effect of marked channel slope changes on mean particle travel velocity revealed. The empirical results lead to the conclusion that particle transport is clearly controlled by local channel slope changes and that these are of fundamental importance in the transport of coarse bed load material.

Coarse bed load transport, channel step, probability of particle entrainment, step length, travel velocity

1. Introduction

It is technically difficult to carry out coarse bed load transport experiments under laboratory conditions. But it is even more difficult to quantify the influences of specific morphological channel characteristics on the transport process of such particles. Thus, different tracer techniques are generally used to analyse the movements of individual stones in natural channels (Bunte & Ergenzinger 1989). Clasts tagged with passive tracers make it possible to determine the total transport distance of the material (Hassan & Church 1990; Gintz & Schmidt 1991). Since the longitudinal channel profile of mountain torrents is very variable, the transport distance of the tracer is influenced, among other controls, by variations in the morphology of the channel bed. The effects of individual channel properties cannot be estimated. It is not surprising, therefore, that attempts to determine relationships between mean travel distances and transport-governing variables have yielded weak correlations (Laronne & Carson 1976; Hassan et al. 1992).

Investigations using the radio tracer technique make the correlation of the transport behaviour of individual particles with different known channel properties possible. The technique provides a high resolution "view" of the transport process, and make it possible to analyse the single step lengths of bed load particles. Einstein (1937) defined the single step length as the distance that a bed load particle travels without resting.

The existence of bottom steps in a channel profile strongly determines the local bed slope conditions. The aim of this paper is to demonstrate the influence of such bottom steps on the movement of particles.

The study site

Our research group has been studying coarse bed load transport in the Lainbach, an alpine mountain torrent, since 1988 (Ergenzinger et al. 1989; Schmidt et al. 1989; Schmidt & Ergenzinger 1992). Most parts of the Lainbach channel are regulated by torrent control structures. The road alongside the torrent is protected by embankments formed of large blocks of parent rock. River bed sediment retention dams, up to 4 m high, hold back great volumes of bed load material. The active channel width upstream of the sediment retention dams (barriers) is increased by raising the channel bed. Downstream of the vertical falls, energy is dissipated by turbulence (hydraulic jump) as the high velocity jet falls into the pool of water.

Since 1982, the barriers downstream of the confluence of the two Lainbach tributaries (Kotlaine and Schmiedlaine, km 6.589) had not been raised further (fig.1). This leads to the conclusion that sediment transport in this river section had attained an equilibrium state. The material transported through this section during floods was collected in sediment traps downstream of the sediment retention dam at km 4.400 and dredged if necessary.

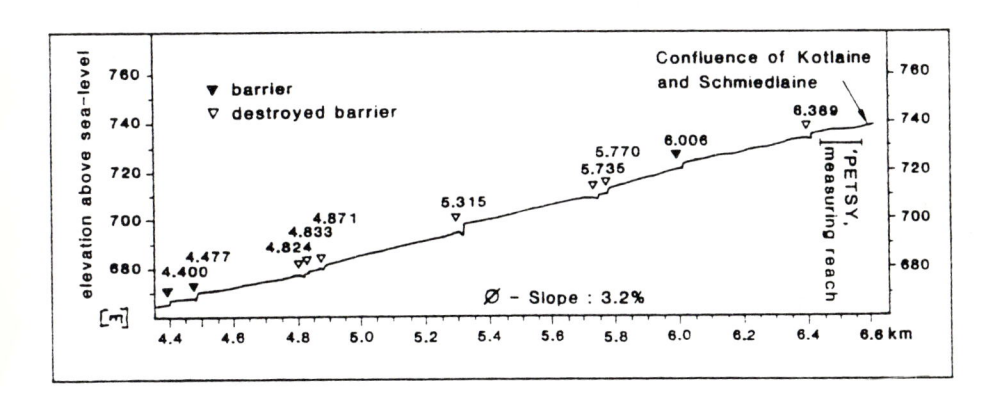

Fig.1: Longitudinal profile of the upper part of the Lainbach (surveyed 1984, vertical exaggeration 7.5 times).

On June 30th, 1990, a heavy thunderstorm occurred in the Lainbach catchment. Pluviometers in the 18.8 km^2 catchment area registered rainfall intensities of up to 52mm/15min (Becht 1991). The thunderstorm led to a large flood in the Lainbach. This discharge event has been classified by the regional water authorities (Wasserwirtschaftsamt Weilheim) as an event with a recurrence interval of more than 100 years.

Seventy percent of the barriers in the river section from 4.400 to 6.600 km were destroyed. Fifteen thousand cubic metres of bed material were eroded and about

23,000 m³ were accumulated in this reach of the river, resulting in a positive mass balance of 8000 m³ (cf. Schmidt, in this volume). Especially in the parts upstream of the destroyed barriers (cf. fig.1) the Lainbach cut into the sediments. This resulted in a markedly increased channel slope in these channel sections. However, the mean channel slope of the reach between the confluence of the two tributaries and the sediment traps remained unchanged.

2. The measuring method

The transport behaviour of individual coarse bed load particles has been analysed by using a variety of passive tracer techniques (Butler 1977, Stelczer 1981, Schmidt et al. 1989, Hassan et al. 1991). The active radio tracer technique 'PETSY' has been specifically developed to provide high spatial and temporal resolution of coarse particle transport in streams (Busskamp & Gintz 1993). The measuring system consists of battery powered radio transmitters which are implanted in test pebbles. The receiving equipment allows the path of the tagged stones to be monitored almost continuously. The measuring accuracy is $\pm$ 1 m when the particle is manually traced over a distance of 10 m. The technique makes it possible to determine the beginning of particle transport, the single step length, the duration of rest period and the cessation of bed load transport (Busskamp & Ergenzinger 1991).

For radio tracer measurements at the Lainbach, 'PETSY' was operated in an automatic mode (Schmidt & Ergenzinger 1992). A stationary antenna system has been installed at a 120 m long test reach (km 6.445-km 6.565). In this section, an antenna switch and a data logger are controlled by computer. The automatic operation of the receiving unit permits the simultaneous monitoring of the motion of up to 8 test pebbles, each having a different transmission frequency. However, this configuration reduces the accuracy of measurement to $\pm$ 2 m, and each radio tracer is monitored in time intervals of 10 seconds.

3. Results

A series of field experiments was undertaken in 1988 and 1989. Six floods were observed during this time. Sixty-four single step lengths of tagged particles were recorded; the hydraulic conditions were known at the beginning of motion. Additional data were obtained after the catastrophic flood in 1991. Nine floods of bed load transporting capacity yielded 66 single step length records together with the respective hydraulic conditions.

Detailed surveys of the test reach in '89 and '91 show strong local changes in bed slope (fig.2). The mean channel slope increased from 2% to 3.5%. Significant bed slope changes appeared in the lower part of the section. This can be explained by breaching of the barrier at km 6.389 during the flood on June 30th, 1990.

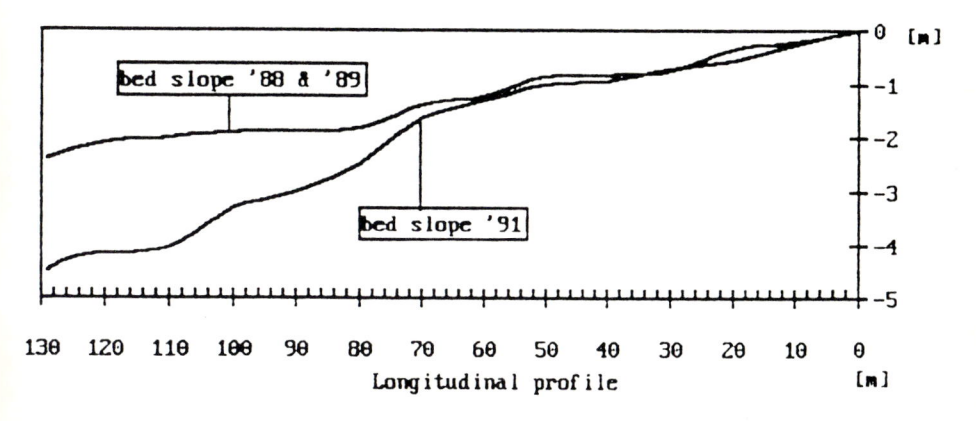

Fig.2: Longitudinal profile of the measuring reach.

Headward erosion of sediments stored behind the barrier was initiated after the destruction of the 2.5 m high sediment retention dam. Owing to the higher bed slope, the forces exerted on the bed surface layer by running water increased. The hydraulic conditions are described by the stream power approach: $\Omega = Q_u S \rho_w$, where Ω is the stream power (kg/m·s), Q_u is the unit discharge (m^2/s), S is the water surface slope (m/m), and ρ_w is the water density (kg/m^3). Higher stream power resulted in an increased bed roughness in the test reach. Fine bed material, previously accumulating upstream of the barrier, is transported through the reach. The increase in roughness was confirmed by direct measurements. As a measure of roughness the k_3-value was used (Ergenzinger & Stüve 1989). The k_3-value is a measure of roughness height obtained by calculating the moving maximum difference between three neighbouring roughness measurements. The roughness height is determined using a frame device (see Ergenzinger 1992). The k_3-roughness increased from a mean value of 5.7 cm to a mean value of 7.2 cm in 1991 (fig.3). This difference is statistically significant at a confidence level of 99%.

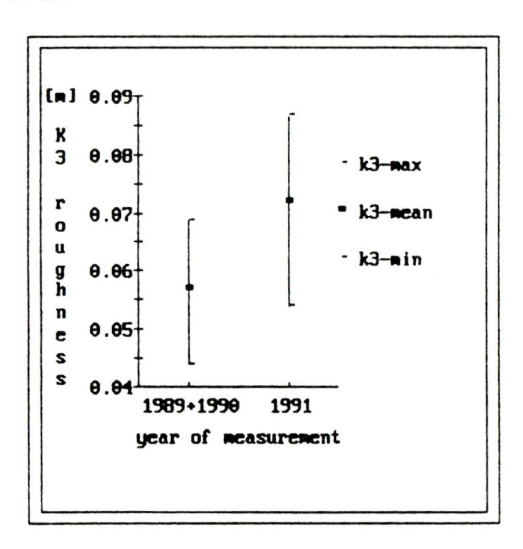

Fig.3: Change of the K_3-roughness (pre flood conditions 1989 + 1990 / post flood conditions 1991).

Probability of particle entrainment

A rise in bed surface roughness has a significant influence on the probability of particle entrainment (erosion) of bed material. Each particle step length starts with the attainment of critical hydraulic conditions. Fig. 4 shows the cumulative distribution of the calculated stream power at the moment and site of initial motion for all recorded particle steps. Thus, the cumulative relative frequency describes the probability of particle entrainment. A comparison of hydraulic conditions at the moment of initial motion of the test pebbles demonstrates the effect of higher bed surface roughness. The curve of '91 has shifted to the right of the results of measurements in '88 and '89 (see fig.4). The probability of erosion decreases at any specified level of stream power.

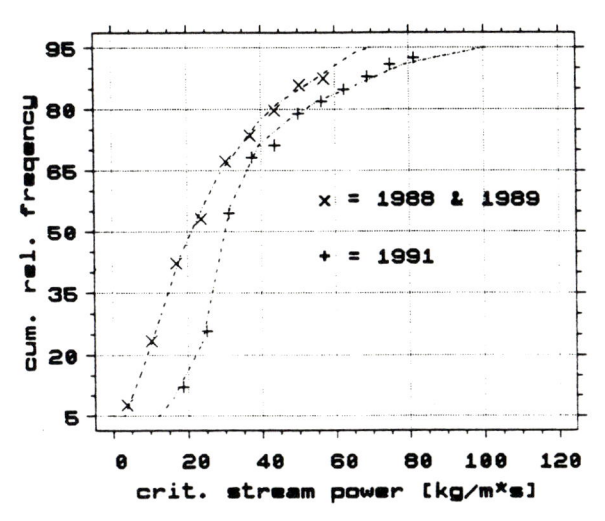

Fig.4: Cumulative frequency of particle entrainment in relation to stream power. (Lines are fitted by eye).

Single step length analysis

Low intensity bed load transport is controlled by series of single step lengths. Fig. 5 shows the results of step length analysis. The empirical values of the cumulative frequency distribution of step lengths are well approximated by a two-parameter gamma function. This result is confirmed by a chi-square (χ^2) test on a significance level of 5% ('88 & '89: $\chi^2=1.1$, d.f.=2, 59% probability of exceeding; 91: $\chi^2=6.4$, d.f.=5, 27% probability of exceeding). The parameters of the gamma distribution are determined by the method of moments.

From the data presented (fig.5) it becomes clear that the step lengths in '91 increase distinctly when compared to the data set of 1988/1989, although the discharge during the measurement periods was very similar in character (fig.6). The increase of single step lengths is examined using Mann/Whitney U-test. With a large sample test statistic Z=2.44, we can conclude that the step lengths in '91 have a significantly higher median.

The mean value of step lengths increases from 16.4 m to 23.7 m. The greater forces acting on the particles not only compensated for the higher roughness of the bed, but also had a positive influence on the length of transport.

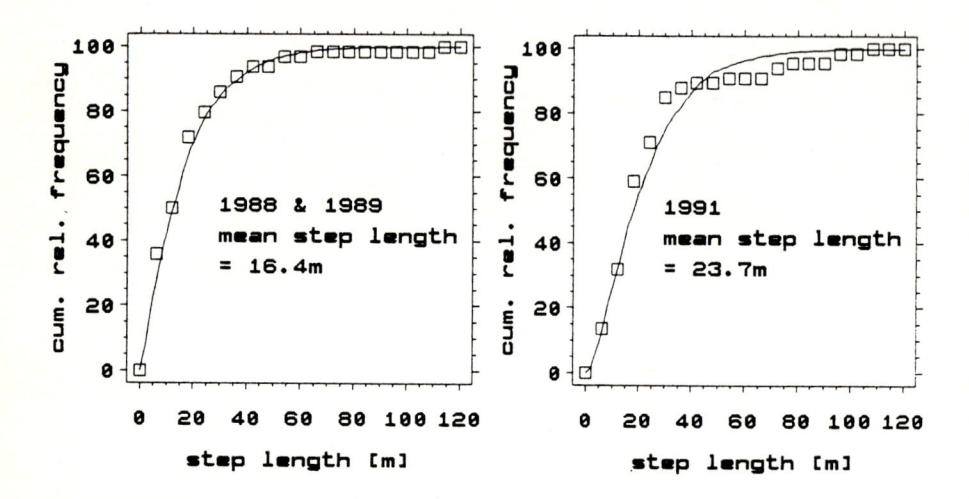

Fig.5: Cumulative relative frequencies of individual step lengths.

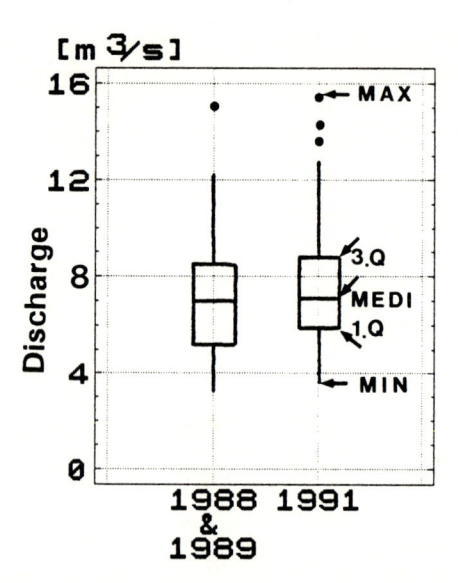

Fig.6: Discharge during the time of measurement. The diagram shows 5 statistical values, (1)minimum, (2)1st quartil, (3)median, (4)3rd quartil and (5)maximum.

4. Discussion

The behaviour of coarse bed load under moderate transport intensities has been analysed using the radio tracer technique 'PETSY'. The results demonstrate the significant influence of local slope changes on step lengths (see fig.5). It is clear that the total transport distance of bed load particles is the sum of all single steps during the periods of bed load movement (fig.7).

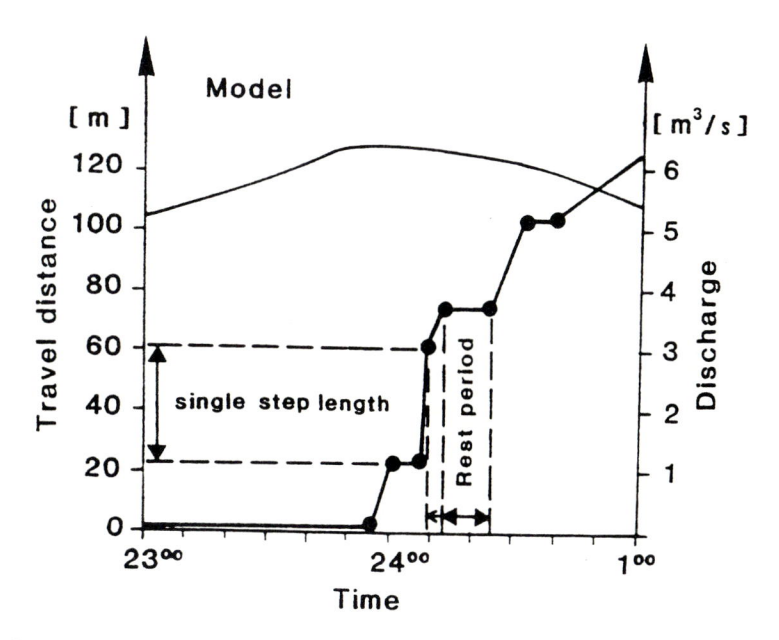

Fig.7: Lagrangian model of bed load movement in a time-travel-diagram.
(Flood on July 24th, 1991)

Bed load transport is divided into phases of movement (single step) and phases of rest (rest period). The duration of rest periods is dependent on the probability of particle entrainment, which is a function of a number of variables (e.g. hydraulic conditions, impulse exchanges by particle collisions, etc). Because the relative number of measurements in the different classes of erosion probabilities are equal and all classes were represented during the experiments (see fig.4), similar duration of rest periods are expected. This theoretical assumption is proved by the measured duration of rest periods. The Mann/Whitney U-test confirms the hypothesis of equal mean duration of rest periods on a confidence level of 99%. Hence, in the Lainbach torrent, there is a rise of total transport distance and, consequently, of the travel velocity of coarse bed load during post-flood events, when compared with pre-flood events of similar discharge. These results document the attenuating effect of bottom steps on bed load transport, and it should not make any difference whether the bottom steps are of man-made (e.g. barriers, weirs) or natural (e.g. step-pool sequences) origin. In addition, the results emphasize the importance of local slope effects, and point out the inherent limitations of using a general (mean) slope approximation. The mean slope value (3.2%) for the middle part of the Lainbach torrent did not change as a result of the destruction of the barriers (cf. fig.1). The travel velocity of particles, however, increases as a result of local slope changes (Laronne & Carson 1976). The mean travel velocity of bed material has a direct influence on the determination of bed load quantities. It is obvious from these investigations that the estimation of transport volumes will be erroneous if the assumed value of bed load travel velocity is determined by simply using average slope values.

ACKNOWLEDGMENTS

The author would like to acknowledge the financial support of the German National Science Associaton (DFG). Special thanks go to Prof. Dr. P. Ergenzinger and Prof. Dr. K.-H. Schmidt for reviewing the draft version. The author would like to thank all co-workers for their help in the field and anonymous reviewers for their comments.

Literature

Becht M (1991) Auswirkungen und Ursachen von Katastrophenhochwassern in kleinen, alpinen Einzugsgebieten. Zeitschrift für Geomorphologie Suppl Bd 89:49-61

Bunte K, Ergenzinger P (1989) New Tracer techniques for particles in gravel bed rivers. Bulletin de la Société Geographique de Liège 25:85-90

Butler PR (1977) Movement of cobbles in a gravel-bed stream during flood season. Geol. Soc. of Am. Bulletin 88:1072-74

Busskamp R, Ergenzinger P (1991) Neue Analysen zum Transport von Grobgeschiebe. Messung Lagrangscher Parameter mit der Radiotracertechnik (PETSY). Deutsche Gewässerkundliche Mitteilungen 35:57-63

Busskamp R, Gintz D (1993) Die Bestimmung des Geschiebetransportes mit Hilfe von Tracern in einem Wildbach (Lainbach / Obb.). In: Barsch D, Mäusbacher R, Pörtge K-H, Schmidt K-H (Hrsg) Messungen in fluvialen Systemen - Feld- und Labormethoden zur Erfassung des Wasser- und Stoffhaushaltes, Berlin [im Druck]

Einstein HA (1937) Der Geschiebetrieb als Wahrscheinlichkeitsproblem. Dissertationsschrift, Mitteilung der Versuchsanstalt für Wasserbau an der Eidgenössischen Technischen Hochschule in Zürich

Ergenzinger P, Schmidt K-H, Busskamp R (1989) The Pebble Transmitter System (PETS): first results of a technique for studying coarse material erosion, transport and deposition. Zeitschrift für Geomorphologie NF 33:503-508

Ergenzinger P, Stüve P (1989) Räumliche und zeitliche Variation der Fließwiderstände in einem Wildbach. Göttinger Geogr Abh 86:61-79

Ergenzinger P (1990) River bed adjustment in a step-pool system: Lainbach, Upper Bavaria. In: Billi R D, Hey R D, Thorne C R, Tacconi P (eds.), Dynamics of Gravel-Bed Rivers, Wiley & Sons

Gintz D, Schmidt K-H (1991) Grobgeschiebetransport in einem Gebirgsbach als Funktion von Gerinnebettform und Geschiebemorphometrie. Zeitschrift für Geomorphologie Suppl Bd 89:63-72

Hassan MA, Church M (1990) The movement of individual grains on the streambed. In: Billi R D, Hey R D, Thorne C R, Tacconi P (eds.), Dynamics of Gravel-Bed Rivers, Wiley & Sons

Hassan MA, Church M, Schick AP (1991) Distance of movement of coarse particles in gravel bed streams. Water Res.Res. 27, 4:503-511

Hassan MA, Church M, Ashworth PJ (1992) Virtual rate and mean distance of travel of individual clasts in gravel-bed channels. Earth Surface Processes and Landforms 17:617-627

Laronne JB, Carson MA (1976) Interrelationships between bed morphology and bed material transport for a small gravel-bed channel. Sedimentology 23:67-85

Schmidt K-H, Bley D, Busskamp R, Gintz D (1989) Die Verwendung von Trübungsmessung, Eisentracern und Radiogeschieben bei der Erfassung des Feststofftransports im Lainbach, Oberbayern. Göttinger Geogr Abh 86:123-135

Schmidt K-H, Ergenzinger P (1992) Bedload entrainment, travel lengths, step lengths, rest periods studied with passive (iron, magnetic) and active (radio) tracer techniques. Earth Surface Processes and Landforms 17:147-165

Schmidt K-H (1993) River channel adjustment and sediment budget in response to a catastrophic flood event (Lainbach Catchment, Southern Bavaria). - In: Comtag-Workshop-Publ., Berlin [in press]

Stelczer K (1981) Bedload transport, theory and practice. Wat. Res. Publ. Fort Collins, Colorado

THE SIGNIFICANCE OF FLUVIAL EROSION, CHANNEL STORAGE AND GRAVITATIONAL PROCESSES IN SEDIMENT PRODUCTION IN A SMALL MOUNTAINOUS CATCHMENT AREA

Karl-Friedrich Wetzel
Lehrstuhl für Physische Geographie
Universität Augsburg
D-86135 Augsburg

Abstract

In an area with highly erodible Pleistocene loose sediments, investigations on the significance of different sediment production processes were carried out. Data obtained during storm events show that channel storage and flushing dominate the sediment loads of the 10.1 ha research basin. Only about 20 % of sediments transported out of the basin are explicable by fluvial erosion on bare erosional scars. Calculations of sediment budgets for two-year periods by means of regression models for hillslope erosion by water and sediment loads of the receiving river, demonstrate that only 50 % of the annual sediment load of the river can be explained by sediment production processes due to rainfall and running water on hillslopes. The rest of the annual sediment load results from gravitational processes, which mainly occur in winter during ablation periods. Therefore an alternation of sediment production by gravitational processes in winter and removal by fluvial processes during summer can be assumed.

1. Introduction

Sediment production in river basins is a consequence of different erosion processes. On slopes, for example, sediment production is the result of splash, sheet, rill and gully erosion during rainfall. Gravitational processes such as rapid mass movement or soil creep also transport solids to the channel. Within the channel sediment is delivered by lateral erosion, downcutting or headwater retreat. Sediment loads which are recorded at the mouth of a catchment basin result from these different processes of sediment production. In many studies it is implied that fluvial erosion on slopes is the dominating sediment delivery process in the catchment area (ROEHL 1962; ROBINSON 1977; BOVIS 1978; SCHRÖDER & THEUNE 1984). Other studies calculate sediment loads of catchment basins by means of test plots or by soil loss equations (WILLIAMS 1977; BRYAN & CAMPBELL 1980; Van VUUREN 1982; HRISSANTHOU, 1988). Here, too, the assumption is made that fluvial erosion is the main sediment-producing process on hillslopes in mountainous terrain. But as SLAYMAKER (1977), DIETRICH & DUNNE (1978), YOUNG (1978), ASHIDA et al. (1981) and BOVIS & DAGG (1988) have shown, gravitational processes and channel storage cannot be neglected in sediment production.

However, little attempt has been made so far to determine the role of different processes in sediment production in mountainous terrain. In a study on the sediment budget of a small catchment area DIETRICH & DUNNE (1978) estimated the relative amounts of different processes by field mapping and process rates from other studies. Examples for field measurements of sediment production and sediment transport processes on different scales of observation were presented by LEHRE (1981), MILLINGTON (1981), CARONI & TROPEANO (1981) and HARVEY (1987). But in none of these studies are sediment delivery ratios given on the basis of storm-events over a period of time.

The current investigation should provide further information about the significance of fluvial erosion on slopes, channel storage and gravitational processes in sediment production in an area with highly erodible Pleistocene loose sediments for a sequence of storm-events. For a period of two years, the role of gravitational processes in sediment production has been estimated by means of models for slope erosion by water and sediment loads of the receiving stream.

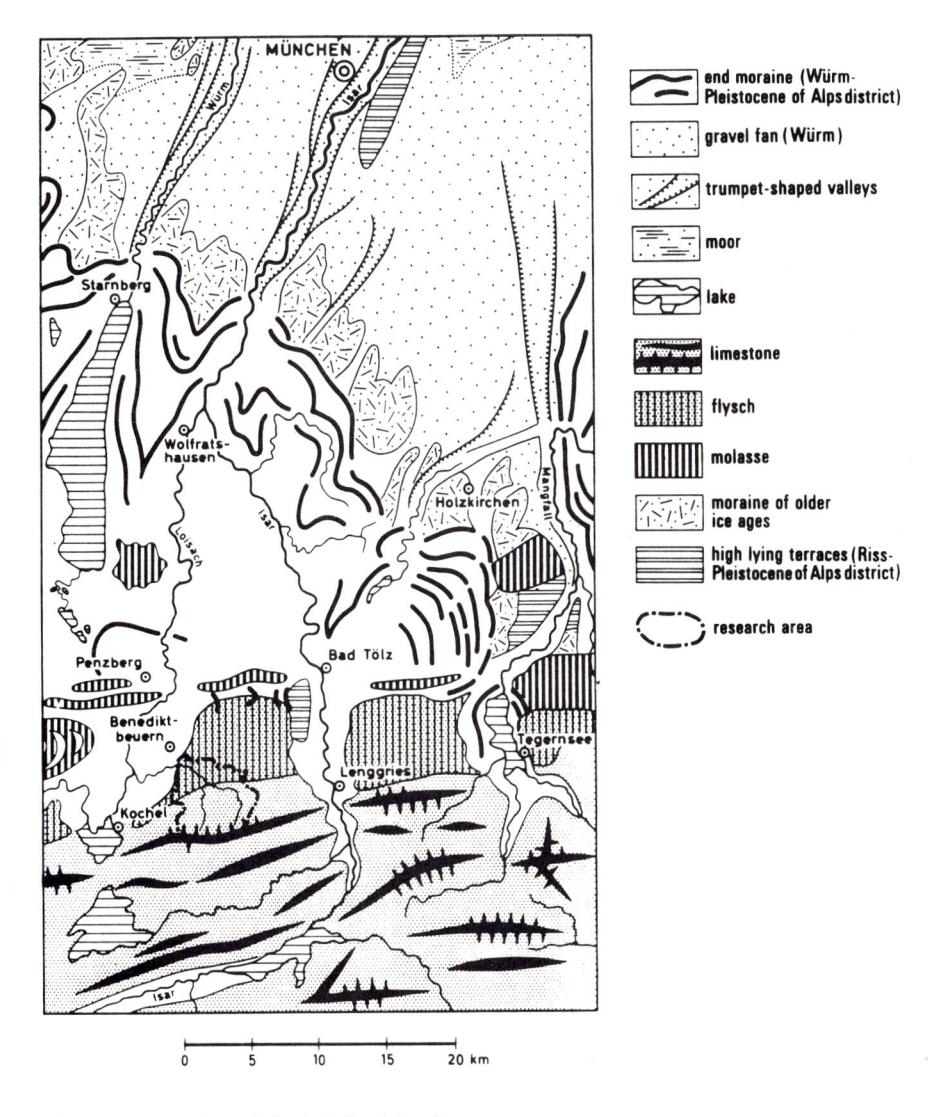

Fig. 1:   Location of the Lainbach basin

Field studies were carried out in the Lainbach basin (18,8 km^2), one of numerous east-west oriented valleys in the Northern Limestone Alps, which were partially blocked or filled by ice masses of large valley glaciers (Inn glacier, Isar-Loisach glacier) during the Pleistocene (fig. 1). A lateral lobe of the Isar-Loisach glacier advanced into the Lainbach valley from the west, depositing till on the valley slopes up to an altitude of about 1020 m (BECHT 1989). Pressure by the overlying ice masses caused the compaction of the sediment.

The mean annual precipitation (12-year record) is about 2160 mm (FELIX 1988). During the summer (May to October) convectional rainfall with many thunderstorms is dominant. Geomorphological processes in the unconsolidated sediments are occurring quite rapidly causing deeply incised erosion scars. Areas with Pleistocene loose sediments are the dominating sediment sources of the Northern Limestone Alps (BUNZA 1989) and investigations on sediment production processes seem to be promising.

2. Measurement methods and study area

The investigation deals with an estimation of the role of gravitational processes in sediment production. Direct measurements of gravitational processes are difficult and time-consuming. By contrast, fluvial erosion on the slopes can be measured comparatively easily by means of sediment traps. Extrapolation of the results of the sediment traps over the whole catchment area makes it possible to estimate the overall erosion by water on slopes. At the outlet of the basin sediment loads can be measured by sampling procedures and runoff recording. The ratio of slope erosion by water to sediment load at the basin outlet is the sediment delivery ratio of fluvial erosion for a single flood-event. For individual flood-events, subtraction of the amount of water erosion on the slopes from the solid load of the stream at the basin outlet allows the amount of channel reservoir change to be assessed, if the other processes of sediment production (downcutting, headwater retreat, lateral erosion) are negligible. Therefore, a research basin within the catchment basin of the Lainbach has been chosen, where torrent control has reduced channel erosion for over a hundred years.

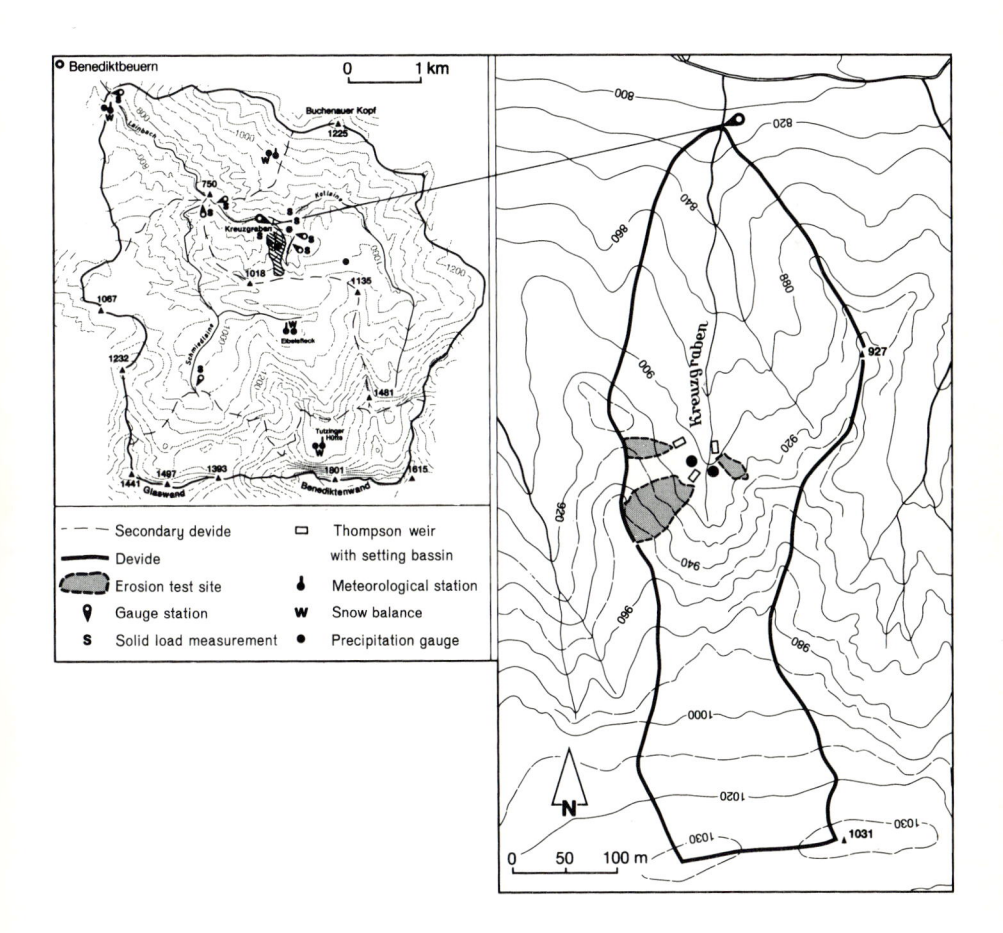

Fig. 2: Location and sites of measurement of the Kreuzgraben catchment.

Because of channel storage and depletion, it is not possible to determine the relative amounts of sediments delivered by gravitational processes on the basis of storm-events using the calculation discussed above. But for a certain period of time, e.g. one year, the significance of gravitational processes can be estimated by calculating the sediment delivery ratio for that period. For this, the assumption is made that most of the sediment produced on slopes by fluvial erosion is transported out of the basin. Under these assumptions, the difference between the sediment load at the basin outlet and the overall erosion by water on slopes is a function of gravitational sediment production.

The sediment delivery ratios were calculated for the Kreuzgraben subcatchment area, which has an area of 10.1 ha (fig. 2). The catchment has a mean altitude of 923 m a.s.l. and the local relief varies by 210.5 m. About 3.2 % of the area is free of vegetation, 13.4 % is covered by sparse grassland, and the remaining 83.4 % is densely vegetated by grassland and forest.

Table 1: Conditions at the four test-sites

	Site-1	Site-2	Site-3	Site-4
Area	793 m^2	243 m^2	1570 m^2	21000 m^2
Local relief	39 m	37 m	59 m	122 m
Slope angle	32 °	50 °	32 °	26 °
Veg. cover	< 10 %	< 25 %	> 85 %	100 %

Within the Kreuzgraben basin four types of slopes (tab. 1) can be differentiated by vegetation cover (WETZEL 1992). Slopes of type 1 are freshly incised erosion scars with bare soil (regosol), sparsely covered by some pioneer plants. On slopes of type 2, the vegetation cover is denser and grass begins to grow. The third stage in the succession comprises slopes densely covered by grass and some individual young trees. Soil development has reached the stage of calcaric regosol. Sites of type four are covered by forest. Deep calcaric cambisols indicate the absence of fluvial erosion.

Fluvial erosion on hillslopes was recorded at four test sites, each representing one type of slope. Divides on sharp-topped crests represent natural boundaries of the investigation sites. The test sites range between 250 m^2 and 1570 m^2 in area, except for site 4 (22,000 m^2). Runoff was recorded using V-shaped weirs via a electronic gauge level recorder (fig. 3). In front of the weirs sedimentation of the solids took place in settling basins during times of discharge. In summer the sediment load was determined by means of a spring balance, ideally after each storm. Meteorological data such as rainfall, air temperature and humidity, as well as soil moisture content were measured within the study area.

Fig. 3: Settling basin with Thomson weir at a test site

Discharge from the Kreuzgraben was recorded in an artificial channel with a gradient of about 5° and an asymmetric V-shaped cross-section. The water level was measured by a nitrogen pressure gauge. Because of the extremely high sediment concentrations and the large range of particle sizes, sampling procedure of the solid load was difficult. Therefore, an event-oriented manual sampling method has been chosen: wide-necked one-litre jars at the end of long rods were drawn repeatedly through the surge at the overspill of the artificial channel to obtain reproducible results. Overspill sampling guarantees a good mixing of water and sediment (LECKACH & SCHICK 1983). Only coarse grained materials with diameters greater than 4 cm passing the jars. Because of the grain size distribution of the Pleistocene sediments (over 90 % < 4 cm in diameter), the error of the sampling method is small.

3. Results

During a two-years period of observation (1989/90) the four test sites show a wide range of sediment loads (tab. 2). The highest amount of sediment was recorded at the test sites with sparse or no vegetation cover. The sites covered with grassland or forest had less than 1 % of the sediment load of the other sites. The vegetated sites showed only little variation. The material derived from these test sites is coarse grained. That indicates an origin of the sediments from the rills between the vegetated areas.

The solid loads from the test sites indicate that sediment production by fluvial processes is derived from barren or sparsely covered ground. Therefore, in the following, only results from the test sites with barren or sparsely covered ground are presented.

Table. 2: Half year's sediment loads [t/ha] of the four test sites.

Period	Site-1	Site-2	Site-3	Site-4
Winter 1988/89	1.76	3.52	0.04	0.0
Summer 1989	55.15	65.70	0.24	0.01
Winter 1989/90	1.37	3.17	0.0	0.0
Summer 1990	53.10	78.57	0.44	0.09

Throughout the year, there is a great contrast in geomorphological processes between the summer and the winter months. In winter fluvial erosion is a consequence of melting snow or advectional precipitation with low intensities. The sediment loads are usually low, because of the high portion of water infiltrating into the ground (HERRMANN 1978). Fluvial erosion in summer produces more than 90 % of the annual sediments collected by sediment traps. Peak loads can be related to convectional rainfall with short duration and high intensities. Figure 4 demonstrates the relation between sediment load and maximum rainfall intensity per 15 min. CARONI & TROPEANO (1981) also found that the amount of erosion on slopes with bare soil is related mainly to maximum rainfall intensity. Correlations of other parameters such as rainfall duration, net rainfall or erosivity of rain with sediment load are less significant, indicating that rain splash is the dominating erosion process. Many of the residuals of the regression line can be explained by different durations of storm precipitation, antecedent soil moisture conditions and a varying soil erodibility throughout the year (WETZEL 1992). In spring erodibility is high, largely owing to ground attacked by frost action. Similar interrelations were found by CARONI & TROPEANO (1981), HARVEY (1987) and YOUNG et al. (1990). During the summer surface sealing processes, exhaustion of sediment deposits and the development of a stone cover leads to lower sediment loads, even if rainfall reaches the same intensities as in spring (WETZEL 1992).

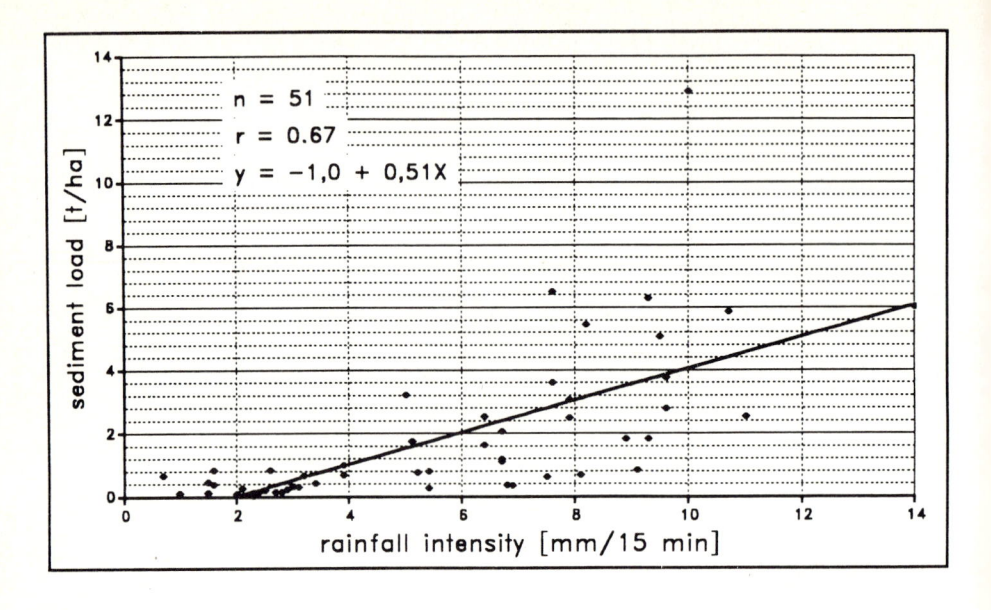

Fig. 4: The relation between maximum rainfall intensity per 15 min and sediment loads at test site-1

[Graph: sediment load [t/ha] vs rainfall intensity [mm/15 min]; n = 51, r = 0.67, y = −1,0 + 0,51X]

Because of the event-oriented manual sampling method only a few floods could be sampled at the gauging station of the Kreuzgraben basin. Although sediment loads of 20 floods were recorded, annual sediment loads of the Kreuzgraben cannot be specified. A comparison of sediment loads resulting from the 20 floods shows that a high sediment production at the test sites is not a condition for high loads of the Kreuzgraben. The sediment delivery ratio shows a wide range of variation from storm to storm. This indicates the dominant influence of channel storage or depletion on sediment loads at the basin outlet. To assess the significance of the different processes of sediment production, a continuous record of sediment loads over a period of at least one year is necessary. During a four-year period of observation (1988-1991) it has been shown that channel reservoir depletion is a consequence of floods with return periods of two years. This indicates a system which is in a state of dynamic equilibrium adjusted to the periodicity of storage and periodic removal (BECHT & WETZEL 1989, WETZEL 1992). In northwest England, HARVEY (1987) found a similar periodicity in a catchment basin with Pleistocene sediments.

4. Discussion

A continuous record of sediment loads for the two-year period of observation is calculated by means of rating curves for slope erosion and sediment transport in the stream. The fluvial erosion on the slopes is estimated by multiple regression models for site-1 and site-2 (WETZEL 1992).

Site-1: $y = 0{,}37 + 0{,}11*I + 0{,}06*R - 0{,}006*E$
           $R(sq) = 0{,}67 \quad SE = 0{,}50 \quad N = 30$

Site-2: $y = 1{,}59 + 0{,}06*R + 0{,}006*E - 0{,}25*M + 0{,}02T$
           $R(sq) = 0{,}62 \quad SE = 0{,}68 \quad N = 30$

I	=	Intensity of precipitation	$[mm*15min^{-1}]$
R	=	Erosivity of precipitation	$[N*h^{-1}]$
E	=	Energy of precipitation	$[kJ*m^{-2}*mm^{-1}]$
T	=	Duration of rainfall	$[h]$
M	=	Point of time	

The regression models require data of rainfall duration (T) and intensity (I), erodibility and energy of rain (R, E) and information about the exact time within the annual erosion cycle (M). The area of the hillsides represented by the test-sites was mapped aerial photographs. Because of the negligible erosion rates on vegetated sites, the overall erosion was calculated on the basis of the areal proportion of sites of type 1 and 2. For the Kreuzgraben site, a rating curve for sediment loads was set up by means of discharge data. Correlation analysis of the data indicates a close relationship between the sediment load and peak discharge of the Kreuzgraben site. Figure 5 shows the rating curve for sediment loads of the Kreuzgraben site. With the two models it is possible to estimate sediment delivery ratios of the Kreuzgraben site for each event on the basis of discharge and meteorological data, respectively.

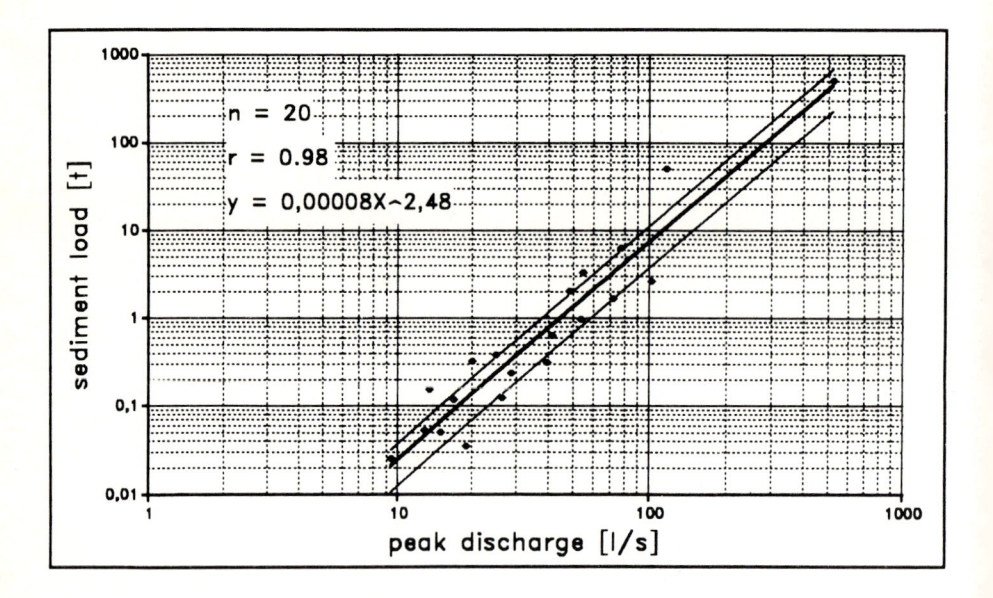

Fig. 5:   The relationship of sediment load and peak discharge at the Kreuzgraben site.

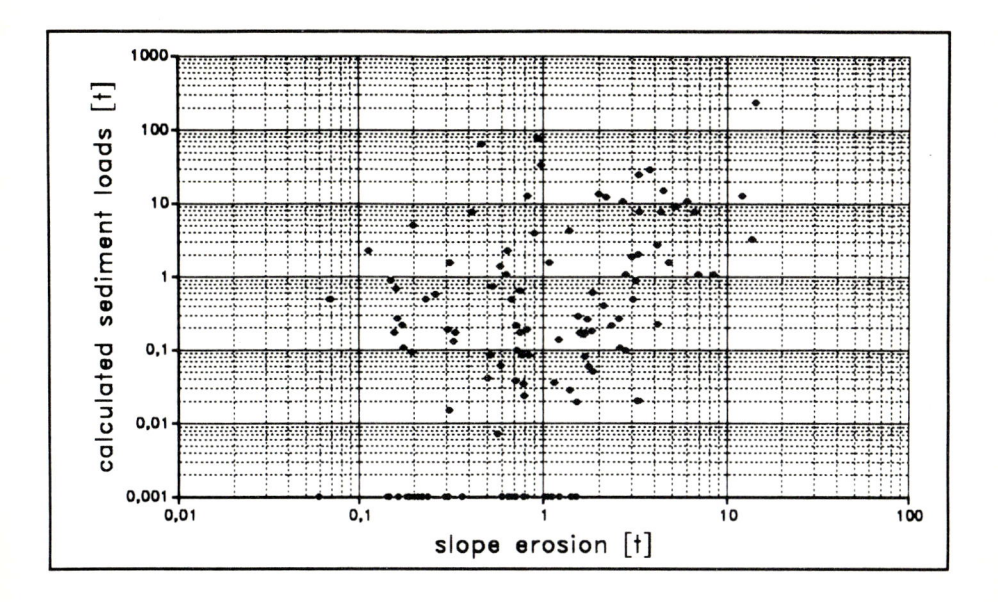

Fig. 6 : The relation between sediment production by fluvial erosion on slopes and solid load at Kreuzgraben gauge station. (No sediment load at Kreuzgraben is equivalent to 0.001 t in the scatter plot)

Focusing on single events a comparison of the computed data demonstrates a poor statistical relationship between slope erosion and sediment load of the torrent (fig. 6). Slope erosion only accounts for about 20 % of the variance of the sediment loads at the mouth of the basin. The reason for the poor correlation must be seen in the different processes producing sediment loads on the slopes and in the stream. On the slopes high loads are the consequence of convectional rainfall with high intensities. Such rainfall usually produces low or medium amounts of discharge, because of the short duration of convectional rainfall and amounts of precipitation less than 20 mm. All floods with sediment loads lower than 0.01 t are the consequence of short duration rainfall with high intensities and rainfall amounts up to 7 mm per event. A high proportion of the sediments eroded on the slopes is deposited in the channel system. On the other hand, continuous rainfall with high amounts of precipitation (> 70 mm/d) produces high peak discharge and sediment loads in the Kreuzgraben stream. But the overall erosion on the slopes is low in general, due to the low erosive effect of rainfall. In those cases, most of the sediments transported out of the basin result from channel reservoir clearance.

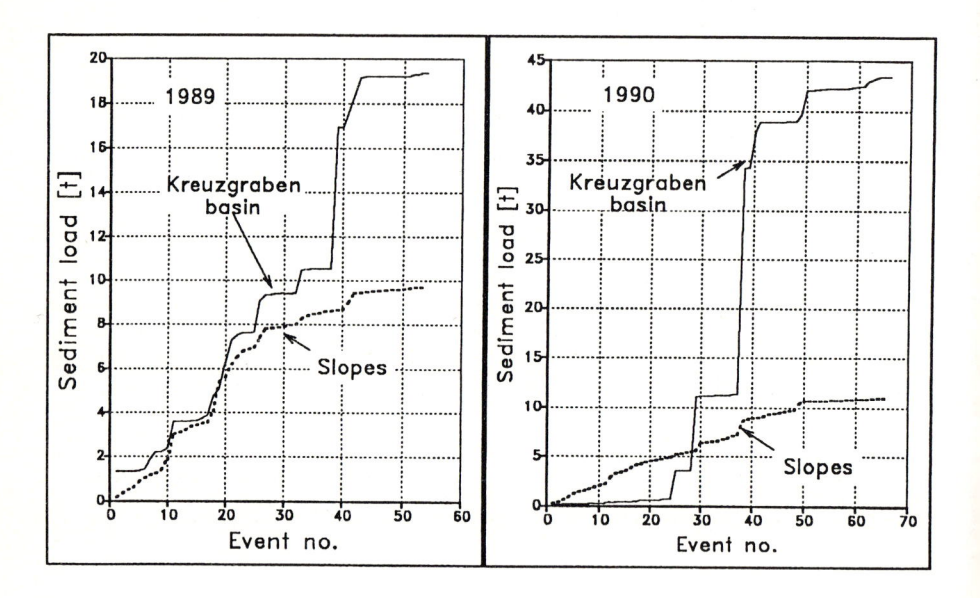

Fig. 7: Annual sequence of sediment loads on hillslopes and at the Kreuzgraben site

Furthermore, sediment budgets for the two years on the basis of computed data are analysed. Figure 7 shows that sediment loads of the Kreuzgraben site are much higher than the loads derived from the slopes by fluvial erosion. An avarage of 33 % of the sediment load of the Kreuzgraben site can be related to fluvial erosion on barren or sparsely vegetated slopes (tab. 3). The slopes with grass cover or forest were not taken into consideration in this calculation (cf. chapter 2). Despite their low erosion rates, about 7 % of the annual erosion may result from vegetated or forested slopes. However, more than 60 % of the Kreuzgraben's sediment load is not explicable by fluvial erosion. Consequently about 60 % of the annual sediment production has to be related to other sediment production processes.

Table 3: Sediment loads of the Kreuzgraben site and sediment production by fluvial erosion on the slopes

	1989	1990
Kreuzgraben	195.7 t	438.8 t
Fluvial erosion on slopes	98.4 t	111.2 t
Difference	97.3 t	327.6 t
Due to slope erosion	50 %	25 %

Because torrent control reduces downcutting and lateral erosion, gravitational processes seem to be the main sediment sources. From other mountainous basins it is known that gravitational processes may be the most important erosional agents (LEHRE 1981, HARVEY 1987). Two kinds of gravitational processes can be differentiated in the Kreuzgraben basin. Firstly spontaneous mass movement as a result of highly saturated soil during ablation periods in winter and secondly soil creep as a long-term uniform movement. Most of the mass wasting products are stored at the slope base and removed gradually by channel erosion.

Measurement of soil creep within a two-year period of observation is quite difficult because of the low rates of movement. DIETRICH & DUNNE (1978) calculated the annual sediment production by soil creep using mean rates of creep velocity as published in the literature. Additionally they took into consideration data about the total length of the hillside and the average thickness of the soil layer. If the relative amount of soil creep resulting in sediment production within the Kreuzgraben basin is estimated by a similar calculation, about 5 % of the annual sediment load is generated by soil creep. This value is in accordance with the sediment loads recorded at the neighbouring forested Waldlaine area (cf. fig. 2), where spontaneous mass movement could not be observed. Sediment loads here are tenfold lower than in the Kreuzgraben basin (BECHT & WETZEL 1992).

Spontaneous mass movement is a common event during ablation periods in winter in the Kreuzgraben basin, even if snow cover is high. The high infiltration rates of meltwater lead to saturation of the soil (HERRMANN 1978), which has already been loosened by frost action. Because of the strata series, a 30 - 50 cm thick loosened and weathered soil layer over compacted impermeable Pleistocene sediment, shallow translational slides result. Often they are followed by mud flows in the lower part of the slides, transporting the solids into the receiving stream. Although slides may also be observed in summer after heavy precipitation such slides are rare and of short extension.

5. Conclusion

The investigations in Pleistocene loose sediments in the Lainbach area show that gravitational processes are dominating sediment production. Slides are most frequent during ablation periods. The alternating climatic conditions in summer and in winter raise the production of sediments, because compacted and stone-covered soil surfaces are loosened and mass wasting products act as a sediment source for floods during summer. Sediment loads of floods in summer are dominated by channel storage behavior.

As a consequence of the results discussed above, rates of erosion or sediment production of catchment basins determined by means of test plots should be interpreted carefully. As a result of channel storage this is also valid for sediment loads of catchments that were calculated on a storm-event basis. Over periods of time erosion rates calculated by means of test plots are misleading, too, because sediment production by gravitational processes is not taken into acount. Therefore, we must assume that erosion has been underestimated. Monitoring of erosion on different scales of observation might be a possible method to improve the results. Gravitational processes at individual points may cause a mean denudation rate that is not typical for comparable drainage basins.

For torrent control, knowledge about erosion processes is needed to increase the effectiveness of control measures. All efforts to reduce water erosion by sowing grass and planting trees are futile if sliding occurs. Nevertheless, because of the dominant role of gravitational processes for sediment production, these should form the focus of future investigations.

References

ASHIDA, K., T. TAKAHASHI & T. SAWADA (1981): Processes of sediment transport in mountain stream channels. - IAHS Publ. no. 132, p 166-178

BECHT, M. (1989): Neue Erkenntnisse zur Entstehung pleistozäner Talverfüllungen. - In: Eiszeitalter u. Gegenwart, Bd. 39, S. 1-19, Öhringen.

BECHT, M. & K.-F. WETZEL (1989): Dynamik des Feststoffaustrages kleiner Wildbäche in den bayerischen Kalkvoralpen. - In: Göttinger Geogr. Abh., Bd. 86, S. 45-52, Göttingen.

BECHT, M. & K.-F. WETZEL (1992): The Lainbach catchment. Its physical landscape and development. - In: Münchener Geogr. Abh. Bd. B 16, S. 15-47

BOVIS, M.J. (1978): Soil loss in the Colorado Front Range: Sampling design and areal variation. - In: Zeitschr. f. Geomorph., Suppl. Bd. 29, S. 10-21, Berlin.

BOVIS, M.J. & B.R. DAGG (1988): A model for debris accumulation and mobilisation in steep mountain streams. - In: Hydrol. Sciences Journ., Vol. 33/6, S. 589-604.

BRYAN, R.B. & I.A. CAMPBELL (1980): Sediment entrainment and transport during local rainstorms in the Steveville Badlands, Alberta. - Catena Vol. 7, p. 51-65, Braunschweig

BUNZA, G. (1989): Abtrag in Wildbächen. - In: Informationsberichte Bayerisches Landesamt f. Wasserwirtschaft, 4/89, S. 81-91, München.

DIETRICH, W.E. & T. DUNNE (1978): Sediment budget for a small catchment in mountainous terrain. - In: Zeitschrift f. Geomorphologie N.F., Suppl. Bd. 29, S. 191-206, Berlin

FELIX, R. (1988): Die Niederschlagsverhältnisse. - In: FELIX, R., K. PRIESMEIER, O. WAGNER, H. VOGT & F: WILHELM (Hrsg.)(1988): Abfluß in Wildbächen. Untersuchungen im Einzugsgebiet des Lainbaches bei Benediktbeuern/Oberbayern. - Münchener Geogr. Abh., Bd. B 6, S. 69-280, München.

HARVEY, A.M. (1987): Sediment supply to upland streams: influence of channel adjustment. - In: THORNE, C.R., BATHURST, J.C. & R.D. HEY (ed.) (1987): Sediment transport in gravel bed rivers. - Chichester / New York.

HERRMANN, A. (1978): Schneehydrologische Untersuchungen in einem randalpinen Niederschlagsgebiet (Lainbachtal bei Benediktbeuern, Oberbayern). - In: Münchener Geogr. Abh., Bd. A 22, München.

HRISSANTHOU, V. (1988): Simulationsmodel zur Berechnung der täglichen Feststofflieferung eines Einzugsgebietes. - Inst. f. Hydrologie u. Wasserwirtschaft univ. Karlsruhe, Bd. 31, 248 S.

LECKACH, J. & A.P. SCHICK (1983): Evidence for transport of bedload in waves: analysis of fluvial sediment samples in a small upland stream channel. - In: Catena, H. 10, S. 267-279, Braunschweig.

ROBINSON, A.R. (1977): Relationship between soil erosion and sediment delivery. - IAHS Publ. no. 122, p. 159-167

ROEHL, J.W. (1962): Sediment source areas, delivery ratios and influencing morphological factors. - In: AIHS Publ. No. 59, S. 202-213, Gentbrugge.

SCHRÖDER, W. & CH. THEUNE (1984): Feststoffabtrag und Stauraumverlandung in Mitteleuropa. - Wasserwirtschaft 74, S. 374-379, Stuttgart

SLAYMAKER, O.(1977): Estimation of sediment yield in temperate alpine environments. - IAHS Publ. no. 122, p. 109-117

Van VUUREN, W.E. (1982): Prediction of sediment yield for mountainous basins in Colombia, South America. - IAHS Publ. no. 137, p. 313-325

WETZEL, K.-F. (1992): Abtragsprozesse an Hängen und Feststofführung der Gewässer. Dargestellt am Beispiel der pleistozänen Lockergesteine des Lainbachgebietes (Benediktbeuern/Obb.) - Münchener Geogr. Abh. Bd. B 17, 176 S.

WILLIAMS, J.R. (1977): Sediment delivery ratios determined with sediment and runoff models. - IAHS Publ. no. 122, p. 168-179

YOUNG, A. (1978): A twelve-year record of soil movement on a slope. - **Zeitschrift f. Geomorphologie** N.F., Suppl. Bd. 29, S. 104-110, Berlin

YOUNG, R.A., M.J.M. RÖMKENS & D.K. McCOOL (1990): Temporal variations in soil erodibility. - Catena Suppl. Bd. 17, S. 41-55, Cremlingen

AN ATTEMPT AT MODELLING SUSPENDED SEDIMENT CONCENTRATION AFTER STORM EVENTS IN AN ALPINE TORRENT

Kazimierz Banasik
Warsaw Agricultural University
Dept. of Hydraulic Structures
ul. Nowoursynowska 166, PL-02-766 Warsaw

Dagmar Bley
Freie Universität Berlin
Inst. für Geographische Wissenschaften
Altensteinstr. 19, D-14195 Berlin

Abstract

The present study is an attempt to apply a lumped parametric model for suspended sediment flow rates to a mountain river. The relationships between rainfall, runoff and suspended sediment are examined. By estimating the model parameters, hydrographs and sediment graphs of the Lainbach river were regenerated for measured rainfalls. Observed and regenerated graphs were compared using the integral square error.
The results show that the model parameters vary from storm event to storm event and a unique set of parameters cannot be applied. While the regeneration of the hydrographs can be classified in all cases as "very good" or better, only 3 out of 6 events show a "good" rating for the sediment graphs. The problem of modelling suspended sediment concentration in small, inhomogeneous watersheds is emphasized.

1 Introduction

Estimates of the temporal distribution of suspended sediment concentrations are essential for the prediction of water pollution in streams and sediment deposits in reservoirs. In the Lainbach catchment in the Bavarian forealps six storm events were examined to simulate the relationships between rainfall, runoff and suspended sediment flow.

The watershed is located in the Northern Limestone and Flysch Alps near Benediktbeuern, Upper Bavaria, about 60 km south of Munich. About 80 % of the catchment is covered by forest. The main sources of the sediments transported during storm events are several non-vegetated slopes in Pleistocene deposits. A detailed description of the area is given by Becht and Wetzel (1992).

The rainfall data used in the investigation were measured at the Melcherreisse gauge, which is close to the main sediment sources. Discharge data belong to a gauging station located below the junction of the two tributaries Schmiedlaine and Kotlaine. The watershed up to this gauge comprises an area of 15.8 km^2. At the same station a continuously recording photoelectric turbidimeter was installed (Bley & Schmidt 1991, Bley 1992). The turbidity readings were calibrated against discrete suspended sediment samples. Highly significant correlations between turbidity and suspended sediment concentration for individual flood events provided continuous and reliable data for the suspended sediment concentration (Schmidt et al. 1989, Schmidt et al. 1992).

The theoretical examinations are based on a lumped parametric model for single rainfall events (Banasik 1990, Banasik & Woodward 1992).

2 Procedure Description

The aim of the investigation was to estimate the parameters of a lumped parametric model for suspended sediment flow rates based on collected data and to use these parameters for "regenerating" the hydrograph and sediment graph for the measured rainfalls. The idea is shown in Fig.1.

The model consists of two parts; a hydrological submodel and a sedimentological submodel. The hydrological submodel is based on the estimated effective rainfall. The unit hydrograph method is used to transform the effective rainfall into the direct hydrograph. The sedimentological submodel uses a sediment yield formula and the unit sediment graph method to transform the sediment yields into sediment flow rates.

```
                              INPUT                      OUTPUT
STEP OF:             ───────▶ TRANSFORMATION ───────▶
```

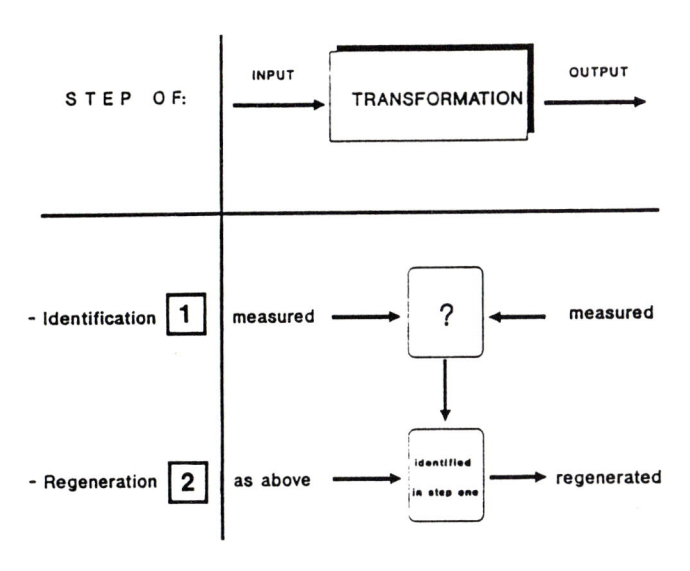

- Identification [1]   measured ──▶ ? ◀── measured
                                      │
                                      ▼
- Regeneration   [2]   as above ──▶ identified ──▶ regenerated
                                     in step one

Fig.1: Schematic sketch of the procedure application

The total effective rainfall, as equal to the direct runoff, and the sediment yield were known from measurements. For the computation the values were distributed into 10-minute time intervals of rainfall duration.

To assess the effective rainfall distribution two methods were examined; the SCS Curve Number method (USDA-SCS 1972) and an exponential formula used by Guillot and Duband (1972). The latter was applied in the further analysis, as it gave a better agreement between the measured and regenerated hydrographs.

The exponential formula for assessing the cumulative effective rainfall, $H_{(t)}$ in mm, is:

$$H_{(t)} = P_{(t)} - b \cdot \left[1 - \exp(-P_{(t)}/b)\right] \qquad (1)$$

where $P_{(t)}$ is the cumulative storm rainfall (in mm) and b is a parameter (in mm) fitted from the P- and H-values of the whole rainfall event. The amount of the effective rainfall in the time intervals was computed from:

$$\Delta H_j = \Delta H_{(t-\Delta t, t)} = H_{(t)} - H_{(t-\Delta t)} \qquad (2)$$
$$\text{for } t = j \cdot \Delta t, \quad j = 1, 2, \ldots n.$$

where Δt is the time interval (10 min) and n is the number of time intervals of rain duration.

The method of moments has been used to estimate the parameters of the model, consisting of N equal reservoirs, each with a storage parameter k, introduced by Nash (1957). The expression for the Instantaneous Unit Hydrograph - IUH of the Nash model is the following:

$$u_{(t)} = \frac{1}{k \cdot \Gamma(N)} \cdot (t/k)^{N-1} \cdot \exp(-t/k) \qquad (3)$$

where $\Gamma(N)$ is a gamma function for N-values as integers ($\Gamma(N) = (N-1)!$).

The Instantaneous Unit Sediment Graph - IUSG was based on the assumption that the source sediment concentration in each time interval varies linearly with the effective rainfall (Williams 1978). So the sediment yield for the j-th time interval (ΔY_j) can be estimated:

$$\Delta Y_j = Y \cdot \Delta H_j^2 / \sum_{j=1}^{n} \Delta H_j^2 \qquad (4)$$

where Y is the total sediment yield of the storm event (in Mg).

The instantaneous dimensionless distribution of the suspended sediment flow can be calculated by:

$$s_{(t)} = \frac{u_{(t)} \cdot c_{(t)}}{\int_0^{\infty} u_{(t)} \cdot c_{(t)} \, dt} \qquad (5)$$

where $c_{(t)}$ is the dimensionless sediment concentration computed from the following exponential formula:

$$c_{(t)} = \exp(-B \cdot t) \qquad (6)$$

where B is a fitted parameter of the equation and t is the time.

3 Computation Results and Concluding Remarks

The parameters estimated in the first step of the investigation (the parameter of the effective rainfall distribution - b, the parameters N and k of the IUH, and B of the IUSG), were used for regenerating the hydrograph, the sediment graph and the time distribution of the suspended sediment concentration (SSC). The values for six flood events are given in Table 1.

The model parameters were found to vary from storm event to storm event. A unique set of parameters was not estimated, since it was obvious that it could not be applied to the watershed.

The regenerated direct hydrograph has been found by convolution of the effective rainfall with the unit hydrograph and the direct sediment graph by convolution of the source sediment yield with the unit sediment graph. The SSC-graph was obtained by dividing the ordinates of the sediment graph by the ordinates of the hydrograph. An example of measured and regenerated graphs is shown in Fig. 2.

Table 1. Events investigated and computed parameters of the model.

Event No	Date	P (mm)	H (mm)	H/P (-)	CN-SCS (-)	b (mm)	N (-)	k (h)	B (1/h)
1	2	3	4	5	6	7	8	9	10
1	8.07.89	26.4	3.52	0.133	82.5	90.1	2.15	1.51	2.00
2	14.07.89	22.4	10.84	0.484	94.3	14.9	1.25	2.60	1.45
3	18.07.89	15.8	2.86	0.181	90.3	38.2	2.55	1.41	0.97
4	27.07.89	9.6	1.88	0.196	94.2	21.2	2.40	1.24	0.32
5	27.06.91	39.4	11.78	0.299	84.5	52.0	1.49	1.98	1.23
6	17.07.91	39.0	9.40	0.241	82.1	67.3	1.21	2.56	2.00
Mean values		25.4	6.71	0.256	88.0	47.3	1.84	1.88	1.33

P = rainfall
H = runoff
H/P = runoff coefficient
CN-SCS = curve number of SCS method
b = parameter of effective rainfall method acc. to Guillot and Duband
N, k = parameters of IUH (N-number of reservoirs, k-retention parameter)
B = parameter of IUSG (sediment routing)

Beside a qualitative visual comparison, e.g. peak reproduction, certain statistical measures, applied by Sarma et al. (1973), were used for the quantitative comparison between observed and regenerated graphs. The integral square error (ISE) of each event can be calculated by:

$$ISE = \frac{\left[\sum_{i=1}^{m}\left[Q_o(i) - Q_c(i)\right]^2\right]^{0.5}}{\sum_{i=1}^{m} Q_o(i)} \cdot 100\% \qquad (7)$$

where $Q_o(i)$ and $Q_c(i)$ are the i-th values of the observed and regenerated graphs, respectively, and m is the number of values in the output series.

The integral square errors of the individual events were then classified according to their regeneration goodness. They are given in Table 2.

Table 2. Evaluation of the sedimentgraph model for Lainbach watershed.

with ISE for SSC-graphs	Number of events		total
	and with ISE for direct hydrographs		
	$0\% \leq ISE \leq 3\%$	$3\% < ISE \leq 6\%$	
$0\% \leq ISE \leq 3\%$	-	-	-
$3\% < ISE \leq 6\%$	-	1	1
$6\% < ISE \leq 10\%$	1	1	2
$10\% < ISE \leq 25\%$	-	2	2
$25\% < ISE$	1	-	1
total	2	4	6

ISE = integral square error
SSC = suspended sediment concentration

The agreement between the SSC-graphs is usually poorer than between the hydrographs. There are, for example, two events with ISE-values for hydrographs not higher than 3% (this means their rating, according to Sarma et al. (1973) is excellent) but none of them shows the same rating for the SSC-graph. One of them is classified as good ($6\% < ISE \leq 10\%$) and the other one as poor ($ISE > 25\%$). In the other group of events with a very good rating ($3\% < ISE \leq 6\%$) for hydrographs, there is one event with a very good rating for the SSC-graph, one with a good rating ($6\% < ISE \leq 10\%$), and two of them have a fair rating ($10\% < ISE \leq 25\%$). On the whole we recieved for three out of six analysed events at least good integral square error ratings for both hydrographs and SSC-graphs.

The relatively poor agreement between measured and regenerated SSC-graphs is probably caused by two major factors, which could not be taken into account in a lumped parametric model:

- the sources of sediment in the Lainbach watershed are of point type (mainly some non-vegetated slopes) and are not uniformly distributed throughout the watershed,
- the amount and intensity of rainfalls taken from only one rain gauge might not be adequate for some of the events.

The simulation might be inproved by combining two separate models for the two subwatersheds, Schmiedlaine and Kotlaine.

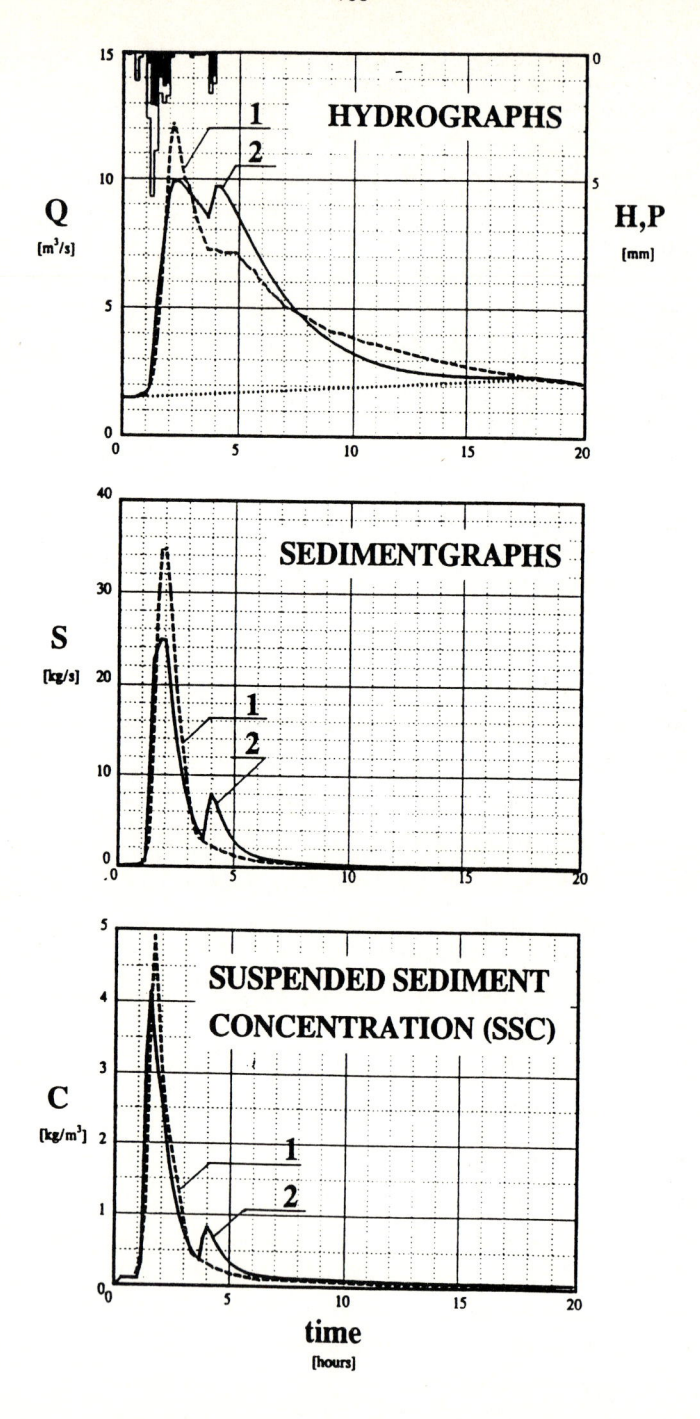

Fig.2: Comparison between observed (1) and regenerated (2) hydrograph, sediment graph and SSC distribution for the flood event of July 14, 1989.

Acknowledgements

The authors would like to thank Dr. M. Becht and his collaborators from the University of Munich for their generosity in supplying rainfall data for the investigation. The assistance of the F.U. Berlin to the first author in the form of a short term research fellowship is gratefully acknowledged.

References

Banasik K. (1990): Sediment graph model for a small watershed. - Proc. Int. Workshop on the Application of Mathematical Models for the Assessment of Changes in Water Quality, UNESCO-AISH Technical University ENIT, Tunis 7-12 May 1990: 302-311.

Banasik K.& Woodward D.E. (1992): Prediction of sediment graph from a small watershed in Poland in changing environment. In: Engman, T. (ed): Irrigation & Drainage Session Proc./Water Forum '92, EE,HY,IR,WR Div/ASCE, Baltimore, MD, 2-6 Aug. 1992: 493-498.

Becht M. & Wetzel K.R. (1992): The Lainbach catchment/Benediktbeuern (Upper Bavaria): Its physical landscape and development. In: Becht M. (ed): Contributions to the excursion during the International Conference "Dynamics and Geomorphology of Mountain Rivers". - Münchner Geographische Abhandlungen, Reihe B, Band B 16: 15-48.

Bley D.& Schmidt K.-H. (1991): Die Bestimmung von repräsentativen Schwebstoff-Konzentrationsgängen - Erfahrungen aus dem Lainbachgebiet/Oberbayern. - Freiburger Geographische Hefte 33: 121-129.

Bley D. (1992): Turbidity measurements and suspended sediment characteristics. In: Bogen J. (ed): Erosion and sediment transport monitoring programmes in river basins.- Poster Contributions, Oslo, Norway, 24-28 August 1992: 18-23.

Guillot P.& Duband D. (1978): Function de transfert pluie-débit sur des bassins versante de l'ordre de 1000 km². Societe Hydrotechnique de France, Paris, Session des 21-22.11.1978 (personal communication).

Nash J.E. (1957): The form of the instantaneous unit hydrograph. - IAHS Publ. 42: 114-118.

Sarma P.B.S., Dulleur J.W. & Rao A.R. (1973): Comparison of rainfall-runoff models for urban areas. - J. Hydrol. 18: 329-347.

Schmidt, K.-H., Bley, D., Busskamp, R. & Gintz, D. (1989): Die Verwendung von Trübungsmessung, Eisentracern und Radiogeschieben bei der Erfassung des Feststofftransports im Lainbach/Obb. - Göttinger Geogr. Abh. 86: 123-135.

Schmidt, K.-H., Bley, D., Busskamp, R., Ergenzinger, P. & Gintz, D. (1992): Feststofftransport und Flußbettdynamik in Wildbachsystemen. Das Beispiel des Lainbachs in Oberbayern. - Die Erde 123: 17-28.

USDA Soil Conservation Service (1972): National Engineering Handbook, Section 4, Hydrology. - Washington, DC.

Williams J.R. (1978): A sediment graph model based on an instantaneous unit sediment graph. - WRR 14/4: 659-664.

INVESTIGATIONS OF SLOPE EROSION IN THE NORTHERN LIMESTONE ALPS

Michael Becht
Institut für Geographie, Universität München
Luisenstraße 37/II, D-80333 München

Abstract: With reference to the Kesselbach valley in the Northern Limestone Alps as an example, the relationship between slope erosion and sediment load yield of the catchment area is presented. Slope erosion is predominated by gravitational processes (debris flows, avalanches). Fluvial erosion prevails only where these processes are not active (forested areas, gentle slopes). Comparing slope erosion with the sediment load yield of the catchment area of the Kesselbach shows a clear predominance of the latter. The solids are eroded in the cuts of the V-shaped valleys. Thus, there is present-day intensive further formation of these valleys, which dominate the scenery.

1. Introduction

Most of the landforms in the Alps have been formed during the Pleistocene and Holocene. Similar geomorphological processes, however, of widely varying intensities, may be at work over a long period of time. For the problems of natural hazards and prevention of damage in the Alps today, it is necessary to get more detailed information about the spatial distribution of the dimensions of present-day morphologically active processes on alpine slopes.

Human impact upon alpine ecosystems may induce an increase in slope erosion. A more intensive use of alpine pastures, for example, enlarges the areas of erosion on the slopes (Mössmer 1985). Only measurements of the amount of erosion and the geomorphological processes on the slopes, however, allow a quantitative estimate of the consequences of human impact on those areas.

The sediment yield of hydrological catchment areas is often measured at gauging stations in sediment basins at the mouths of the streams. The mean denudation rate of the slopes in the catchment area is derived from the measured sediment yield (Peters-Kümmerly 1973, Karl et al. 1975, Fatorelli 1988). It is assumed that temporary sedimentation in parts of the streams and at lower parts of the slopes is in balance with the transport from these sedimentation areas to the receiving streams, although many areas of sedimentation in alpine torrents may be mobilized during extreme flood events. There are no measurements of the relationship between slope erosion and sediment load yields of catchment areas nor investigations of the development of temporary deposits of sediments. Investigations of the sediment supply of torrents over a long monitoring period may provide more information about the sediment balance than short-term monitoring because of the possibility of extreme flood events which may erode stored sediments or accumulate new sediments in the stream. However, if there has been an increase in sediment deposition in parts of the catchment area during the Holocene (Strunk 1988), for instance, slope erosion will have been underestimated.

The spatial differentiation of slope erosion in the catchment areas cannot be determined by the measurement of the sediment yield of the torrents. According to Bogardi (1974) and Beschta (1981), sediment transport in streams of the humid regions is not limited by the hydraulic conditions or the amount of run-off but mainly influenced by the availability of erodible sediments in the catchment area. Large deposits of Pleistocene sediments dominate the sediment yield of catchment areas in the Northern Limestone Alps (Becht 1989). Preventive measures are only possible on the basis of knowledge about the spatial and temporal distribution of slope erosion and sediment load yield of the catchment areas.

2. Investigations in the catchment area of the Kesselbach

Between 1990 and 1992 the relative amounts of slope erosion and sediment load yield of the receiving stream in the catchment area of the Kesselbach were measured. The Kesselbach is located near the Dürrach in the Karwendel Mountains, which are a part of the Northern Limestone Alps (fig. 1).

Fig. 1: Location of the catchment area of the Kesselbach in the Karwendel Mountains

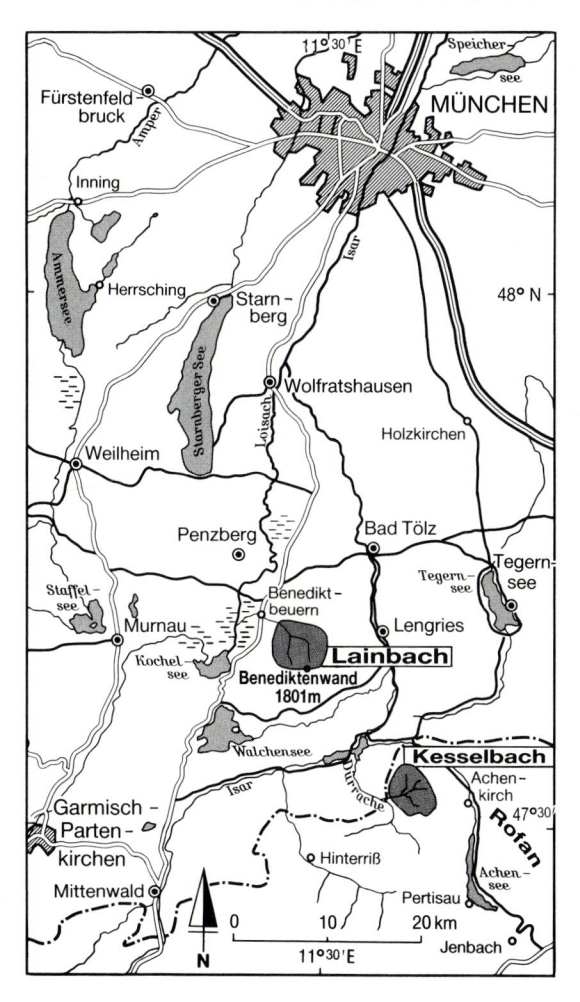

Slope erosion in the valley of the Kesselbach has been investigated at 13 test sites which were installed in small natural drainage basins with areas ranging from < 1 ha up to about 5 ha (fig. 2). The selection of the test sites was the result of a geomorphological, geological and vegetation mapping of the catchment area of the Kesselbach. They were chosen to represent the main geomorphological processes that occur in different altitudes of the drainage basin. The areas of the test sites have to be small enough to differentiate between the influence of individual geomorphological processes (e.g. fluvial erosion, debris flows, avalanches) on the slopes.

Fig. 2: The test sites in the catchment area of the Kesselbach

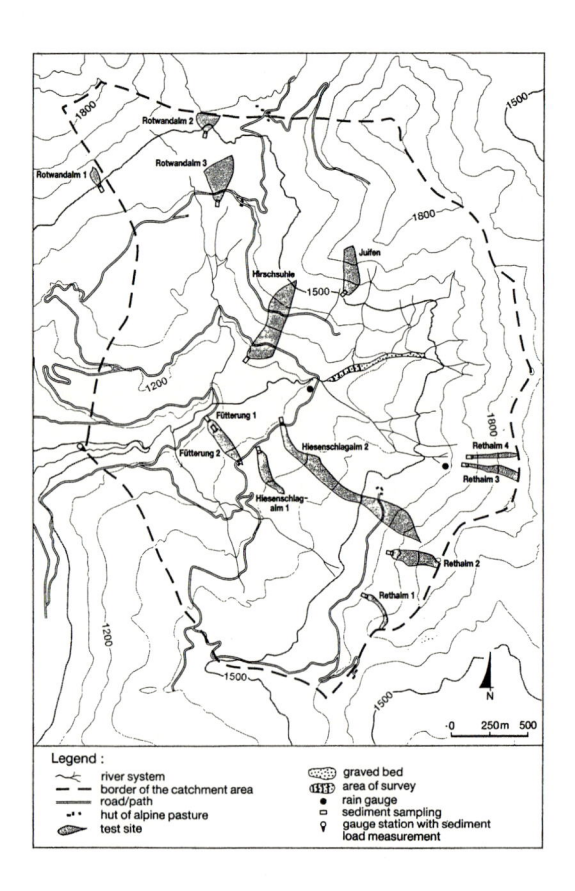

Table 1: Description of the test sites on slopes in the catchment area of the Kesselbach (geology according to Miller 1992)

area	size (m²)	altitude a.s.l. (m)	geology		geomorpho-dynamic processes	vegetation		mean slope (%)	aspect	land use
Rethalm 1	8375	1600-1775	80%	hearthstone	avalanches	66%	meadow	47.4	NW	alpine pasture
			20%	Allgäu beds	cattle tracks	34%	forest			
Rethalm 2	22950	1640-1900	84%	massive Aptych.	avalanches cattle tracks	9% 89%	meadow forest	69.3	W	alpine pasture
			9%	radiolarite		2%	barren of vegetation			
			7%	hearthstone						
Rethalm 3	11400	1640-1875	73%	marly Aptych.	Blaiken cattle tracks	57% 43%	meadow with Blaiken	63.2	W	alpine pasture
			27%	massive Aptych.						
Rethalm 4	6450	1655-1865	51%	marly Aptych.	Blaiken cattle tracks	36% 64%	meadow with Blaiken	65.8	W	alpine pasture
			49%	massive Aptych.						
Rotwand-alm 1	5700	1480-1580	89%	Haupt-dolomit	avalanches	52%	mountain pine	69.8	SE	game
			11%	platy limestone		48%	barren of vegetation			
Rotwand-alm 2	17700	1490-1515	100%	Kössen schist	cattle tracks creeping of soil	35% 65%	trees meadow	16.6	S	alpine pasture
Rotwand-alm 3	33200	1380-1480	100%	Kössen schist	cattle tracks	78% 22%	meadow trees	34.0	S	alpine pasture
Juifen	26400	1510-1760	100%	massive Aptych.	Blaiken avalanches	91% 9%	meadow with Blaiken	69.9	S	game
Hiesen-schlagalm 1	13850	1225-1340	100%	Kössen schist	cattle tracks creeping of soil	89% 11%	meadow forest	33.7	NW	alpine pasture
Hiesen-schlagalm 2	100375	1210-1780	37%	Rotkalke	cattle tracks	34%	meadow	49.2	NW	alpine pasture
			22%	Kössen schist	creeping of soil	33%	trees			
			18%	hearthstone	earth slips	23%	forest			
			15%	radiolarite	Blaiken	10%	barren of vegetation			
			8%	massive Aptych.						

area	size (m²)	altitude a.s.l. (m)	geology		geomorpho-dynamic processes	vegetation		mean slope (%)	aspect	land use
Hirsch-suhle	57057	1195-1530	42%	Rotkalke	creeping of soil	64%	forest	54.0	SW	game
			24%	Allgäu beds	channel	19%	trees			
			21%	hearthstone	erosion	13%	meadow			
			13%	moraine		4%	barren of vegetation			
Fütterung 1	19575	1095-1247	100%	Kössen schist	creeping of soil	60%	forest	39.2	NW	forest
					earth slips	10%	trees			
						26%	meadow			
						4%	barren of vegetation			
Fütterung 2	15325	1122-1247	100%	Kössen schist	creeping of soil	62%	forest	39.2	NW	forest
						11%	trees			
						27%	meadow			

notes: - Only geomorphological processes which, besides fluvial erosion, are active on the test areas are listed.
- trees = very sparse forest (tops covering less than about 20% of the ground)
- Blaiken = local name for rotational erosional features in the soil

Artificially limited test sites are not used because of the errors due to border effects. In most cases, the hydrological catchment area is not known, interflow from the upper parts of the slopes may lead to processes of fluvial erosion at the borders.

There are three main processes which influence the distribution and intensity of erosion on the slopes: run-off, avalanches, and debris flows. The sediment yield by fluvial transport was measured weekly during the summer season (time without snow cover) at small sediment basins with a volume of 55 l or 70 l. Samples were taken to analyse the eroded material in the laboratory (weight, grain size, organic matter).

Sediment transport by avalanches may occur only in spring, when ground avalanches originate from a heavy snow accumulation during winter. The sediment load of the avalanches can be well estimated by taking samples from the surface (0.5 m²), as there was no solid matter mixed inside the snow.

For debris flows, the amount of debris was estimated soon after the event by measuring and mapping the area and volume of accumulation of debris flow material.

The sediment load yield of the catchment area of the Kesselbach was measured at a gauging station by the Tyrolean Waterpower Company (TIWAG) using a "Tyrolean weir". With this equipment it is possible to record the bed load transport of the Kesselbach continuously (Hofer & Klein 1992, Hofer 1985). In addition to that, the suspended sediment load was recorded by a turbidity meter and the transport of dissolved loads was measured by the electric conductivity.

3. Slope erosion processes in the catchment area of the Kesselbach
3.1 Fluvial erosion
3.1.1 Erosion from test areas

The intensity of fluvial erosion on the slopes of the Kesselbach valley shows considerable variations (fig. 3). Under forest cover, higher sediment transports only occur in small V-shaped valleys (Hirschsuhle) where run-off concentrates. Outside these drainage channels there is very little erosion, as can be seen from the little basinlike valley of the measuring site Fütterung 2. The eroded material mainly derives from the channels. The relative amounts of organic matter decrease with growing erosion and following channel formation in V-shaped valleys. The mean contents are 17% at the test area Fütterung 2 and decrease to less than 5% at Hirschsuhle.

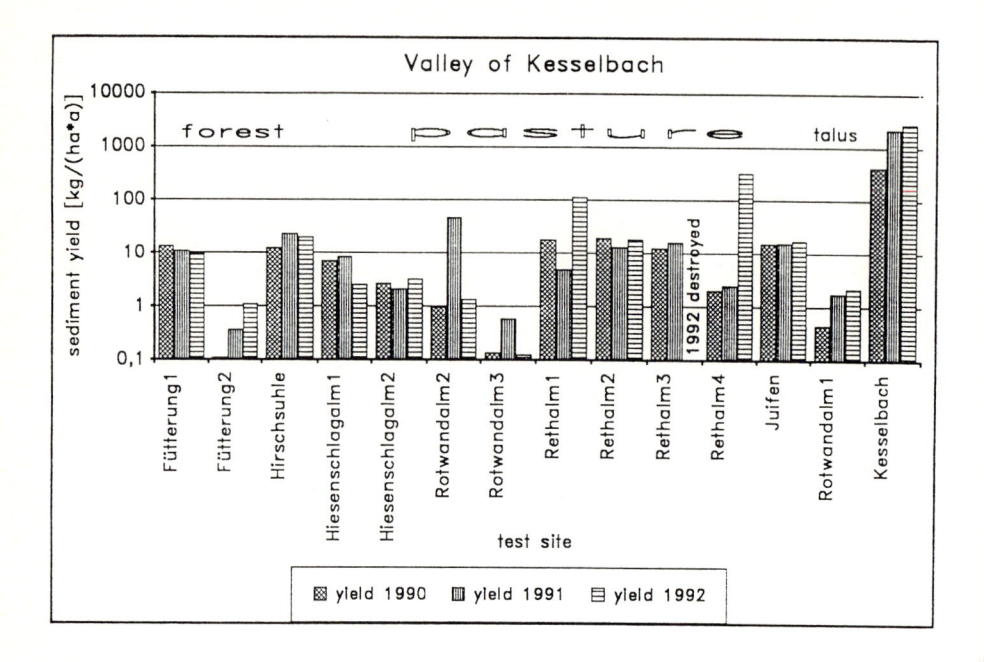

Fig. 3: Sediment yield of test sites in the Kesselbach valley in the years 1990-1992

The measuring site at Fütterung 1 represents a special situation which differs from the higher-lying area Fütterung 2 because of its more intense erosion (cf. fig. 2). Here the extension of the net of forestry roads led to an undercutting of the slope, which resulted in a translation slide. The nearly perennial run-off (0.2-0.3 l/s) at the measuring site Fütterung 1 proves that here the slope water horizon has been cut.

Erosion from pastures varies widely between the test areas. On all alpine pastures, the number of animals is high (Blechschmidt 1990); thus the intensity of erosion is mainly determined by the differing substrata. Least erosion occurs on the Kössener beds which are hardly permeable to water (Rotwandalmen). This seems contradictory, because direct run-offs should be high here. Compared to the steep slopes of the Rethalmen (cf. tab. 1) situated on permeable limestones, the slopes of the Rotwandalmen are much gentler; thus the smaller relief intensities can explain the lower rates of erosion. Higher transport of sediment occurred on the area Rotwandalm 2 in 1991 due to episodic crumbling away of the banks of cattle tracks. Therefore, statistical analyses show no significant correlations between fluvial slope erosion and the mean inclination of the slopes in the test areas.

There is little fluvial erosion from rock-covered areas, which have only a limited extension in the Kesselbach valley because of the low altitudes of the peaks (highest elevation Juifen 1988 m a.s.l.). The test area Rotwandalm 1 lies at the foot of a steep rock-covered slope mainly of Hauptdolomit (cf. tab. 1).

3.1.2 The influence of precipitation on fluvial slope erosion

The slopes in the Kesselbach valley show differing reactions of erosion to precipitation. A distinction has to be made between short precipitation events (rainfall) of high intensities and those of long durations, high amounts but lower intensities. The correlations for the intervals of maintenance at the test areas are shown in table 2.

Table 2: Correlations between yields of solids from test areas and precipitation parameters in the Kesselbach valley

test areas	number of observ. (n)	correlat. for precipitation sum	significance [%]	correlat. for precipitation event	significance [%]	correlat. for prec. intensity/ 30 min.	significance [%]
forest areas:							
Fütterung 1	55	0.60	0.1	0.32	5	0.14	–
Fütterung 2	55	0.40	1	0.52	0.1	0.60	0.1
Hirschsuhle	55	0.52	0.1	0.47	0.1	0.50	0.1
meadow and pasture areas:							
Hiesenschl. 1	44	0.66	0.1	0.26	10	0.10	–
Hiesenschl. 2	55	0.46	0.1	0.52	0.1	0.67	0.1
Rotwandalm 2	55	0.40	1	0.07	–	0.09	–
Rotwandalm 3	53	0.04	–	0.04	–	0.05	–

test areas	number of observ. (n)	correlat. for precipitation sum	significance [%]	correlat. for precipitation event	significance [%]	correlat. for prec. intensity 30 min.	significance [%]
Rethalm 2	44	0.59	0.1	0.40	1	0.48	0.1
Rethalm 3	39	0.83	0.1	0.54	0.1	0.09	--
Rethalm 4	47	0.23	--	0.39	1	0.66	0.1
Juifen rocks and talusses:	36	0.67	0.1	0.36	5	0.07	--
Rotwandalm 1	53	0.03	--	0.02	--	0.05	--

precipitation sum = precipitation sum in the observation period
precipitation event = highest precipitation sum of a single precipitation event
precipitation intensity = highest precipitation intensity in the observation period

Erosion from areas entirely covered by forest vegetation with dense thickets (Fütterung 2) shows especially good correlation with precipitation intensity, although there can be only little splash effect (Bork 1988, Poesen 1987) here. Precipitation of lower intensities completely infiltrates into the forest soil; thus only higher rainfall intensities can cause direct run-off and erosion. Infiltration rates measured by double-ring infiltrometers amount to 3-7 mm/min.

On the other hand, the Hirschsuhle measuring site shows that high precipitation totals, too, may lead to intensive transport of sediment, as the run-off originates from a bog about 150 m above the measuring site, which contributes to an increased run-off in channels in the case of longer lasting precipitation. The alpine pasture areas Hiesenschlagalm 1 and Rotwandalm 2 with little bogs acting like sponges in their areas also show positive correlations with precipitation totals.

On very compacted soils of alpine pastures over Pleistocene sediments (Hiesenschlagalm 2), all precipitation events lead to erosion. Here infiltrometer measurements show no infiltration either in spring or in summer. This is supported by the results of measurements of the direct run-off by small channels which catch the soil water flow parallel to the slope close to the surface. On the area Hiesenschlagalm 2, run-off in the upper soil (depth 0-3 cm) was collected after every precipitation event. Often it even exceeded the capacity of the measuring instruments. Comparative measurements in the forest (near test area Hirschsuhle) testify to only limited interflow there.

Differing reactions of test areas on karstified slopes (Rethalmen, Juifen) are due to the influence of spring water, which can be detected by the significantly higher electric conductivities in the run-off of the areas Rethalm 3 (mean value 197 µS/cm) and Juifen (mean value 169 µS/cm). Compared to this, the mean electric conductivities in the run-off of the test areas Rethalm 1 and 4 measure only 65 and 87 µS/cm respectively. Because of the delayed and damped beginning of the run-off of episodical springs, correlations with rainfall of long duration and high totals are especially good at the measuring sites Rethalm 3 and Juifen, whereas the test areas Rethalm 1 and 4 show high correlations with precipitation intensities.

Further differences exist in the origin and composition of the eroded material. In channels originating from karst springs there is mainly transport of coarse sediment from the channel bed itself. The relative amounts of clay and silt are lower than 5%, and there are hardly any organic substances (<1%). From the areas Rethalm 1 and 4, on the other hand, an average of 37% and 14% respectively of fine material and 4-7% of organic substances are eroded. Though there is altogether less (less frequent) erosion here (cf. fig. 3), more soil material is eroded. These results are proved at the areas Hiesenschlagalm 1 and 2 which also correlate well with precipitation intensities. The contents of fine-grained material here even amount to 51% and 82% respectively on the average due to the composition of the initial substratum (Pleistocene loose sediments), the contents of organic substances are 12% and 6% respectively.

3.2 Erosion by avalanches

After only small snow-slab avalanches, which hardly transported any sediments, occurred in the years 1990 and 1991, there were frequent ground avalanches (de Quervain 1966, Wilhelm

1975) in spring 1992. In contrast to the two preceding winters, there was considerable snow accumulation in the winter of 1991/92. In the nearby Lainbach area (cf. fig. 1) records of the snow cover exist since 1971. The snow accumulation in spring 1992 exceeded the maximum measured before.

Whereas loose snow avalanches in midwinter transport no sediments according to the observations, ground avalanches at Zotenjoch in the NW Kesselbach valley and predominantly in the area of the Rethalmen caused intensive erosion in 1992. The spatial distribution of the avalanches was analysed by means of a GIS. The areas where avalanches occurred are characterized by their steepness (29-34°), their location outside of dense forests and the predominance of karstified rocks (Hauptdolomit, Aptychus beds) in the subsoil. Geomorphologically soft rocks (Kössener beds) are free of avalanches because of their gentler slopes.

Accumulation of sediment at Zotenjoch amounted to about 18.4 t on an area of 3500 m² (=5.26 kg/m²). For the entire slope (24.5 ha) this corresponds to an erosion of 75g/m² (= 0.03 mm for a rock density of 2.6 g/cm³).

At the Rethalmen, snow accumulation and following avalanche activity was higher. Only limited sampling of the areas was possible due to the extemely high risk of avalanches; thus the determined sediment loads are estimated values. According to these, the test area Rethalm 2 shows an accumulation of 750 t of solids, the test areas Rethalm 3 and 4 even 7500 t of solids. For the entire area of the Rethalmen the sampling and measuring results lead to an estimated erosion in a magnitude of 5 mm.

The stronger erosive effect at the Rethalmen is due to:
- the use as alpine pastures,
- the thicker soil cover, that means greater availability of loose material,
- the small area covered by trees and bushes, and
- the aspect.

While on the S-oriented slopes at the Zotenjoch the accumulated snow is alredy reduced by melting during winter, the snow cover is preserved until spring on the W- and NW-facing slopes. The greater thickness of the snow cover increases the erosive effect of the ground avalanches.

3.3 Debris flows

Debris flows occur on the slopes of alpine valleys after precipitation events of high intensities. In the valleys, on the other hand, debris flows are mainly caused by high precipitation totals (Rickenmann & Zimmermann 1992); here it is difficult to distinguish between debris flows and strong bed-load transport of a torrent is difficult. The precipitation intensity causing debris flows on slopes depends on the mean meteorological conditions in the catchment areas. In dry regions in the Central Alps intensities of about 20 mm/30 min. may lead to debris flows, whereas in the Kesselbach valley no debris flows can be observed up to at least 30 mm/30 min.. Here an excessive precipitation event with a maximum measured intensity of 39.9 mm/30 min., which caused debris flows on the slopes, occurred in the area of the Rethalmen and the adjoining catchment area in the SW on 1 Aug. 1992. In the wetter marginal areas of the Alps, on the other hand, only an even stronger thunderstorm (75 mm/30 min.) led to debris flows in the Lainbach area.

In the Kesselbach valley, 750 m^3 of loose material were deposited on an area of altogether 2500 m^2 at the Rethalmen (about 1350 t with a compactness of 2.0 g/cm^3). On the test area Rethalm 2, 17.8 m^3 were eroded (= 13.9 t/ha = 0.5 mm), on the test area Rethalm 4 20.6 m^3 (= 57.5 t/ha = 2.2 mm). Calculation of the solids eroded by debris flows for the entire Rethalm area yields a mean erosion of 0.7 mm.

Analysing areas of present-day debris flows and fossil debris flow deposits at Juifen and Zotenjoch by the GIS showed that areas of debris flows are characterized by:

- a steepness of more than about 27°.
- a location on highly permeable limestones.
- only episodic run-off of the channels, and
- a low drainage density.

Thus there is no formation of debris flows on hardly permeable rocks, as high direct run-off has a strong diluting effect on the suspension; thus sediments are transported by the run-off. Debris flows originating from high-lying areas lose their debris flow characteristics both with increasing deposition of debris and with dilution of the suspension by affluents.

4. Slope erosion and sediment load yield in the Kesselbach valley

The sediment load yield of the catchment area of the Kesselbach varies in the years 1990 to 1992 (tab. 3). This is mainly due to the amount of bed-load transport, which begins to have an intense effect with run-off peaks of 7 m^3/s. The distribution of the yields of suspended load, bed-load and dissolved load over the year (fig. 4) clearly shows the variations of the individual values between the months.

Table 3: Sediment load yield of the Kesselbach valley

area	year	load of solids [t a]	total load [t/a]	yield of solids [kg/(ha a)]	total yield [kg/(ha a)]	erosion of solids [mm/a]	total erosion [mm/a]	percent. dissolved mat.
Kessel-	1990	341	1649	426	2061	0.016	0.079	80%
bach	1991	1766	3138	2208	3923	0.085	0.151	44%
	1992	2010	3435	2513	4294	0.097	0.165	41%

Fig. 4: Seasonal distribution of sediment load yield in the Kesselbach valley in the year 1990

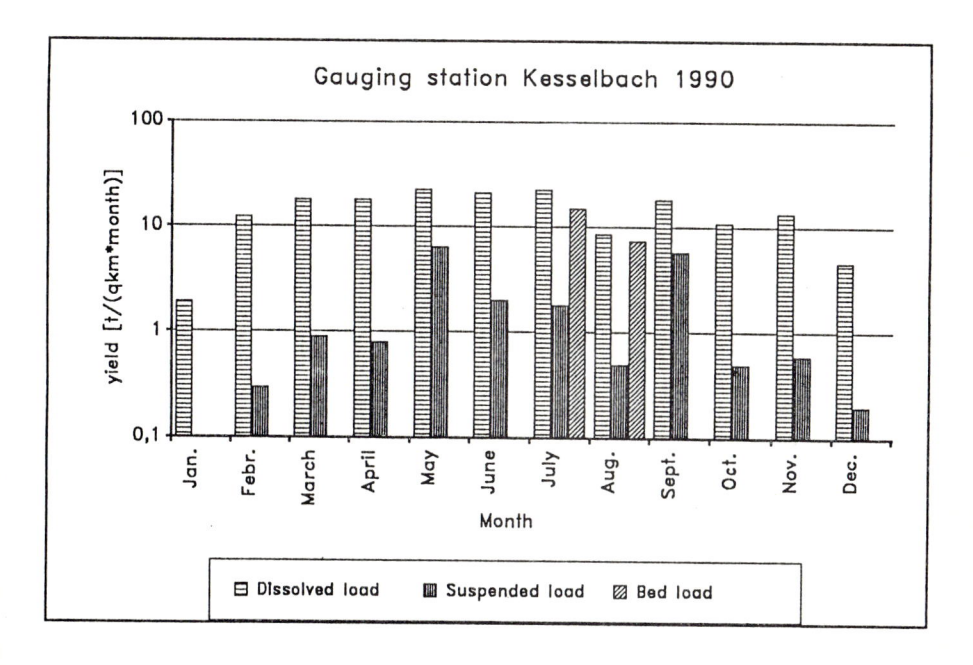

[Bar chart: Gauging station Kesselbach 1990, showing monthly yield in t/(qkm·month) on a logarithmic scale from 0.1 to 100 for Dissolved load, Suspended load, and Bed load for each month January through December.]

The yield of dissolved load shows the smallest range of both intra- and interannual variation. It accounts for about 50% of the total sediment load yield and thus must not be neglected, as has been the case in other investigations with reference to the high relief intensity and erosion in high mountains (Jäckli 1957, Karl et al. 1975, Vorndran 1979). The load of dissolved matter depends on the solubility of the rocks but mainly on the annual sums of precipitation and run-off respectively (Walling & Webb 1983). Therefore, erosion of dissolved material, which lies in the magnitude of 0.066 mm/a in the Kesselbach valley, increases further by about 50% towards the northernmost areas of the Alps (Lainbach valley) (0.11 mm/a, Becht et al. 1989). Values for the Tiroler Ache (Marquartstein), the Isar (Mittenwald) and the Halbammer also range between 0.09 and 0.15 mm/a (Prösl 1985). For the Kreidenbach, which has a similar situation to the marginal regions of the Alps as the Kesselbach valley, Prösl, too, gives a significantly lower value for the erosion of dissolved material of 0.039 mm/a.

Over the year, the yield of dissolved load predominates in winter (fig. 4). During and after snowmelt transport of solids occurs in connection with rainfall, which, in other years, may clearly exceed the yield of dissolved load. The maximum in transport of suspended load per month occurs mostly in spring, when the soil is loosened by the winterly frost action; thus the first rainfall events in spring erode more loose material than in summer or autumn. Similar results exist for the Lainbach area (Wetzel 1992). The peak in suspended load in September (fig. 4) must be caused by problems of measuring with the suspension probe, as neither corresponding precipitation nor bed-load transport were measured.

The relative amount of suspended load decreases with growing total sediment yield per year (tab. 3). The mean percentage is about 20% and hence much less than in other valleys of the Northern Limestone Alps (Becht 1989, Sommer 1980). Similar run-off leads to 5-10 times higher concentrations of suspended load in the Lainbach. This is mainly due to the relatively coarse initial material and the limited occurrence of Pleistocene loose rocks in the Kesselbach valley. In the adjoining catchment area of the Dürrach, the percentage is 33% according to Sommer (1980) and thus significantly higher.

The erosion of the Kesselbach during the study period can be put in a long-term perspective by means of the series of observations of the Dürrach (fig. 5). Unfortunately, often only mean values for several years of the sediment yield at the reservoir are available here. Nevertheless, the data confirm that below-average erosion in 1990 was followed by an average sediment yield in 1991 and 1992. Compared to the Dürrach, the sediment yield of the Kesselbach area is less, which is also due to a more limited occurrence of loose rocks.

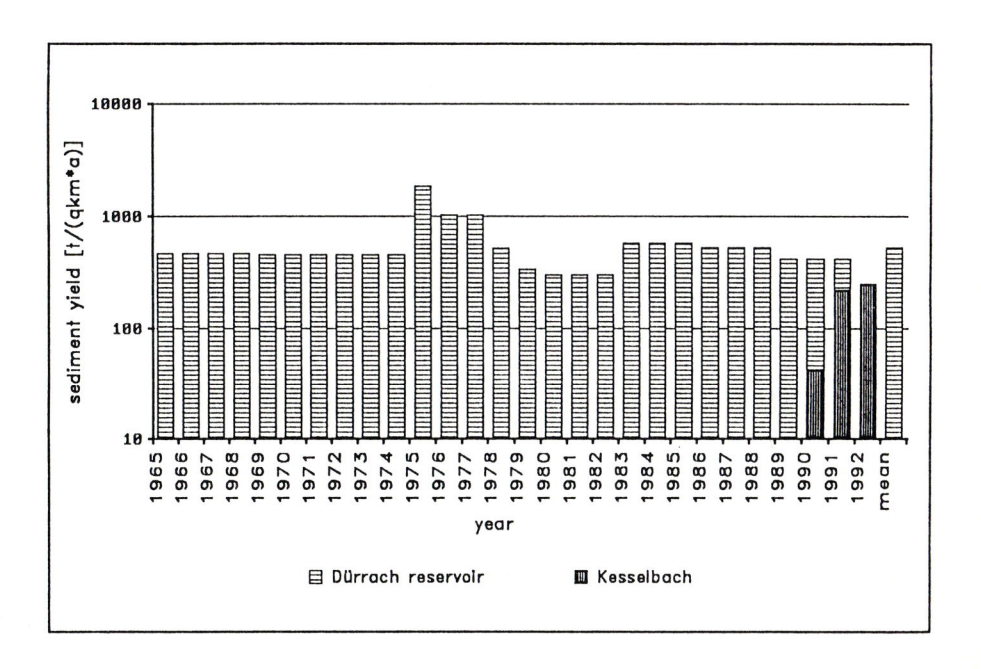

Fig. 5: Annual sediment yield of the catchment area of the Dürrach since 1965

The fluvial sediment yield of the Kesselbach valley is more than ten times greater than the fluvial erosion on the slopes (cf. fig 3). In winter the slopes are covered with snow; therefore, there is no erosion (Becht 1990), whereas in the receiving streams transport of sediments occurs along with higher run-offs. Mainly, however, the higher sediment load yield of the catchment area is a consequence of the spatial distribution of erosion in summer. Samples taken during a flood flow on 17 Jun. 1991 prove that the transport of solids from the high-lying areas into the streams is less than their sediment load yield (fig. 6). The greatest increase of sediment load occurs in the course of the stream in the lower part of the catchment area (below 1200 m a.s.l.). This part of the course of the Kesselbach lies in a V-shaped valley with very high and steep slopes.

Fig. 6: Sediment erosion on slopes and sediment load yield of the catchment area of the Kesselbach from 17 Jun. to 24 Jun. 1991

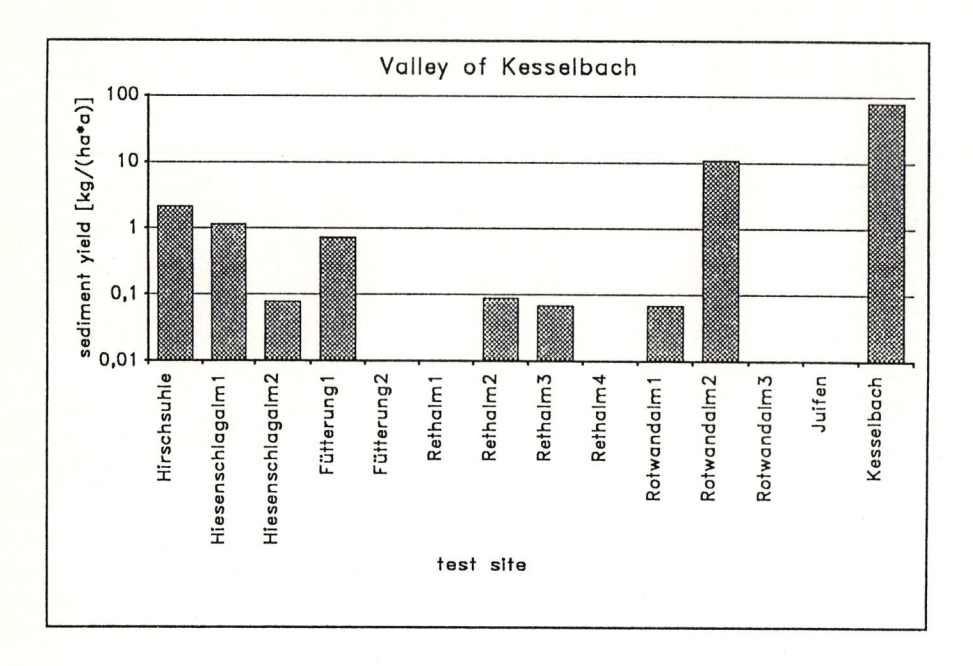

During the sampling intervals, fluvial erosion from the test areas shows an especially high correlation with the sediment load yield of the catchment area where slope erosion correlates well with high precipitation totals (tab. 4). On the other hand, slopes on which fluvial erosion mainly depends on precipitation intensities (Rethalm 4) show no correlation with the sediment load yield of the area. Therefore, slope erosion and sediment load yield do not run synchronously on many test areas; thus the solids eroded here are temporarily re-deposited within the catchment area. This holds true both for single events and for the annual sediment balances.

Table 4: Correlations between sediment erosion from test areas and sediment load yields of the Kesselbach valley during the sampling intervals 1990 and 1991 (in comparison α-values for the correlations between precipitation and sediment load yields of the test areas)

test area	number of observations	Kesselbach suspended load		Kesselbach total sediment load		test areas	
						prec. int.	prec. sum
		r	sign. [%]	r	sign. [%]	sign. [%]	sign. [%]
Hirschsuhle	55	0.41	1	0.56	0.1	0.1	0.1
Hiesenschlagalm 1	44	0.38	1	0.40	1	–	0.1
Hiesenschlagalm 2	55	0.41	1	0.19	–	0.1	0.1
Fütterung 2	55	0.21	–	0.34	5	0.1	0.1
Rotwandalm 1	53	0.05	–	0.35	5	–	–
Rotwandalm 3	53	0.00	–	0.00	–	–	–
Rethalm 1	48	-0.06	–	0.05	–	–	–
Rethalm 3	39	0.39	1	0.19	–	–	0.1
Rethalm 4	47	0.01	–	0.08	–	0.1	–

prec. sum = precipitation sum in the observation period
prec. int. = highest precititation intensity in the observation period

5. Outline of present-day relief formation in the Kesselbach valley

Relief formation in the Kesselbach valley is characterized by intensive downward erosion of the streams. It can be seen not only in a measurably much higher sediment load yield of the catchment area compared to slope erosion but also in the deeply incised V-shaped valleys in the scenery. These are currently under intensive further formation. Thus high erosion rates under forest cover occur also in the little valley at the test area Hirschsuhle (cf. fig. 3 and 6). Locally there is intense gravitational erosion by avalanches and debris flows. For the intervals of re-occorrence of these processes there exist only rough estimates on the basis of experiences from other areas and the reports of the Alpine herdsmen. On the assumption of a recurrence interval of 10-20 years, these processes of formation predominate slope erosion on the most affected slopes of the Rethalm area. Here it lies in the magnitude of the average sediment load yield of the catchment area as calculated from the sediment load of the Kesselbach.

The environmental changes induced by human activities and the clearing of the steep slopes at the Rethalmen have increased slope erosion. However, this is not mainly due to the use as alpine pastures, but a consequence of an altered system of geomorphological processes. The high slope erosion at the Rethalmen is predominated by avalanches and debris flows, which do not occur in the forest. Use as alpine pastures alone, as it is also found at the Rotwandalmen, does not lead to increased slope erosion there, as only fluvial processes are active (cf. fig. 3).

Especially the Rethalmen show a combination of different processes and transports. Only a small fraction of the currently fluvially transported material is newly eroded from the soil or the regolith cover. Usually it is redeposited material. This means that fluvial erosion, too, is increased by gravitational processes of sediment transport. On the other hand, it is difficult to speak only of erosion rates here, as in many cases material - sometimes after a very long period of time - is only taken from previousdeposits and then transported further. But since the collected and measured sediments do not give any indication of their origin, the rate of erosion is determined from the transported solids in this investigation.

The effects of exteme flood flows on the sediment budget in alpine catchment areas is often overestimated. During the excessive precipitation event on 1 Aug. 1992, only about 75 m^3/km^2 of sediments were transported out of the Kesselbach. The total amount of about 600 m^3 was less than the annual sediment load yield (cf. tab. 3). According to the Tyrolean Waterpower Company, 35000-40000 m^3 of sediments were transported out of the most affected adjoining catchment area of the upper Dürrach (= 650 m^3/km^2). This is equivalent to two to three times the sediment load yield of the long-term mean (cf. fig. 5). Less frequent extreme flood events have greater effects in the areas. Old channels are widened and new ones are opened; thus, the redeposited bed-load in the catchment area of the Dürrach is estimated at about 300.000-350.000 m^3 according to measurements of the TIWAG, which is equivalent to ten times the amount of the sediment load yield.

The present-day development of the V-shaped valleys can also be confirmed for the Lainbach valley (Becht 1993). Here, the high precipitation sums are of decisive importance for the geomorphological formation, as they cause intensive fluvial erosion and thus a predominance of the sediment load yield of the catchment area. In the dryer Central Alps, on the other hand, gravitational processes of slope erosion prevail. This does not hold true for extensively glaciated catchment areas, where strong meltwater run-off increases fluvial erosion (Becht 1993).

References

BECHT, M. (1989): Suspended sediment load yield of a small alpine drainage basin in Upper Bavaria.- In: Catena, Suppl.-Bd. 15: 329-342, Cremlingen.

BECHT, M. (1990): Auswirkungen der Schneeschmelze auf die Schwebstoffführung von Wildbächen.- In: Beitr. z. Hydrologie, Bd. 11, H. 2: 1-27, Kirchzarten.

BECHT, M., FÜSSL, M., WETZEL, K.-F. & WILHELM, F. (1989): Das Verhältnis von Feststoff- und Lösungsaustrag aus Einzugsgebieten mit carbonatreichen pleistozänen Lockergesteinen der Bayerischen Kalkvoralpen.- In: Göttinger Geogr. Abh., Bd. 86: 33-43, Göttingen.

BECHT, M. (1993): Untersuchungen zur aktuellen Reliefentwicklung in alpinen Einzugsgebieten. - Habilitationsschrift LMU München.

BESCHTA, R.L. (1981): Patterns of sediment and organic-matter transport in Oregon Coast Range Streams. - IAHS No. 132: 179-189, Christchurch.

BLECHSCHMIDT, G. (1989): Ursachen und Ausmaß der Blaikenerosion im Karwendel. - Diss. Fak. f. Wirtschafts- und Sozialwiss. der TU München, München.

BOGARDI, J. (1974): Sediment transport in alluvial streams. - Budapest.

BORK, H.-R. (1988): Bodenerosion und Umwelt - Verlauf, Ursachen und Folgen der mittelalterlichen und neuzeitlichen Bodenerosion. Bodenerosionsprozesse, Modelle und Simulationen. - In: Landschaftsgenese u. Landschaftsökologie, H. 13, Braunschweig.

FATORELLI, S., LENZI, M., MARCHI, L. & KELLER, H.M. (1988): An experimental station for the automatic recording of water and sediment discharge in a small alpine watershed. - In: Hydrol. Sciences Journal. Vol 33, No. 6: 607-617.

HOFER, B. (1985): Der Feststofftransport von Hochgebirgsbächen. - Diss. Inst. f. konstr. Wasserbau und Tunnelbau, Univ. Innsbruck.

HOFER, B. & KLEIN, W. (1992): The diversion system Bächental barrage/captation Kesselbach of the Achensee power plant.- In: Münchener Geogr. Abh.,Bd. B16: 49-58, München.

JÄCKLI, H. (1957): Gegenwartsgeologie des bündnerischen Rheingebietes. - Beiträge z. Geologie d. Schweiz, Geotechn. Ser., Lieferung 36, Zürich.

KARL, J., SCHEURMANN K. & MANGELSDORF, J. (1975): Der Geschiebehaushalt eines Wildbachsystems, dargestellt am Beispiel der oberen Ammer. - In: Deutsche Gewässerkundl. Mitteilungen, Jg. 19, H. 5: 121-132, Koblenz.

MILLER, H. (1992): Geologische Karte des Kesselbachgebietes. Zusammenfassung von Diplomkartierungen am Inst. f. Allgemeine und Angewandte Geologie der LMU, München, unveröff.

MÖSSMER, E.-M. (1985): Blaikenbildung auf beweideten und unbeweideten Almen.- Jb. d. Ver. z. Schutz d. Bergwelt 59: 79-94, München.

PETERS-KÜMMERLY, B. (1973): Untersuchungen über die Zusammensetzung und den Transport von Schwebstoff in einigen Schweizer Flüssen. - Geogr. Helv., Jg. 28, H. 3: 137-151.

POESEN, J. (1987): Transport of rock fragments by rill flow - a field study. - In: Catena, Suppl.-Bd. 8: 35-54, Cremlingen.

PRÖSL, K.-H. (1985): Dissolved Load of Alpine Creeks and Rivers.- In: Beiträge z. Hydrol., Sonderh. 5.1: 235-244, Kirchzarten.

de QUERVAIN, M. (1966): On avalanche classification: a further contribution. - In: Int. Ass. of Scient. Hydr., Publ. No. 69: 410-417.

RICKENMANN, D. & ZIMMERMANN, M (1992): Beurteilungen von Murgängen in der Schweiz: Meteorologische Ursachen und charakteristische Parameter zum Ablauf. - In: Int. Symp. Interpraevent, Bd. 2: 153-164, Bern.

SOMMER, N. (1980): Untersuchungen über die Geschiebe- und Schwebstofführung und den Transport von gelösten Stoffen in Gebirgsbächen.- In: Intern. Symp. Interpraevent, Bd. 2: 69- 94, Bad Ischl.

STRUNK, H. (1988): Episodische Murschübe in den Pragser Dolomiten semiquantitative Erfassung von Frequenz und Transportmengen. - In: Zeitschr. f. Geomorph. N.F., Suppl.-Bd. 70: 163-186, Berlin, Stuttgart.

VORNDRAN, G. (1979): Geomorphologische Massenbilanzen. - Augsburger Geogr. Hefte, Nr. 1, Augsburg.

WALLING, D. E. & WEBB, B. W. (1983): The Dissolved Loads of Rivers: a Global Overview. - In: Int. Ass. Hydr. Sciences, Publ. No. 141: 3-20.

WETZEL, K.-F. (1992): Abtragungsprozesse an Hängen und Feststofführung der Gewässer. Dargestellt am Beispiel der pleistozänen Lockergesteine des Lainbachgebietes (Benediktbeuern/Obb.). - Münchener Geogr. Abh., Bd. B17, München.

WILHELM, F. (1975): Schnee- und Gletscherkunde. - Berlin.

Examples from areas outside the Alps

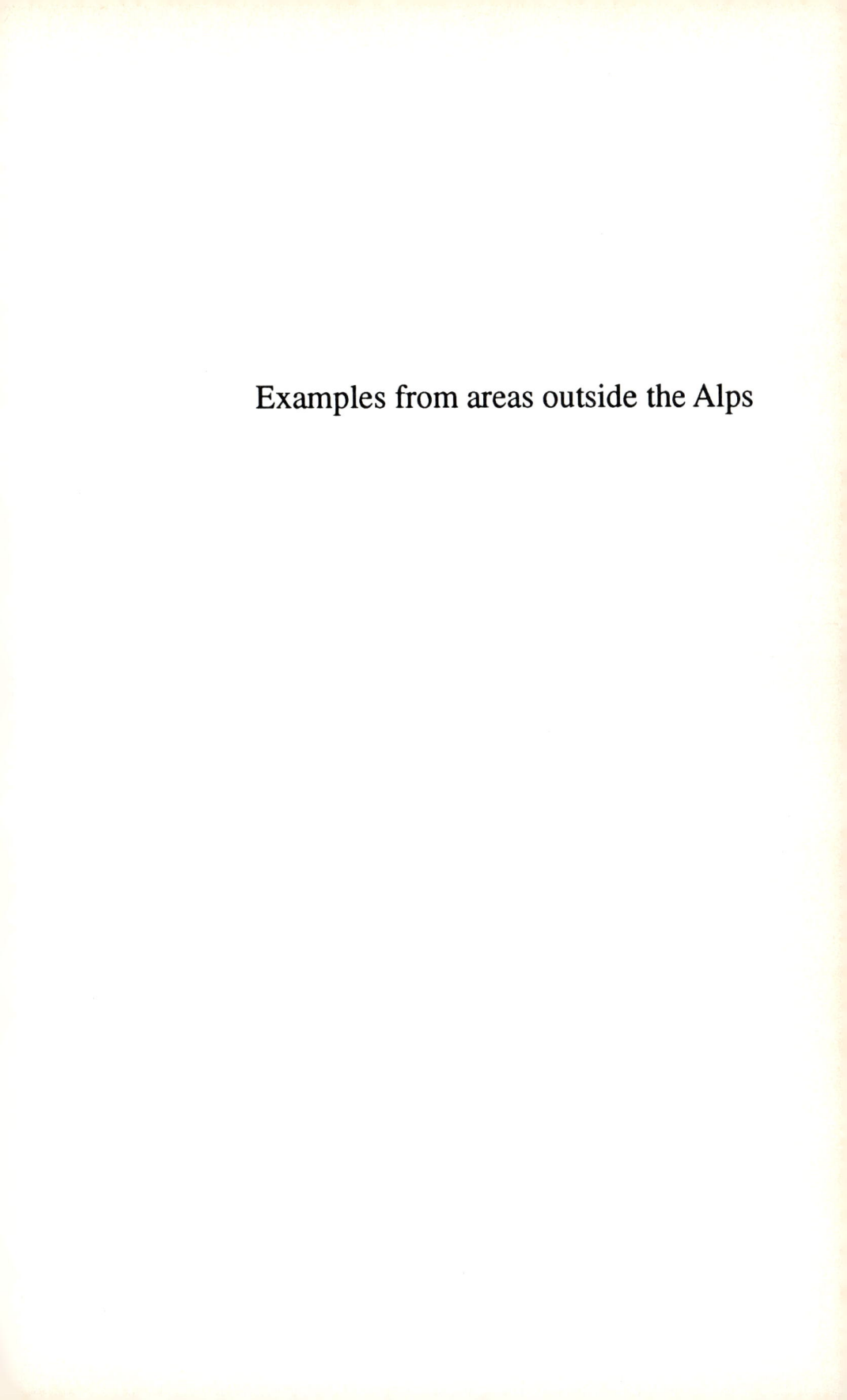

STREAMBED DYNAMICS AND GRAIN-SIZE CHARACTERISTICS OF TWO GRAVEL RIVERS OF THE NORTHERN APENNINES, ITALY

Paolo Billi
Dipartimento di Ingegneria Civile, Università di Firenze
Via S. Marta 3, 50139 Firenze, Italia

ABSTRACT

Better sampling strategies are needed to understand gravel-bed river dynamics. The Leopold transect-line pebble count method needs to be extended to finer material. A new procedure which extends the method into the fine sand range is presented. Sampling underwater bed material with the proposed device is easier and more reliable. Such procedure was applied to two gravel-bed rivers of the Northern Apennines where 13 and 12 sampling sites were selected respectively. The data gathered point out the great variability of grain size and much weaker than expected bar surface armoring. To investigate this result, replicate sub-surface sampling was carried out on two selected bars. Analysis of different characteristic diameters, including D50 of truncated grading curves, was performed as well. The apparent lack of bar surface armoring is discussed in terms of equal mobility and streamwise sorting. The downstream distribution of sand was analyzed since large quantities of fine material, entrapped in a few pools, were observed in sparse reaches. Possible mechanisms for local sediment accumulation may include bar winnowing, dissection or wash out as well as increased pool trapping efficiency are discussed.

INTRODUCTION

The morphology and grain-size distribution of the main physiographic units making up the bed of a gravel river represent the ultimate products of many complex processes which are most active during floods. How, and to what extent, a single flood affects the stream bed is not straightforward. The duration of the preceding base flow and the intensity of the previous flood can also be important. The dynamics of a river, a coarse-grained river in particular, can be viewed as a series of more or less relevant changes in thalweg morphology and grain size.

Several papers dealing with the channel geometry of gravel rivers and their transitions have been published. They commonly report overall changes of stream pattern, development and migration of channel bars or longitudinal profile changes (e.g., McGowen & Gardner 1970; Jackson, 1976; Lewin, 1976; Gregory,1977; Hein & Walker, 1977; Shumm, 1977; Bluck, 1979; Church, 1983; Ferguson & Werritty, 1983; Harvey, 1987; Ergenzinger, 1992). By contrast very few papers deal with time and space variation of bed material grain size in gravel rivers (Keller, 1971; Bluck, 1982; Mosley & Tindale, 1985; Billi, 1990; Lisle & Madej, 1992). Notwithstanding it is relevant data for many practical purposes. For instance, many of the commonly used bedload formulas require a characteristic bed sediment size (see Gomez & Church, 1989, for a broad review). This poverty of information is due to many factors such as a lack of standard sampling methodologies, difficulty in obtaining reliable underwater samples, and the wide range in grain-size for sedimentary and geomorphological units making up the streambed.

With the twofold aim of contributing to the solution of these problems and investigating the relation between river dynamics and bed material characteristics, a sediment survey campaign was carried out on two gravel-bed streams of the Northern Apennines.

STUDY AREA

The two streams under study are the Orcia and Cecina Rivers. They are in southern and mid-west Tuscany, respectively (Fig. 1). Both rivers drain the middle Tuscany range, the inner mountain belt of the Northern Apennines arch.

The Orcia is located 40 Km south of Siena. It is the main left tributary of the Ombrone, the second largest river in Tuscany. The area of the Orcia River catchment is 860 Km^2 and the main channel length is about 60 Km. The average annual precipitation is about 800 mm. The monthly mean discharge, measured at Mt Amiata Station (where the upstream basin is about 580 Km^2),

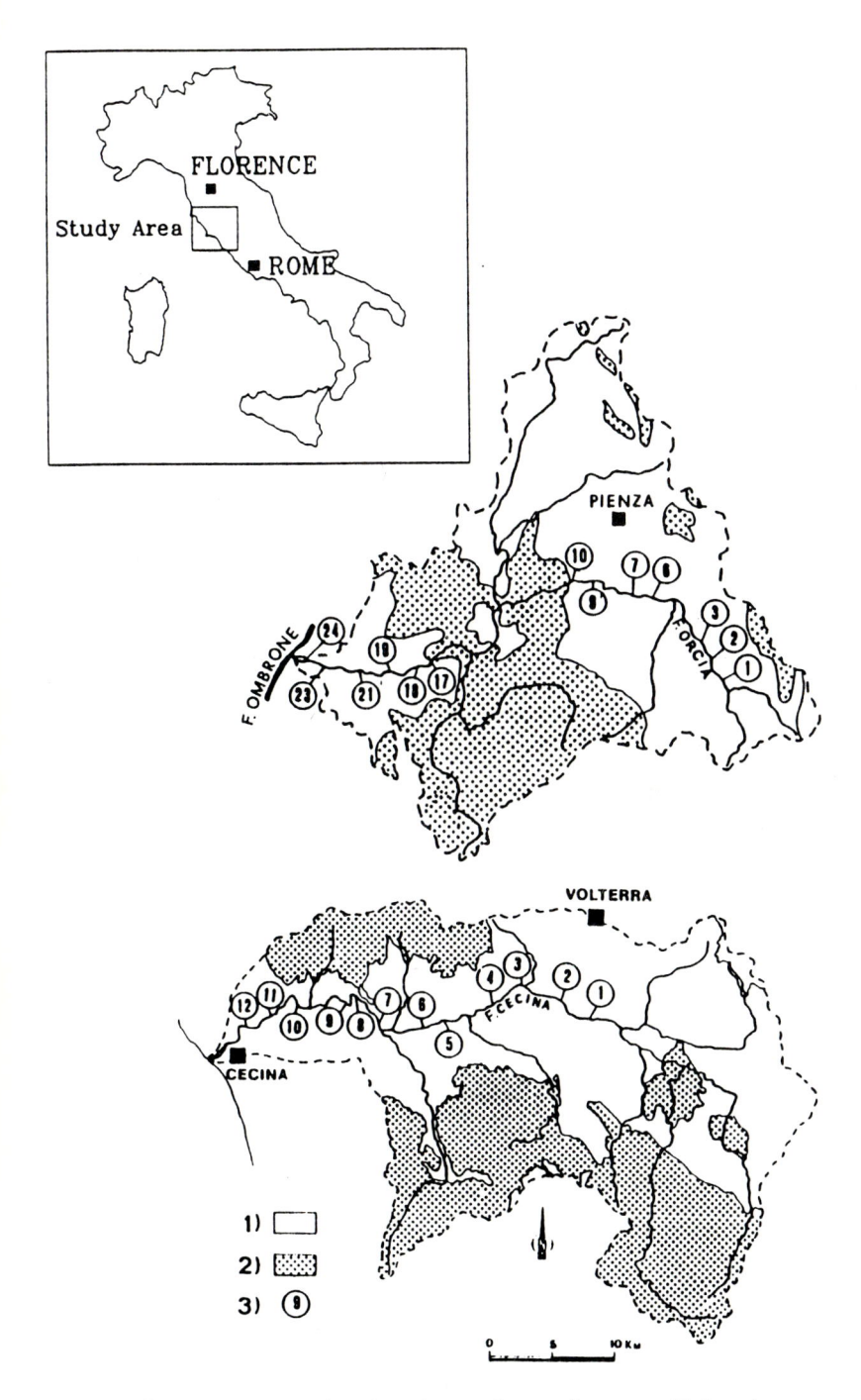

Fig. 1 - *Location map showing the study catchments: 1) incoherent and semi-coherent rocks; 2) bedrock; 3) sampling site.*

ranges from 1 to 10 m^3/s, while bankfull discharges approach 500 m^3/s.

Half the watershed is underlain by incoherent and semi-coherent rocks. These being Miocene to Holocene marine and fluvio-lacustrine gravel, sand and clay, outcropping in the basin head and in its most downstream reach. The center of the basin is dominated by bedrock outcrops, mainly Apenninic Sequences, into which the Orcia has cut a deep and narrow gorge.

The Cecina River is west of Siena (Fig. 1). After a course of about 70 Km, it outflows into the Tyrrhenian Sea, a few Km south of Leghorn. The catchment area is about 900 Km^2. The average annual precipitation is around 900 mm. The geological setting is very similar to that of the Orcia.

A flow gauge is located 18 Km upstream of the outlet and monitors a watershed of 634 Km^2. Here, bankfull discharge is around 600 m^3/s, while the highest flow was recorded in November 1966 with a peak discharge of about 1000 m^3/s.

Both the Orcia and the Cecina Rivers have a gravel bed. They generally show a straight channel morphology, with alternate lateral bars, with a bias toward "wandering" (Church & Jones, 1982) in the most downstream reaches.

SAMPLING METHODOLOGY

The thalweg of a gravel-bed river commonly consists of many morphological and sedimentary units which have been described in detail by several authors (e.g. McGowen & Garner 1970; Jackson, 1976; Hein & Walker, 1977; Bluck, 1979; 1982; Church & Jones, 1982). Three main, larger scale, physiographic units can be schematically distinguished however. These are bar (surface and sub-surface), riffle and pool(Fig. 2). Such units were considered as suitable sampling locations. Along the Orcia 24 (later reduced to 13) sites were selected (Fig. 1). Along the Cecina 12 sites were selected (Fig.1).

The grid-by-number method (Leopold, 1970) is considered the most reliable procedure in describing grain-size distribution for coarse bed material (Kellerhals & Bray, 1971; Hey & Thorne, 1983; Mosley & Tindale, 1985; Church et al., 1987; Billi, 1989; Wolcott & Church, 1991). Unfortunately, the main physiographic units considered in this paper cannot be sampled this way due to many practical problems such as the underwater position of the channel units and the inherent difficulty of application to sub-surface sediment.

In the present study, the bar surface was sampled by the transect line method (Leopold, 1970). Bulk samples of about 30 kg were taken from bar sub-surface, riffle and pool. Bar sub-surface material was sampled after the armor layer (as thick as D90) was removed. Riffle and pool bulk samples include both surface and sub-surface material. According to Kellerhals and Bray (1971), grid-by-number samples are equivalent to sieve-by-weight bulk samples. Therefore, the grain-size data of different units are comparable. This sampling strategy has been adopted in other bed sediment studies (e.g. Mosley & Tindale, 1985; Church et al., 1987; Lisle & Madej, 1992). Still there is a practical limitation of the pebble count method due to the difficulty in measuring particles finer than 8 mm. To overcome this restriction and to collect bulk underwater samples, a visual fine-grain comparator and a modified scoop-type sampler were used (Billi, 1990; 1992).

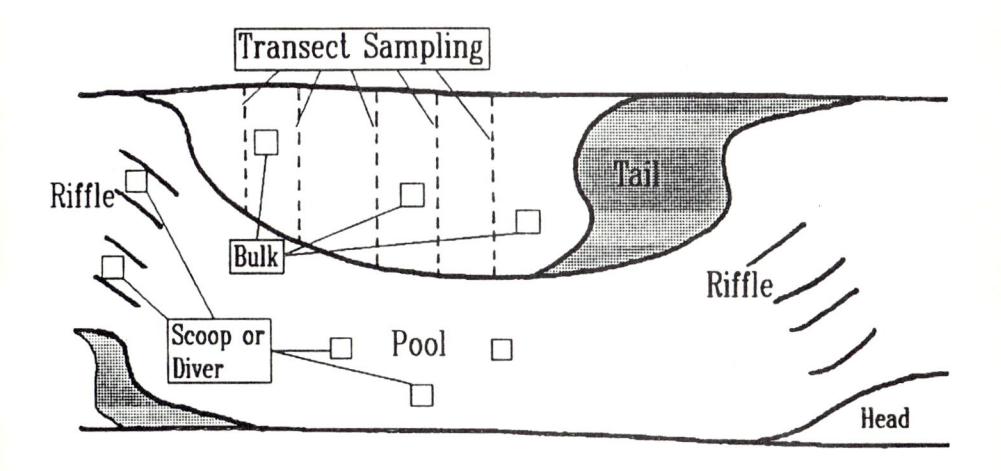

Fig. 2 - *Sketch of the main physiographic units with the sampling methods adopted.*

The visual comparator consists of a twofold tablet to which natural sediment specimens, ranging in size from 16 to 0.063 mm and arranged according to 1/2 phi scale, are attached. The operator compares bed sediment with that on the tablet and assigns the closest phi size. Laboratory experiments indicate 70% agreement between actual size and that inferred by the visual comparator. A bias of 0.5 phi occurs in only 25 % of the observations (Billi, in preparation).

The scoop sampler consists of a rectangular frame (50X30 cm) to which a bag, of the type used for the Helley-Smith sampler (Helley & Smith, 1971), is connected (Billi, 1990; 1992). A downstream operator holds the sampler firmly on the bed, another upstream, shovels a prefixed volume of bed material into the sampler. Almost all the fines are collected as the size of the bag net opening is 0.1 mm. In deep water reaches a diver adopted the same sampling procedure simply by swimming to the bed and filling the scoop from upstream. Each bulk sample consisted of three sub-samples with a total weight of more than 30 kg. The sampling campaigns on the Orcia and Cecina rivers were carried out during spring 1990 and 1991 respectively.

DATA ANALYSIS AND DISCUSSION

The downstream grain-size data are widely scattered and do not at any significant level follow a simple correlation model (Fig. 3). Averaging the data of each sampling site and interpolating over distance for the Orcia (Fig. 3 a, dashed line) shows a downstream increase in grain size. Downstream reaches having a higher gradient than those upstream. By contrast, the Cecina data follow a more ordinary pattern, with a slight downstream decline (Fig. 3 b, dashed line).

In terms of grain size, the different morphologic units seem to vary independently. The studied streams show different behaviors probably related to flood history. Particularly surprising is the low correlation between bar surface and bar sub-surface D50. Figures 3 and 4 show that the study streambeds are only locally armored and that armoring seems to be a phenomenon less common than expected. This is even more surprising if we consider that both the Orcia and the Cecina riverbeds have undergone a remarkable degradation due to past extensive gravel exploitation.

In the literature, armoring is commonly described as a typical feature of gravel-bed rivers. Sub-surface D50 is reported to be to 1/2 - 1/4 surface D50 (Parker & Klingeman, 1982; Andrews & Parker, 1985). Obviously, extending the pebble count down to 4 phi makes surface material grading curves and D50 closer to those of sub-surface sediment. Results similar to those in figure 4 were found by Mosley & Tindale (1985), Church et alii (1987) and Lisle & Madej (1992). In the first two papers the authors truncated the grain-size distributions of both surface and sub-surface material at 8 mm (smaller grains are difficult to handle with the pebble count method). Lisle & Madej

(a)

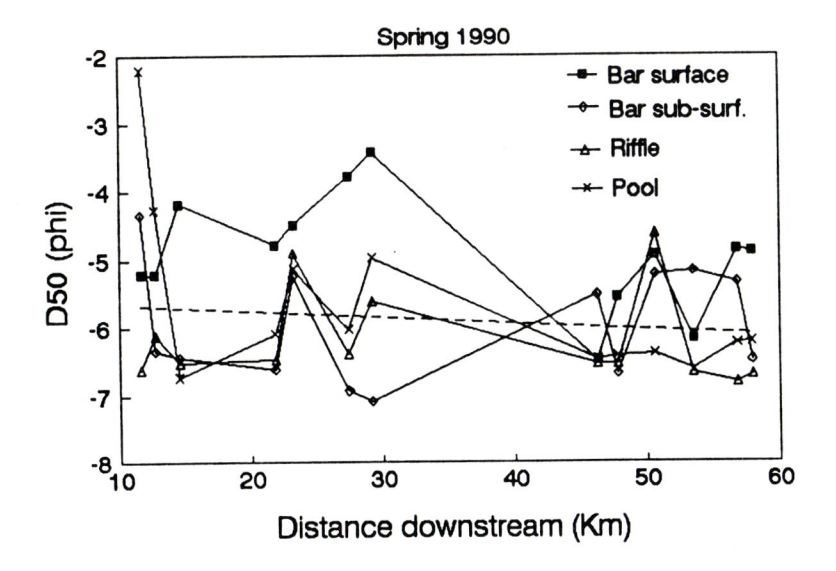

(b)

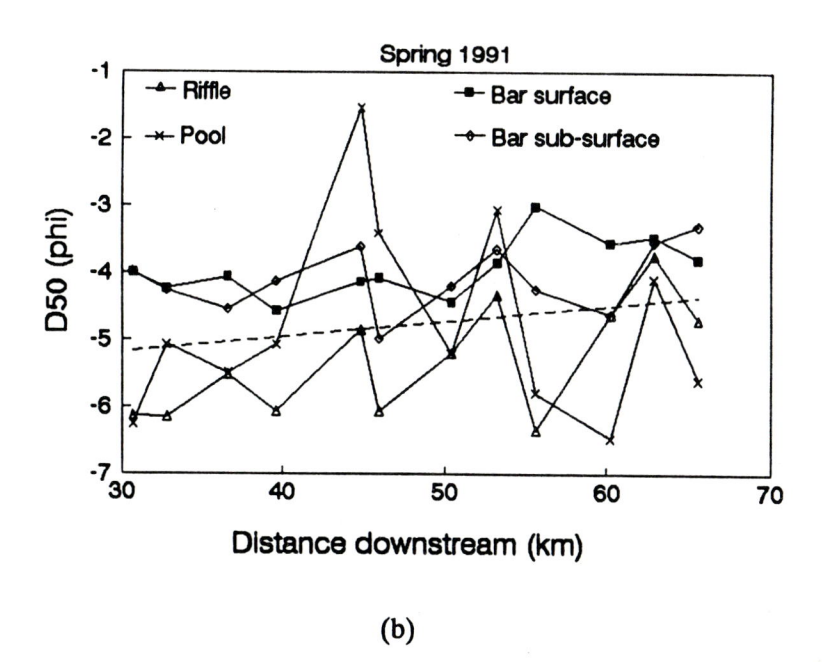

Fig. 3 - *Downstream variation of D50 for the main physiographic units: a) Orcia River; b) Cecina River. The dashed lines indicate the general trend of D50.*

limited the pebble count to 4 mm, but estimated the finer fraction using the less than 4 mm size distribution of the sub-surface material. Both procedures, imply artificially modified grading curves. The modifications have the advantage of making the two sample populations comparable as does the use of the visual comparator technique developed in this study (presumably with higher precision). In the study streams, the lack of evident bed Armour could be accounted for by the small number of sub-surface samples collected, although Mosley & Tindale (1985) found the same situation with a larger number of samples. In order to investigate the absence of armor as a due to the limited number of samples, replicate sub-surface sampling was carried out before and after a large bankful flood on two bars of the Cecina (one in the upstream reach, on site 2, one in the downstream reach, on site 10, Fig. 1). A total of 40 bulk samples were therefore collected.

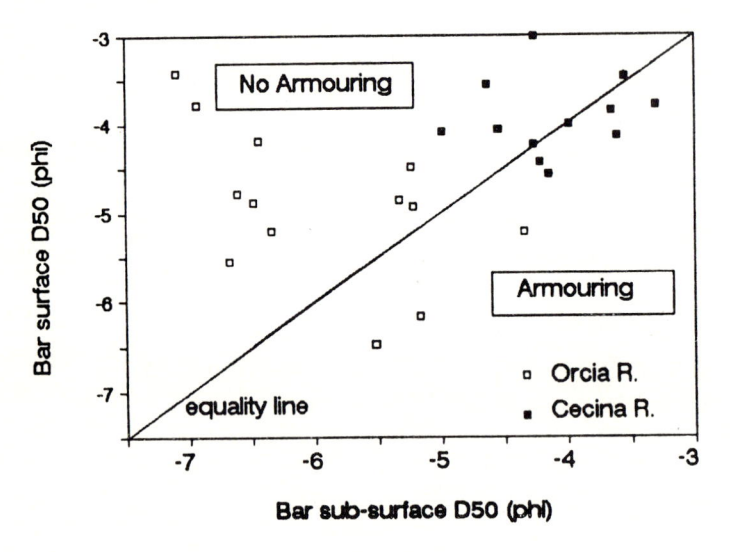

Fig. 4 - *Plot of bar surface vs sub-surface D50.*

The D50 of site 2 bar sub-surface samples is almost constant around -4.3 phi. In the bar of site 10 it varies widely ranging from -1 to -6.6 phi. The large variation was found to be closely related to the fine material (sand) content. Apparently, as the sand percentage increases from head to tail of the bar on site 10, the sorting processes, less effective in the upstream reaches, become more and more important downstream. This leads to explicit sediment segregation commonly associated with bar migration and accretion, as observed on the bar of site 10.

Further detailed sediment studies are necessary to investigate this point and to develop an effective strategy for bar sub-surface sampling.

To further investigate the lack of armoring, and eliminate the effect of sand content, surface and sub-surface grain-size distributions of the Cecina were truncated at 2 mm, the new D50s were recalculated. The gravel D50s were well correlated ($R = 0.86$), but the data again showed a no armoring for most of samples (Fig. 5). Only the finer samples lay in the armoring field (Fig. 5). Because the D50 of the gravel fraction can approximate the D84 of a nontruncated sample, a plot of surface D84 to subsurface D84 can be constructed (Fig.6). Figure 6 is very similar to figure 5 with a good correlation coefficient ($R = 0.85$). The same four points in the armoring region. Use of

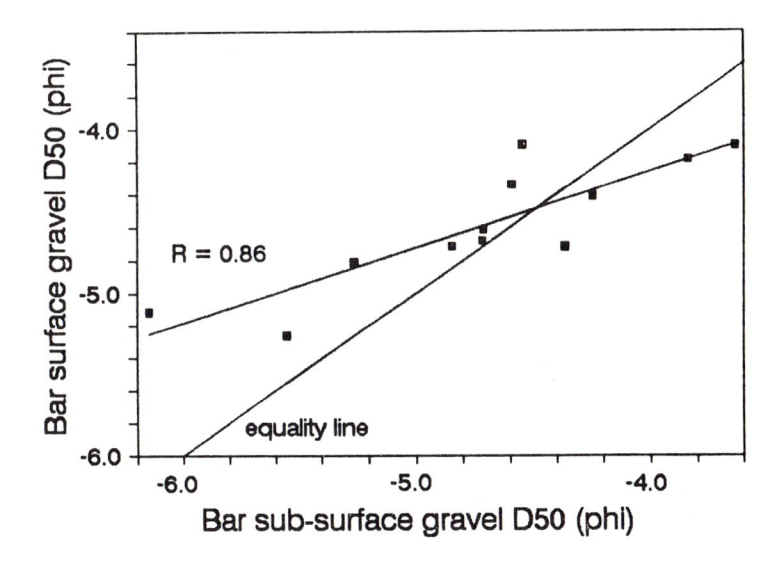

Fig. 5 - *Cecina River: plot of bar surface vs sub-surface D50, grain-size distributions truncated at 2 mm. R is the correlation coefficient.*

D16 of the gravel fraction to make a similar kind of plot produced no significant correlation.

These results support the conclusion reported by Lisle & Madej (1992), although they came to it through different reasoning, that equal mobility (Parker et al., 1982) can coexist with streamwise selective transport. The lack or armoring suggests equal mobility while the armoring may be related also to selective transport. Thus, both equal mobility and selective transport exist, but at different position along the stream bed. One can speculate that in the upstream reaches, where bed material is typically coarser, the local flow conditions are able to entrain particles of all sizes which are deposited after a short distance. Here, no sediment segregation occurs and equal mobility prevails over selective transport, although the latter may operate also. Down-

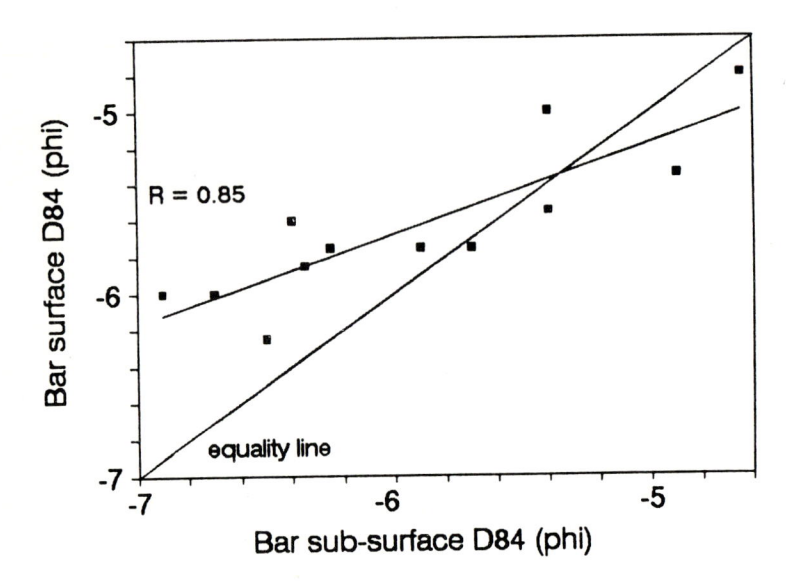

Fig. 6 - *Cecina River: plot of bar surface vs sub-surface D84 of actual grain-size distributions. R is the correlation coefficient.*

stream, on the bars bed material becomes finer and sorting processes are more and more effective, leading to selective transport more favorable to armoring.

From analysis of data gathered, D84 has to be the preferred characteristic diameter, for describing bed material of gravel rivers. The D84 is affected by content variations of the finer fraction to a lesser extent than D50. In fact, the occurrence of fines in general, and sand in particular, on a gravel stream bed is only partly related to the flood flow characteristics (peak discharge, duration, etc.). Winnowing processes, which are very active during receding flow, may play a very important role in eroding or depositing large masses of fine sediment.

In figure 7 the downstream variations of sand content for the main physiographic units of the study rivers are reported. As previously observed for D50, no significant correlation exists among the units and the two rivers show different behaviors. In the Orcia the overall sand percentage tends to decrease

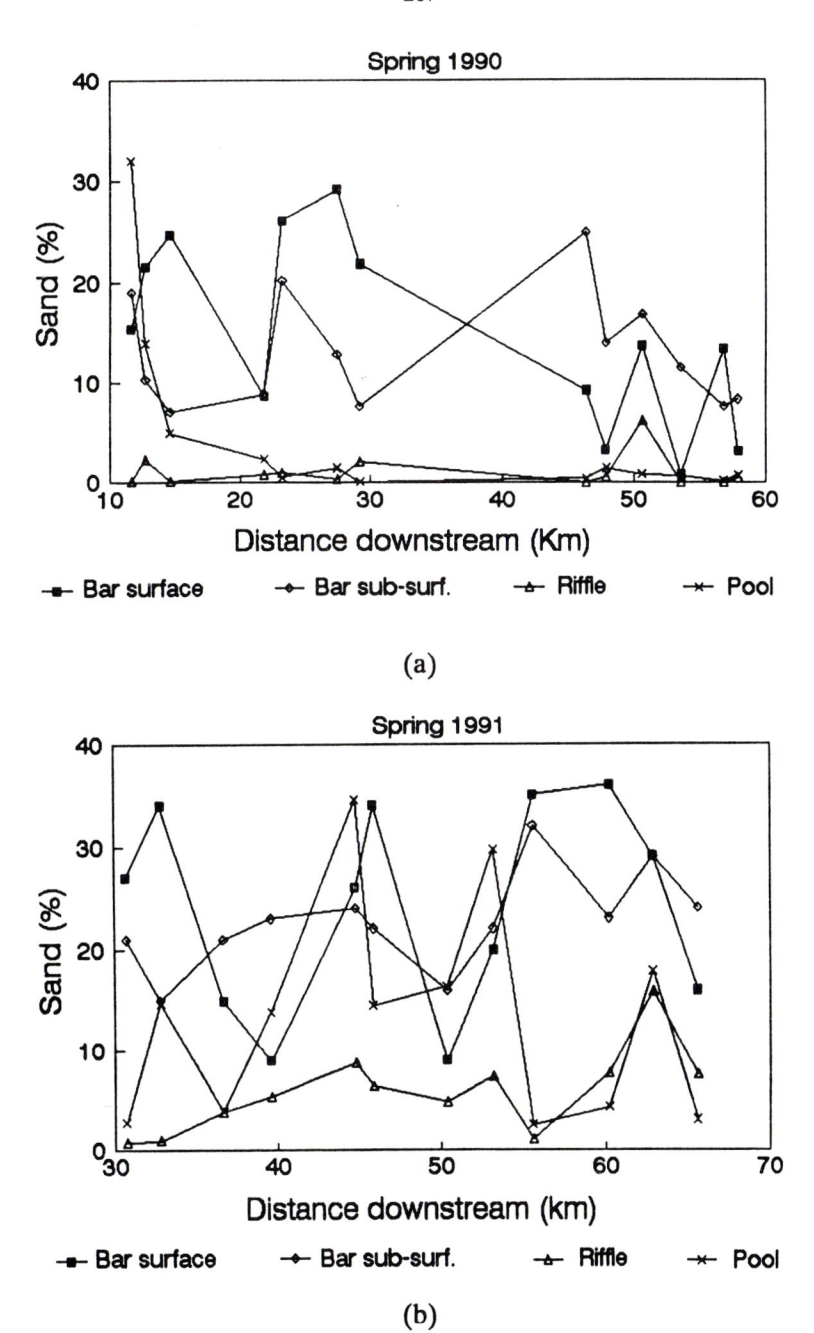

Fig. 7 - *Downstream variation in sand content of the main physiographic units: a) Orcia River; b) Cecina River.*

in a downstream direction (Fig. 7 a). This, as the unusual downstream increase of D50 (Fig. 3 a), is probably due to the higher gradient of downstream reaches. The Cecina shows a more regular pattern. The two rivers are generally comparable in terms of sand content but for the pools (Fig. 7 a and b). In fact, the sand content of the Cecina pools is higher than those of the Orcia. The poverty of sand in the Orcia pools may be due to low intensity flood flows which characterized a long period before the sampling campaign. They were probably able to wash out only the finer bed material.

Riffles of both rivers had the lowest and least variable sand content. The Cecina river pools had the widest variation. During the sampling campaign, it was noticed that a few of the Cecina pools were filled with a large quantity of very fine, unconsolidated sediment, while others displayed an imbricated, coarse particle bed. The former correspond to the peaks of figure 7 b. The latter have a sand content close to that of riffles (Fig. 7 b).

The data suggest that pools may act as fine sediment traps.
Since there were also pools without much sand, it can be argued that sine-sediment trapping depends on local factors such as trapping capability, bar dissection, winnowing and upstream sediment supply. Unfortunately, data on pool geometry is not presently available for the study streams. Since the trapping efficiency of a pool is related to its geometry, this fundamental factor deserves more detailed field investigation.

The study rivers have a predominantly straight channel with alternate lateral bars. Their morphology and downstream migration match the model described by Bluck (1979), so they are commonly attached to a bank and have low angle slip faces dipping either downstream or transversely toward the adjacent pool. At or near bankfull discharge the flow is capable of shaping the bed as a continuum. During the receding stages, however, many processes such as bar dissection, development of transverse channels, progradation of microdeltas and overall bar washout may yield quantities of fine sediment on the static (likely coarse grained) bed of the adjoining pool. Here the flow conditions are close to or below the transport threshold (Keller, 1971). Therefore, during the late flood stages, bars and pools may act as an interrelated system in which the morphological evolution of the former deeply affects the latter. This statement would seem to be proved by the diagrams of figure 7 b, where low values for the bar surface are approximately matched by high sand percentages in the pools and vice versa. This aspect deserves further investigation.

Finally, a third, more speculative explanation for the sand accumulation in a few Cecina pools may be the occurrence of scattered large masses of fine sediment, moving downstream, mainly during the receding flood flow, on a static pebbly bed. These sand waves may match the kinematic wave model of Leopold (Leopold et al., 1964), as they probably exchange much of their mass with the fine sediment in transit while slowly moving downstream. As the flow wanes, the mass exchange ends and the sand is temporary stored in one or a few adjacent pools due to their higher trapping efficiency (or because the sand wave was just there). A second field survey, carried out on the Cecina soon after a large bankfull flood, seems to confirm this hypothesis (Billi, unpublished data).

CONCLUSIONS

With very simple equipment, exposed and underwater units of gravel-bed rivers can be reliably sampled and the pebble count method can be extended down to 4 phi.

This improved procedure was applied to two gravel-bed rivers of the Northern Apennines. No significant correlation was found among the D50s of the sedimentary units considered in the paper (i.e. bar surface, bar sub-surface, riffle and pool). Therefore, the inference of underwater unit D50s from exposed bar D50s is not straightforward. Some correction factors related to flood history and hydraulic characteristics, not provided in this paper, need further investigation.

Gathered data for the study rivers show that exposed parts of the bed (bars) are generally not armored. This unexpected result can be partly explained by the limited number of sub-surface samples. In two experimental bars replicate sampling was carried out. In the bar of the upstream reach D50 and sand content were almost constant. In the bar of the downstream reach the sand percentage and, as a consequence, D50 varied greatly, becoming finer from bar head to bar tail. This can be accounted for by the sorting processes being more effective in the downstream than in the upstream reaches.

The fine fraction (sand) adds a sort of background noise which makes the relationship between bedload processes and bed material more difficult to understand. The bar surface and sub-surface grain-size distributions were truncated at 2 mm and the D50s of the new curves recalculated. The gravel D50s resulting exhibited high correlation, but a general armoring condition was still lacking. Only a few finer samples show an armoring ratio (surface D50/sub-surface D50) slightly higher than one. Identical results were obtained using D84 of the actual grading curves.

In the upstream reaches, where bed material is typically coarser, sorting processes are scarcely active and equal mobility prevails. In the downstream reaches, selective transport becomes more important giving rise to favorable conditions for development of Armour on bar surfaces. Therefore, equal mobility and selective transport can coexist in gravel-bed rivers with streamwise sorting like those considered in this paper.

D84 seems to be a more stable characteristic diameter than D50 since the former is affected less by the sand fraction amount. To investigate the interactions between bar sediment and flow characteristics, D84 is probably a more appropriate diameter.

Downstream variation in sand content for the physiographic units considered in this paper was investigated. In the Cecina, a few pools displayed fine sediment filling. This is connected with riffle winnowing, bar dissection and bar washout during the receding flood flow. Pool geometry or its trapping efficiency, however, may be important as well. For these reasons the grain size of pool bed material is unpredictably variable and in places unrelated to the main hydraulic characteristics of the reach. Therefore, much care must be taken to properly select pool sampling sites for field investigation of bed armoring processes.

ACKNOWLEDGMENTS

The author is grateful to an unknown referee that greatly improved the style and the clarity of the manuscript. F. Vannacci, A. Bettazzi and C. Manieri are acknowledged for their help in collecting the field data and for the laboratory analysis of the samples.

This research was financially supported by National Research Council (CNR) contracts n. 90.02707.CT05 and 91.00737.CT05.

REFERENCES

ANDREWS, E.D., PARKER, G., 1985, *The coarse surface layer as a response to gravel mobility.* In THORNE, C.R., BATHURST, J.C. and HEY R.D. (eds), Sediment Transport in Gravel-Bed Rivers, Wiley Chichester, 269-325.

BILLI, P., 1989, *Sediment survey of alluvial channels,* Proceedings of the International Congress on Geoengineering, Torino, 27-30 Settembre, 3, 1525-1530.

BILLI, P., 1990, *Sediment dynamics studies in the Orcia River,* 3rd International Workshop on Gravel-Bed Rivers Field Excursion Guide, 1-14.

BILLI, P., 1992, *Variazione areale delle granulometrie e dinamica degli alvei ghiaiosi: metodologie di campionamento ed analisi dei primi risultati*, Atti del Convegno su: Erosione ed Alluvionamenti in aree caratterizzate da subsidenza o da tettonica, Universitá di Ancona, 14-15 Ottobre 1991, in press.

BLUCK, B.J., 1979, *Structure of coarse braided stream alluvium*, Trans. Royal Soc. Edimburg., 70, 181-221.

BLUCK, B.J., 1982, *Texture of gravel bars in braided streams*. In HEY, R.D., BATHURST, R.D., THORNE, C.R., (eds), Gravel-Bed Rivers, Wiley, Chichester, 339-355.

CHURCH, M., 1983, *Pattern of instability in a wandering gravel bed channel*, I.A.S. Spec. Publ. 6, 169-180.

CHURCH, M., JONES, D, 1982, *Channel bars in gravel-bed rivers*. In HEY, R.D., BATHURST, R.D., THORNE, C.R., (eds), Gravel-Bed Rivers, Wiley, Chichester, 291-324.

CHURCH, M.A., McLEAN, D.G., WOLCOTT, J.F., 1987, *River bed gravels: sampling and analysis*. In THORNE, C.R., BATHURST, J.C. and HEY R.D. (eds), Sediment Transport in Gravel-Bed Rivers, Wiley Chichester, 43-88.

ERGENZINGER, P., 1992, *Riverbed adjustments in a step-pool system:, Lainbach, Upper Bavaria*. In BILLI, P., HEY, R.D., THORNE, R.C. & TACCONI, P., (eds.), Dynamics of Gravel-Bed Rivers, Wiley, Chichester, 415-430.

FERGUSON, R.I. WERRITTY, A., 1983, *Bar development and channel change in the gravelly River Feshie, Scotland*, I.A.S. Spec. Publ., 6, 181-193.

GREGORY, K.J. (ed.), 1977, *River channel changes*, Wiley, Chichester, 448 pp.

GOMEZ, B., CHURCH, M., 1989, *An assessment of bed load sediment transport formulae for gravel bed rivers*, Water Resour. Res., 25(6): 1161-1186.

HARVEY, A.M., 1987, *Sediment supply to upland streams: influence on channel adjustment*. In THORNE, C.R., BATHURST, J.C. and HEY R.D. (eds), Sediment Transport in Gravel-Bed Rivers, Wiley Chichester, 121-146.

HEIN, F.J., WALKER, R.G., 1977, *Bar evolution and development of stratification in the gravelly, braided, Kicking Horse River, British Columbia*, Can. J. Earth Sci., 14, 562-570.

HELLEY, E.J., SMITH, W., 1971, *Development and calibration of a pressure-difference bedload sampler*, U.S. Geol. Surv., Open-File Report, 18 pp.

HEY, R.D., THORNE, C.R., 1983, *Accuracy of surface samples from gravel bed material*, J. Hydraul. Eng., ASCE, 109, 842-851.

KELLER, E.A., 1971, *Areal sorting of bed-load material: the hypothesis of velocity reversal*, Geol. Soc. Am. Bull., 82: 753-756.

KELLERHALS, R., BRAY, D.L., 1971, *Sampling procedures for coarse fluvial sediments*, Proc. ASCE, J. Hydr. Div., 97: 1165-1179.

JACKSON, R.G. II, 1976, *Depositional model of point bars in the lower Wabash River*, Jour. Sedim. Petrology, 46: 579-594.

LEOPOLD, L.B., 1970, *An improved method for size distribution of stream bed gravel*, Water Resour. Res., 6(5): 1357-1366.

LEOPOLD, L.B., WOLMAN, M.G., MILLER, JP., 1964, *Fluvial processes in geomorphology*, Freeman, San Francisco, 521 pp.

LEWIN, J., 1976, *Initiation of bedforms and meanders in coarse grained sediment*, Geol. Soc. Am. Bull., 87: 281-285.

LISLE, T.E., MADEJ, M.A., 1992, *Spatial variation in armouring in a channel with high sediment supply*. In In BILLI, P., HEY, R.D., THORNE, R.C. & TACCONI, P., (eds.), Dynamics of Gravel-Bed Rivers, Wiley, Chichester, 277-293.

McGOWEN, J.H., GARNER, L.E., 1970, *Physiographic features and stratification types of coarse-grained point bars; modern and ancient examples*, Sedimentology, 14: 77-112.

MOSLEY, M.P., TINDALE, D.S., 1985, *Sediment variability and bed material sampling in gravel-bed rivers*, Earth Surf. Proces. and Landforms, 10: 465-482.

PARKER, G., KLINGEMAN, P.C., 1982, *On why gravel bed streams are paved*, Water Resour. Researches, 18: 1409-1423.

PARKER, G., KLINGEMAN, P.C., McLEAN, D.G., 1982, *Bedload and size distribution in paved gravel-bed streams*, ASCE, J. Hydraulics Div., 108(HY4),544-571

SCHUMM, S.A., 1977, *The fluvial system*, Wiley, New York, 330 pp.

WOLCOTT, J., CHURCH, M.A., 1991, *Strategies for sampling spatially heterogeneous phenomena: the example of river gravels*, Jour. Sedim. Petrology, 61(4): 534-543.

Discharge and fluvial sediment transport in a semi-arid high mountain catchment, Agua Negra, San Juan, Argentina

Dietrich Barsch, Hans Happoldt, Roland Mäusbacher,
Lothar Schrott & Gerd Schukraft

Department of Geography, University of Heidelberg, INF 348,
D-69120 Heidelberg, Germany

Abstract

The measurements of discharge and fluvial sediment transport carried out in a semi-arid catchment of the High Andes of Cuyo show that the concentration of suspended sediment increases as the catchment area becomes larger, whereas the discharge shows only a minimal increase further downstream or even a decrease in certain parts, due to the high rates of seepage and evaporation. The discharge and suspended sediment yield of a tributary basin (covering 57 km²) was measured over a period of 5 months, from November 1990 until April 1991. With mean values of discharge of 0.35 m^3/s and peaks of up to 1.5 m^3/s the total suspended sediment yield was 249 tons, i.e. 4.4 g/m². During one week in January 1991 the Agua Negra river showed suspended sediment yields ranging between 17.9 t in its upper course (57 km²) and 263.6 t in its lower course (617 km²). Compared with the suspended sediment yield which is clearly dominant in sediment transport the bedload is very low. However the analysis of the morphology and stratigraphy of some terraces indicates flood events in the past with sediment transports much higher than today's.

1. Introduction

Very little attention is paid to the complexity of streams in high mountain areas. Data series about discharge and sediment supply are normally not available for proglacial catchments in high altitudes (cf. GURNELL 1987 a,b). Obvious differences in the geomorphic processes involved and variations in precipitation in time and space make extrapolation from lowland river data impossible. Thus, even short data series from mountain rivers enlarge our restricted knowledge. This is especially true for mountains where up to now hardly any measurements have been made. In addition, most of our information is restricted to mid-latitude mountains. Subtropical and tropical mountain systems have often experienced only a marginal interest. Therefore, the studies reported have been carried out in the semi-arid Andes of San Juan (Argentina) at latitude 30° S. In this part of the Andean region, detailed analyses of the river system become very important because of the water scarcity and the increase of desertification. Agriculture, for example, totally depends on irrigation, mainly drawing upon the meltwater of the mountain rivers (MINETTI & SIERRA 1989). A broad range of problems is caused by abrupt snowmelts or high rates of fluvial sediment transport (damage to roads and railways, blocked irrigation pipes). The purpose of this paper is to present some initial information about the varying quantities of discharge and sediment supply for a typical proglacial mountain stream with a nival glacial regime under semi-arid conditions.

2. Study area

The investigation was carried out in a catchment area of the river system Rio Jachal in the province of San Juan.

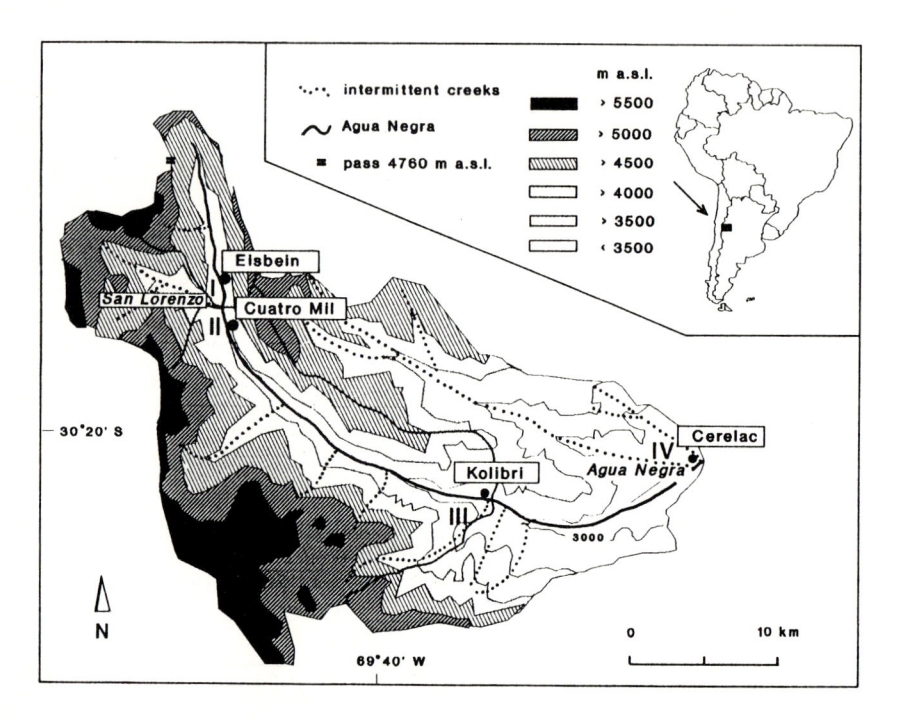

Fig. 1. Map of the Agua Negra basin showing the location of the gauging stations

One advantage of the study area was its good accessibility due to the mountain pass to Chile, another the existence of topographical maps with a high resolution (scale 1:10.000) as well as aerial photographs. The studied catchment area has a surface of 617 km², its vertical extension being 3500 m (see Fig. 1). The area mainly faces SSE. The water of the Agua Negra is channelled into pipes below the Cerelac station and is used for irrigation of the oases.

Three further gauging stations were installed along the longitudinal section of the Agua Negra (see Fig.1). Only a small part (about 8.1 %) of the basin is covered by glaciers as a result of its semi-arid climate and high solar radiation (SCHROTT 1991). The two stations situated in the upper course of the river - Eisbein (4150 m a.s.l.) and Cuatro Mil (4000 m) - draw from tributary basins of 57 km² and 113 km² respectively. Large parts are underlain by high mountain permafrost, the presence of which is indicated by the active rock glaciers that appear above 4000 m (HAPPOLDT & SCHROTT 1992). The two stations situated in the lower course of the Agua Negra , Kolibri (3150 m) and Cerelac (2650 m), draw from much larger areas, covering 365 and 617 km² respectively.

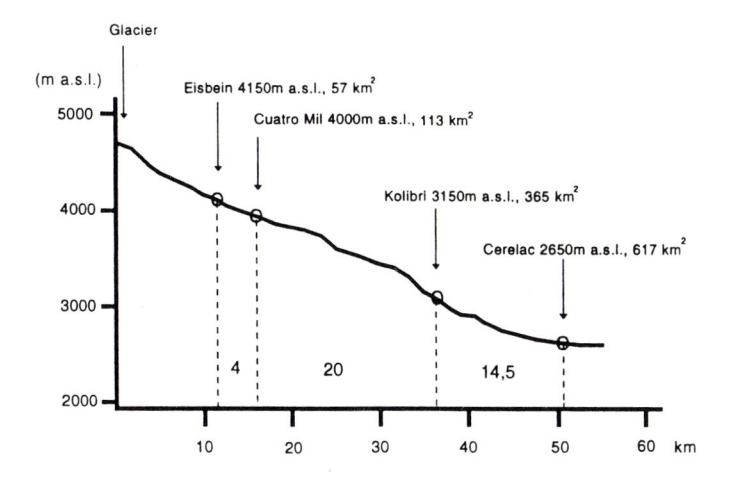

Fig. 2. Longitudinal section of the Agua Negra river showing the elevation and area covered by the gauging stations

3. Methods

The measurement stations were equipped with dataloggers for continuous recording of changes in the water level, air and water temperature as well as electrical conductivity. A total of 88 measurements of runoff, which were carried out by means of a current meter and a NaCl tracer at different water levels, form the basis of the rating curves. As a result of the ever changing cross-section and high turbulence the results produced by the NaCl tracer method were more exact. In order to determine the relationship between discharge and suspended sediment concentration water samples of 1 l were taken manually from the fastest running part of the stream at regular intervals (7, 14, 21 h) and additionally during floods with an obvious increase in suspension concentration. The water samples were passed through a 0.2 μm membrane filter.

The bedload was established by means of sediment baskets and pebble and cobble tracers (see also BARSCH et al. 1992). The basket is 80 cm long, 50 cm wide, and 40 cm high. The mesh size is 15 mm. The pebble and cobble tracers, which had been marked, measured and weighed beforehand, were laid out in different cross-sections. Indirect methods such as grain size analysis on terraces were applied in order to obtain information about possible competences and sediment load during flood events in the past.

4. Diurnal and seasonal stream discharge variation

At the Eisbein station (4150 m) the discharge was measured throughout the whole ablation period of 1990/91 with the exception of the minimal quantities - due to snow and ice -

during the winter months. The nival-glacial river regime at the Eisbein station reflects a chronological sequence of snow, glacier- and permafrost-melt. The highest daily mean of discharge as well as the maximum was reached in December with 731 l/s and 1344 l/s, respectively. During the ablation period of 1990/91 the mean discharge was 324 l/s. For the tributary basin I (see Fig. 1) this corresponds to a total discharge of $5.07 \cdot 10^6$ m^3/s.

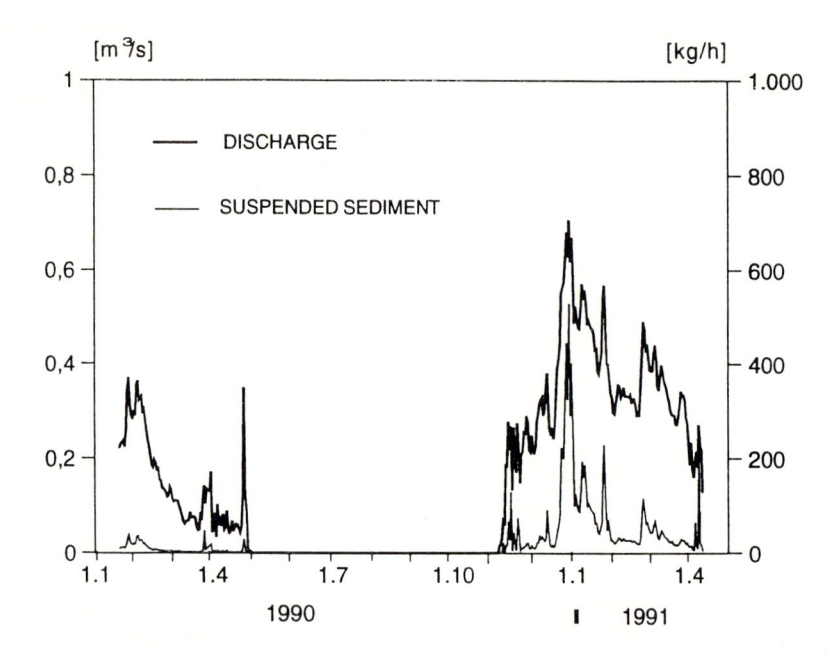

Fig. 3. Discharge and suspended sediment transport rate at the Eisbein gauging station (4150 m)

The diurnal and annual variability of runoff in the catchment is heavily influenced by the intensity of solar radiation as well as sensible heat flux. Precipitation is mainly concentrated during the winter months, therefore its influence on the discharge volume is - within the scope of this study - negligible. Runoff maxima are observed in different time lags from radiation maxima depending on the time of the year and the changes in the drainage system of the glacier. Up to the end of December the discharge is largely determined by snowmelt. Parallel discharge measurements that were carried out near the glacier tongue and at the Eisbein gauging station provide information about the proportions of glacier- and permafrost-melt in January 1991. It must be remembered that the lag time of the water-wave from the glacier tongue to the Eisbein gauging station is almost 4 h over a distance of 9.5 km (flow velocity 0.7 m/s). The one third increase in discharge volume at the gauging station gives the volume of permafrost-melt in the area between the glacier and the gauging station. When, after the snowmelt, the albedo is lower and the drainage system is better developed, strong variations in radiation or temperature are in general immmediately reflected by the runoff.

Figure 4 shows that with continuously high solar radiation, taking account of a certain time lag, changing temperatures are also reflected by an increase or decrease of discharge.

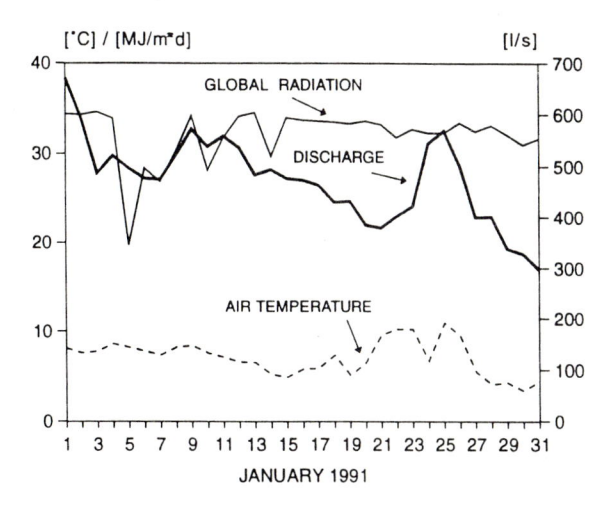

Fig. 4. Global radiation, air temperature and discharge at the Eisbein gauging station (4150 m)

Short-term variations in temperature like the decrease on January 24 are compensated by high global radiation.

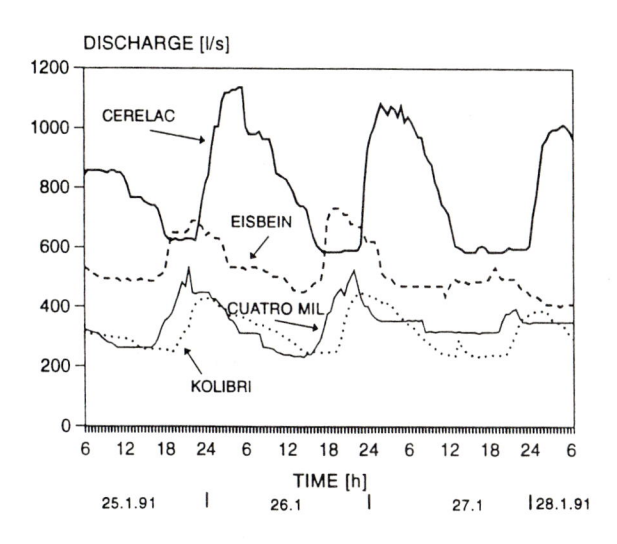

Fig. 5. Diurnal discharge variations at the four gauging stations

During January 1990/91 the diurnal fluctuation of discharge was measured at three other gauging stations in the lower course of the river. Figure 5 illustrates the development in the longitudinal section during a characteristic part of this period. At all four gauging stations a diurnal variation is clearly visible. According to the size of the tributary basins, the highest diurnal mean of discharge is measured at the Cerelac station (2650 m) with 785 l/s, whereas at the highest station, Eisbein (4150 m), in the upper course of the Agua Negra the mean was 341 l/s. At the Cuatro Mil and Kolibri stations, however, the situation is completely different. Although Cuatro Mil is only about 4 km below Eisbein (see Fig. 1), with another creek (San Lorenzo) flowing into the Agua Negra 1 km above the station, the discharge is considerably lower. Another 14.5 km down the Agua Negra river at the Kolibri gauging station the discharge is hardly any higher in spite of the big increase in the size of the basin. This means that the total discharge decreases considerably between the Eisbein gauging station, situated relatively near to the glacier, and the stations Cuatro Mil and Kolibri further downstream. There are two reasons for this phenomenon: Firstly a certain amount of water infiltrates into the extensive fluvio-glacial deposits and secondly water is further lost due to evaporation. These types of water losses have been observed in other high mountain catchments (cf. MAIZELS 1983).

5. Discharge and suspended sediment yield

The calculations of the total suspended sediment supply are based on the specific relationship between discharge and sediment concentrations at each gauging station (see Fig. 6). In spite of a few cases in which concentration of sediment decreases and discharge increases (hysteresis), which are caused for example by water outbursts from the glacier, a relationship between discharge and suspended concentration can be observed. The transport rates which are calculated on the basis of the moderate correlation should be seen as a specific and not a universal case. The difficulty of the terrain makes obtaining long-term data unfeasible.

SUSPENDED SEDIMENT CONCENTRATION [g/l]

$Suspension = 0.0032\ e^{0.0058\ Q}$

$n = 25$

$r = 0.69$

DISCHARGE [l/s]

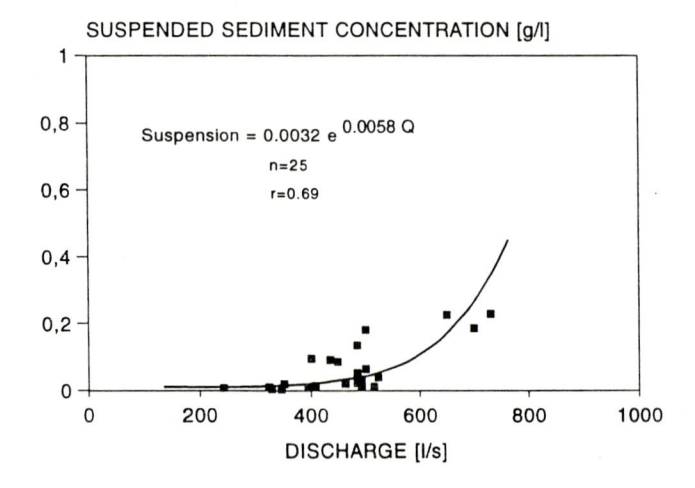

Fig. 6. Assumed relationship between discharge and suspended sediment concentration at the Eisbein gauging station (4150 m)

The variations of suspended sediment concentration with unchanging discharge can be explained by the different available amounts of loose material (GURNELL 1987 a,b). Quantities of unconsolidated material are caused by the extreme climatic conditions in the area and by the rock properties. Aridity, high solar radiation and the consequent extreme heating of the rock surface are important factors within the weathering process. Thus, talus is abundant, small grain sizes (silt) are very frequent in all talus deposits and at their surface. This material is mobilized by mudflows from melting snow patches and by the outflow from intermittent springs. The latter was generally observed during days with high solar radiation. Part of the suspended material was directly washed into the river. Other material is temporarily deposited in the great alluvial fans.

The observations and measurements of suspended sediment concentration near the glacier tongue suggest that the most important sediment source is the proglacial area, where sediments are remobilized by fluvial processes.

At all four stations not only were regular water samples taken 3 to 5 times a day, but also a complete diurnal course with a 3 hours interval was made in order to be able to differentiate more clearly the relationship between discharge and suspended sediment concentration. Normally days without events reflect a good correlation of suspended sediment concentration and discharge.

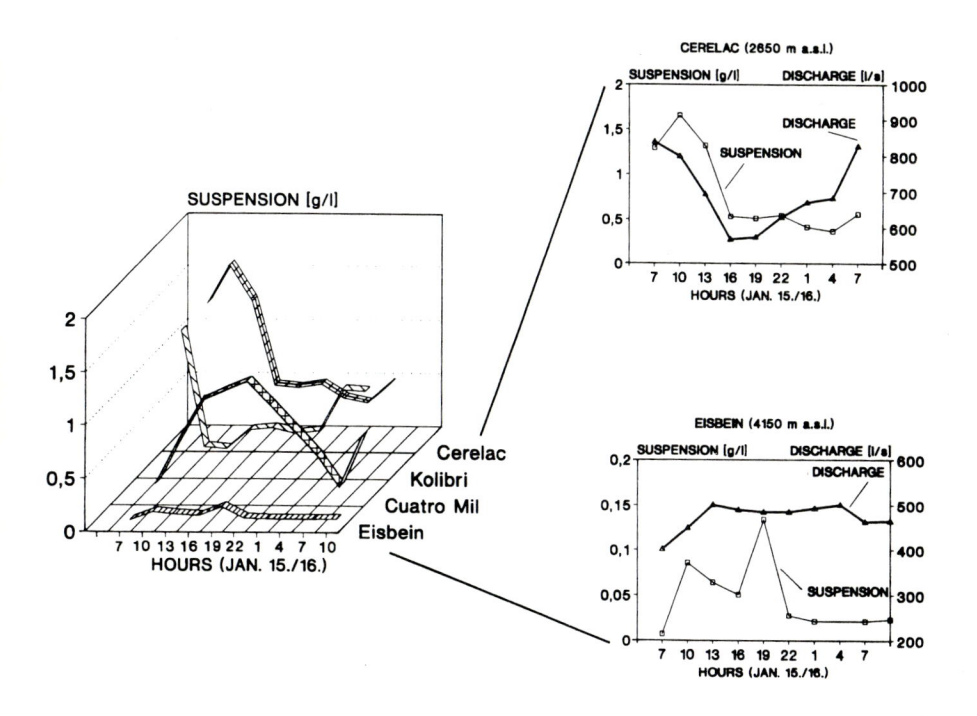

Fig. 7. Diurnal variation of suspended sediment concentration and discharge

By means of the rating-curve the suspended sediment yield at the highest station, Eisbein, can be calculated for the measuring period (see Fig. 3). The mean suspended transport rate during the ablation period of 1990/91 was 57.4 kg/h. This equals a total suspended sediment yield of 249.15 tons or 4.368 g/m². Two thirds of the suspended sediment were transported in December and January with a mean of 112.8 kg/h and short peaks of more than 1000 kg/h. According to these calculations a maximum of 1345 kg/h was reached on December 27. Owing to the size of tributary basin I and widespread permafrost at 4000 m a.s.l. the lowest suspended sediment yield was measured at the Eisbein station. At the Cerelac station, about 40 km further downstream, the concentrations of suspended sediment are considerably higher and show well marked diurnal variations. The mean suspended sediment transport rate is about 1500 kg/h for the period of January 1991. It can, however, triple within short periods of time (Fig. 8).

The diurnal fluctuations of discharge and suspended sediment concentration are a result of the fluctuations of air temperature and solar radiation. During the day advective heat and high solar radiation produce an intensification of water flow from snow patches, glacier- and permafrost areas. Owing to the large catchment area minimum and maximum discharge are subject to significant time lags. This means that at the Cerelac gauging station maximum discharge occurs approximately 12 h later than the radiation readings.

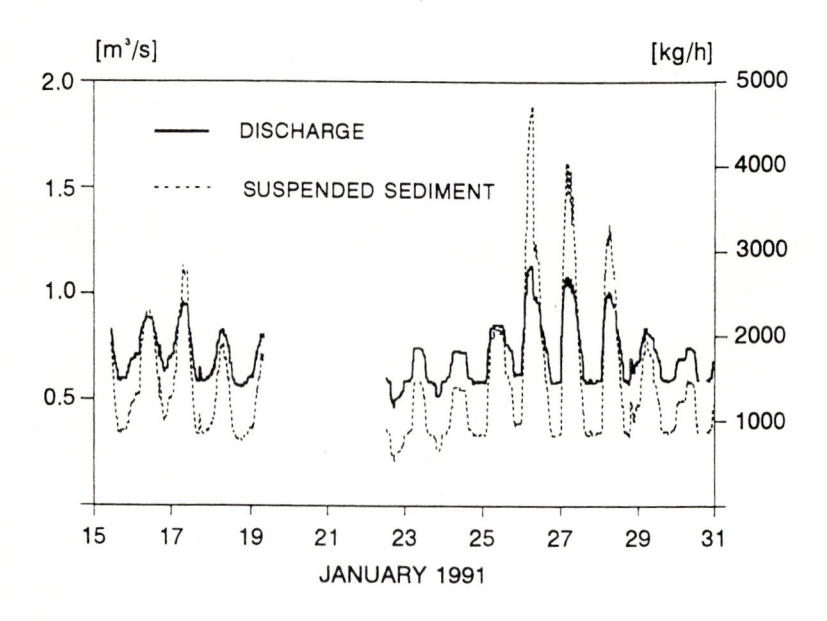

Fig. 8. Discharge and suspended sediment transport rate at the Cerelac gauging station (2650 m)

The parallel recordings at all four gauging stations make it possible to compare the different suspended sediment loads. The budget of the week from 24 to 30 January shows a continuous increase in the suspended sediment yield from the highest to the lowest gauging station.

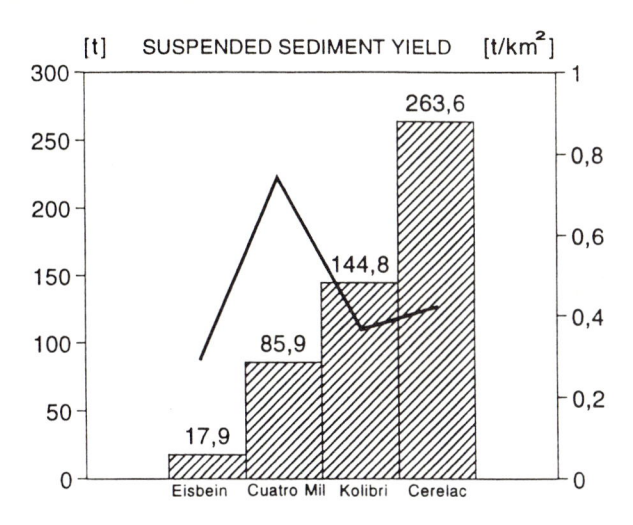

[Chart: SUSPENDED SEDIMENT YIELD showing values – Eisbein: 17.9, Cuatro Mil: 85.9, Kolibri: 144.8, Cerelac: 263.6]

Fig. 9. A comparison of the sediment yield in the longitudinal section of the Agua Negra from January 24 to January 30.

6. Bedload

The measurements relating to bedload in a high-energy-system are influenced by the amounts of available material and various factors such as magnitude of floods, flow velocity, turbulence, vegetation and grain size. As these parameters, as well as the course of the river, are subject to variation, the calculation of bedload transport is very problematic and subject to various errors (GOMEZ 1987).

Errors occur as a result of slower currents caused by resistance from the sampler and as a result of the loss of fine bedload particles through the sampler mesh. Therefore basket sampler measurements are not representive of the absolute value of bedload charge (cf. NANSON 1974). The investigation is further complicated by technological and logistical problems in regions with extreme conditions. Therefore two different methods for analysing the bedload transport were chosen, methods which can even be used in high mountain areas. On the one hand, the transported material was caught with the help of sediment baskets; on the other hand, previously marked and measured pebble and cobble tracers weighing between 0.15 - 3.1 kg were laid out at different sites of the Agua Negra river. The intention was to determine possible distances of transport of bigger stone tracers in correlation to size, shape and weight (SCHMIDT & ERGENZINGER 1992). Unfortunately the recovery rate of pebble and cobble tracers was only 50 % (35 out of 70).

Distances between the study sites made it impossible to check the baskets sufficiently frequently to give any more than a limited estimate of bedload. However, these preliminary measurements in the High Andes still show some interesting tendencies:

1. Bedload transport was quantified using basket measurements only at the upper stations Eisbein and Cuatro Mil (see Tab. 1). The method did not work for the two lower stations Kolibri and Cerelac because of vegetation, lower turbulence and decreasing slope in these areas.

2. Although the hydrograph at the Cuatro Mil station showed a lower discharge its bedload was always higher than that of the Eisbein station.
3. Most of the pebbles caught in the baskets were between 2 and 6.3 cm in size.
4. The use of the marked pebbles and cobbles did however prove the occurrence of bedload transport at the lower stations. There was no consistent difference in the distance over which light rounded stones and bigger angular stones were transported.

Therefore it seems as if, as well as shape and weight, there are other more significant factors, such as turbulence and shear stress - based on the flow rate - determining distance of transport (RICKENMANN 1990, GINTZ & SCHMIDT 1991). On average the pebbles and cobbles were transported over a distance of 0.1 to 2 m within 3 months. Maximum distances of pebble transport were observed at the Cuatro Mil station (11 and 23 m).

Tab. 1. Bedload at the Eisbein (4150 m) and Cuatro Mil (4000m) gauging stations

Eisbein station (bedload)[1]

date	time [hours]	grain size [cm] >6,3	>2,0	>0,63	total load	load / hour
8.1 - 9.1	20	-	3,75	4,0	7,75	0,387
9.1 - 10.1	30	-	11,0	6,5	17,5	0,590
10.1 - 15.1	114	1,1	15,9	9,0	26,0	0,228
15.1 - 18.1	71	0,48	0,92	0,3	1,7	0,024
Σ8.1 - 18.1	236	1,58	31,57	19,8	52,59	0,224

Cuatro Mil station (bedload)[1]

date	time [hours]	grain size [cm] >6,3	>2,0	>0,63	total load	load / hour
22.1 - 24.1	45	9,27	12,7	1,9	23,9	0,531
24.1 - 25.1	33	4,5	22,5	22,5	49,5	1,5
25.1 - 26.1	21	2,75	26,5	12,0	41,3	1,96
26.1 - 28.1	48	6,25	33,0	12,5	51,8	1,08
Σ22.1 - 28.1	148	22,78	94,7	48,93	166,4	1,124

[1] values = [kg]

Owing to technical problems a relation between bedload transport rate and stream velocity could not be established.

7. Information on previous flood events

In addition the terrraces of the Agua Negra were analysed with respect to the type of deposit and the grain size distribution (cf. BAKER et al. 1983, BAKER 1989). The lack of vegetation in the flood plain at the Kolibri and Cerelac station can be interpreted as an indication of regular flooding of the valley floor. It has to be assumed that the discharge during these events is many times larger than the discharge measured in January 1991. During these previous events great parts of the flood plain were reworked and considerable amounts of sediment and rocks were transported.

On the upper terraces boulders between 20 and 40 cm were found. Grain size analysis was therefore carried out at different levels in order to obtain criteria for determining the magnitude of flood events. It showed that grain sizes of more than 6.3 cm are more frequent on the upper than on the lower terraces.

8. Conclusions

The hydrological measurements carried out at different altitudes in the High Andes of Cuyo display the present diurnal and annual range of fluctuations of discharge and suspended sediment transport:

- Whereas the concentration of suspended sediment increases in the lower course of the Agua Negra river, the discharge in some parts downstream decreases, owing to the high rate of seepage in certain sections of fluvio-glacial outwash deposits and to evaporation.
- In contrast to the Alps, relatively low flow rates produce quite high suspended sediment yields. An important prerequisite for this phenomenon is the availability of unconsolidated material with a high content of silt.
- In comparison to the transport in suspension, the bedload is in general insignificant. It is probably restricted to flood events after winters extraordinarily rich in snow. During the field measurements in January 1991 bed load of some importance was only measured at the highest stations (Eisbein at 4150 m and Cuatro Mil at 4000 m a.s.l.).

Acknowledgements

The work was funded in part by a scholarship from the Gottlieb Daimler and Karl Benz Stiftung (Ladenburg) to L. Schrott. The Hiehle Stiftung, the DAAD and the University of Heidelberg provided further financial support. The authors would like to thank Dr. A. Corte (Mendoza), Dr. J. L. Minetti (San Juan, now Tucuman) and all students taken part in the excursion "Argentina 1990/91" for their cooperation and field assistance. We also thank M. Wulf for translating this paper into English.

References

BAKER, V. R. (1989): Magnitude and frequency of paleofloods.- In: BEVEN, K. and CARLING, P. (eds.): Floods: Hydrological, sedimentological and geomorphological implications: 171-183.

BAKER, V. R., KOCHEL, C. R., PATTON, P. C. and PICKUP, G. (1983): Paleohydrologic analysis of Holocene flood slack-water sediments.- In: COLLINSON, D. J. and LEWIN, J. (eds.): Modern and ancient fluvial systems, International association of Sedimentologists, Special publication 6: 229-239.

BARSCH, D., GUDE, M., MÄUSBACHER, R., SCHUKRAFT, G. and SCHULTE, A.(1992): Untersuchungen zur aktuellen fluvialen Dynamik im Bereich des Liefdefjorden in NW-Spitzbergen.- In: Stuttgarter Geographische Studien, Bd. 117: 217-252.

GINTZ, D. and SCHMIDT, K.-H. (1991): Grobgechiebetransport in einem Gebirgsbach als Funktion von Gerinnebettform und Geschiebemorphometrie.- In: Zeitschrift für Geomorphologie, Supplement-Band 89: 63-72.

GOMEZ, B. (1987): Bedload.- In: GURNELL, A. M. and CLARK, M. G. (eds.): Glacio-fluvial sediment transfer: 355-368. GURNELL, A. M. (1987a): Suspended Sediment.- In: GURNELL, A. M. & CLARK, M. G. (eds.): Glacio-fluvial sediment transfer: 355-366.

GURNELL, A. M. (1987a): Suspended sediment.- In: GURNELL, A. M. and CLARK, M. G. (eds.): Glacio-fluvial sediment transfer: 305-345.

GURNELL, A. M. (1987b): Fluvial sediment yield from alpine, glacierized catchments.- In: GURNELL, A. M. and CLARK, M. G. (eds.): Glacio-fluvial sediment transfer: 415-418.

HAPPOLDT, H. and SCHROTT, L. (1992): A note on ground thermal regimes and global solar radiation at 4720 m a.s.l., High Andes of Argentina.- In: Permafrost and Periglacial Processes, Vol 3: 241-245.

MAIZELS, J. K. (1983): Proglacial channel systems: channel and thresholds for change over long, intermediate and short timescales.- In: COLLINSON, D. J. and LEWIN, J. (eds.): Modern and ancient fluvial systems, International Association of Sedimentologists, Special publication 6: 251-266.

MINETTI, J. L. and SIERRA, E. M. (1989): The influence of general circulation patterns on humid and dry years in the Cuyo Andean Region of Argentina.- In: International Journal of Climatology, Vol.9: 55-68.

NANSON, G. C. (1974): Bedload and suspended-load transport in a small, steep, mountain stream.- In: American Journal of Science, Vol 274: 471-486.

RICKENMANN, D. (1990): Bedload transport capacity of slurry flows at steep slopes. Mitt. der Versuchsanstalt für Wasserbau, Hydrologie und Glaziologie der ETH Zürich, 103: 249pp.

SCHMIDT, K.-H. and ERGENZINGER, P. (1992): Bedload entrainment, travel lengths, step lengths, rest periods-studied with passive (iron, magnetic) and active (radio) tracer techniques.- In: Earth Surface Processes and Landforms, Vol. 17: 147-165.

SCHROTT, L. (1991): Global solar radiation, soil temperature and permafrost in the Central Andes, Argentina: a progress report.- In: Permafrost and Periglacial Processes, Vol 2: 59-66.

SEDIMENT TRANSPORT AND DISCHARGE IN A HIGH ARCTIC CATCHMENT
(LIEFDEFJORDEN, NW SPITSBERGEN)

Dietrich Barsch, Martin Gude, Roland Mäusbacher, Gerd Schukraft, Achim Schulte
(Department of Geography, University of Heidelberg,
Im Neuenheimer Feld 348, D-69120 Heidelberg, Germany)

Abstract

In a joint geoscientific project in northwestern Spitsbergen at latitude 80°, the fluvial dynamics and transports in three small catchments (approx. 5 km^2 each) including different proportions of glacierized areas were studied in 1990 and 1991. Only during the short summer are runoff and sediment transport possible. Therefore, the studies started before the beginning of the snowmelt. In the study area, transport in suspension is most effective, transport in solution and especially bedload are less important. Both summers are similar regarding the total discharge (Kvikkåa catchment: 2.8 to 2.4*10^6 m^3). But in 1990 flood peaks were higher by 40% (2200 l/s vs. 1600 l/s). Therefore, suspension and bedload transports were higher by a factor of about 2 (237 t vs. 124 t for suspension and 5 t vs. 2.2 t for bedload).

1. Introduction

The investigation of fluvial transports in the area of Liefdefjorden (NW Spitsbergen, cf. fig. 1) is part of the geoscientific project "Land-Sea Sediment Transport in Polar Geoecosystems". Detailed complex studies of different sections of the ecosystem are supposed to lead to a primarily quantitative characterization of energy and mass fluxes, i. e. of the active process complex (BLÜMEL et al. 1988, BLÜMEL ed. 1992). The investigation of the fluvial subsystem concerning the transport processes and sediment quantities represents the link between the terrestrial and the marine environments. The aim of the investigations is therefore to quantify the active processes in the rivers, related to both discharge and sediment transport.

This paper is based on the analysis of the data of two campaigns in 1990 and 1991. This enables a comparison of the two years' data, especially as far as yield of sediment transport is concerned (BARSCH et al. 1992).

Limited accessibility as well as restricted knowledge of the investigation area require a concept of data acquisition techniques that is flexible in view of both the installation in the field and the operation during measurements. Since the rivers are characterized by rapid changes of discharge and extremely turbulent flow, robust measurement tools need to be installed in protected positions.

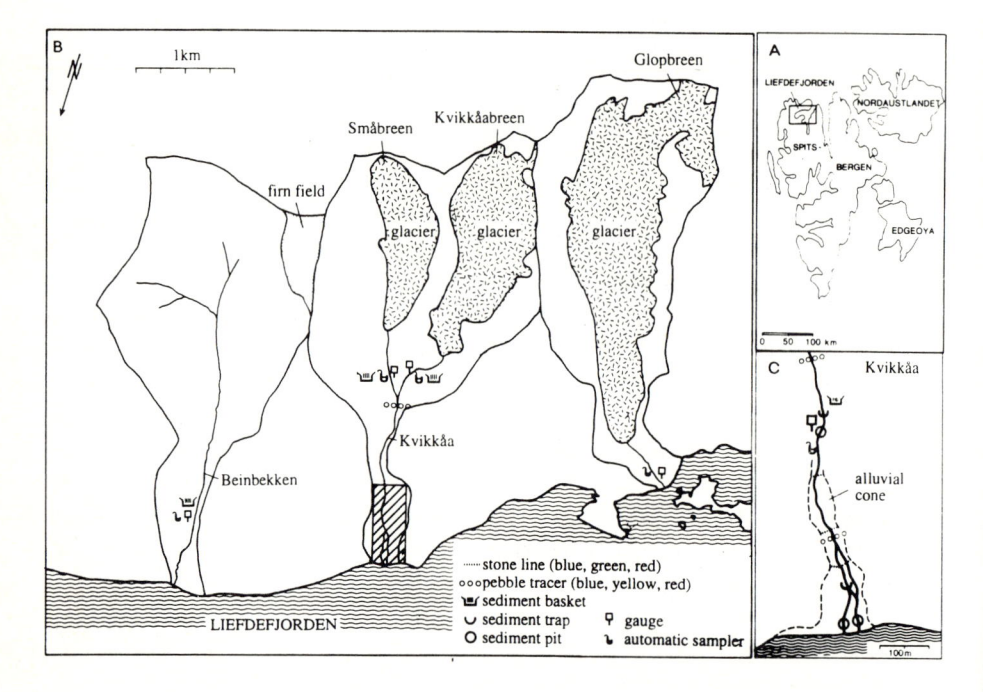

Fig. 1: Map of the investigation area (A), the catchments of the Beinbekken, the Kvikkåa and the Glopbreen (B), and the alluvial fan of the Kvikkåa (C)

2. Investigation Area and Measurement Methods

The investigated catchments were selected by analysis of vertical aerial photographs. The decisive parameters were river regimes, the shape of the channel bed, the size of catchment area, and the accessibility from the base camp.

Three catchments of approximately 5 km^2 each, containing different proportions of glacier-covered area, are investigated. Despite their alpine-like shape the highest elevations are at approx. 750 m a. s. l. and the rivers drain into the Liefdefjorden. The catchments are characterized in the upper parts by steep taluses and in the middle parts by areas with moderate gradients which lead into the alluvial fans. The geology is dominated by Devonian metamorphic rocks in the Kvikkåa and Devonian sandstones in the Beinbekken. The vegetation is typical high arctic with mosses and lichens, and, in lower areas, a large variety of vascular plants.

The Beinbekken represents a purely nival regime, whereas the Glopbreen river is dominated by glacial meltwater. A nival/glacial regime characterizes the Kvikkåa. Three continuously registering gauge stations including sediment sampling facilities (cf. Fig. 2) are installed at different reaches of the Kvikkåa. At the gauge station at the head of the alluvial fan the gradient of the channel is approx. 5-6°, whereas in the lower sections near the mouth it decreases to 1-2°. At the Beinbekken, only the total output is measured. During the campaign in 1990, only the Beinbekken and the Kvikkåa were investigated, because a layer of aufeis with a thickness of approximately 150 cm prevented the installation of measurement tools in the Glopbreen river.

Discharge and sediment transport measurements are required to start with the break-up of the channels. As the channels were difficult to find owing to the wind-drifted snow, discharge data of this period refer only to repeated measurements of cross and velocity profiles in the permanently changing channels. After installation of the gauge stations discharge is measured by means of NaCl-tracers; occasionally current meters are used in surveyed cross sections. The current meters also provide data concerning the sediment transport process, since the velocity in the cross section of the channel and the shear stress at the bed are controlling factors for the transport of bed load.

The study of sediment transports (suspension and bed load) is carried out using common methods (GOMEZ 1987, GURNELL 1987a; cf. Fig. 2). Water samples containing the suspended load are automatically taken at constant intervals (3.5 hours, during floods 1 hour). Additionally, samples from different parts of the cross section are taken in case of visible increase of suspension concentration. The latter is measured in the camp by filtration (0.2 μm) of the samples.

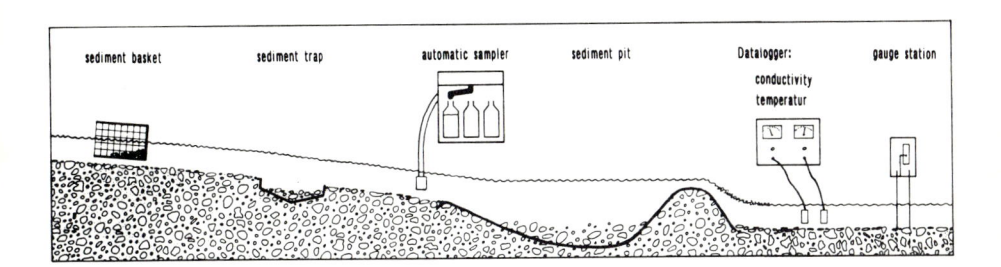

Fig. 2: Measurement tools installed in the rivers

Measurement of bed load is even more complicated than that of suspension load (ERGENZINGER & SCHMIDT 1990, GOMEZ 1987). This is true especially for arctic areas, where measurement tools have to be installed in the catchments without technical aid. Therefore, a combination of different measurement tools has been chosen for the investigations (cf. Fig. 2). They were installed at the beginning of the fluvial activity and sampled constantly.

Despite their limited efficiency, sediment baskets were used since they are easy to carry and can be installed in any part of the river. They have an opening of 0.4*0.5 m and are provided with a fairly course meshscreen (14 mm mesh) to prevent extreme backwater while filling rapidly. The baskets covered approx. 1/10 of the river width. They were emptied either frequently or, during floods, whenever they were filled up.

In order to take into account the resulting problems concerning grain size distribution and efficiency, two different types of sediment pits are installed in the channel floor. The first is an arrangement of open boxes placed along a cross profile (cf. Fig. 2: sediment trap). They can be emptied after floods and therefore data about different periods are supplied. On the other hand, sediment pits (cf. Fig. 2: sediment pit), emptied only once at the end of the campaign, give information about the total amount of bed load.

The measurement concept is completed by pebble tracers and coloured lines. The latter are painted on cross profiles, which cover the pebbles on the alluvial fan including the active channels. The change in these lines allows the sediment transport during the break-up period to be estimated. The marked pebble tracers placed in the channel at different sites are transported specific distances especially during floods, depending on their size and shape. The pebbles are then tracked after the floods and at the end of the campaign in order to record the distances of transport. The efficiency of this method is limited owing to the fact that few of the pebbles can be located after transport (GINTZ & SCHMIDT 1991).

3. Discharge and sediment transport during break-up

Both the Beinbekken and the Kvikkåa are characterized by discharge-initiating slush-flows, running sometimes beneath the snow-covered channels. Discharge of slush-flows is difficult to investigate because of its high snow and ice content and the high annual variability of the tracks (CLARK & SEPPÄLÄ 1988).

Slush-flows also initiate the sediment transport. On the Kvikkåa alluvial fan the coloured lines show the areas affected by mobilization and transportation during these floods. In 1991 pebbles with a grain size up to 60 cm were moved. Because of this, the investigation facilities reveal no possibility of directly measuring the sediment transport. Rough estimations of the sediment yields can be derived from the study of the accumulation after snowmelt (RAPP 1960). These problems are dealt with in a separate publication (BARSCH et al. 1993).

4. Discharge dynamics of the three catchments

The parallel investigation of three similar sized catchments, defined by different proportions of glacier-covered area, makes it possible to compare discharge dynamics (FLÜGEL 1981). The Beinbekken, the Kvikkåa, and the Glopbreen catchments are characterized by their typical pattern of snow, glacier, and permafrost melting and the resulting hydrographs. Similar differentiations are investigated in alpine rivers (e. g. RÖTHLISBERGER & LANG 1987). Rainfall is of minor importance (in 1990 and 91 less than 20 mm during the summer period, respectively). The discharge increases rapidly after break-up in the snowmelt-dominated Beinbekken and Kvikkåa (cf. Fig. 3). Opposed to this, the Glopbreen river hydrograph shows a retarded and gentle increase, but high discharges at the end of August (since the analysis of sediment transports in the Glopbreen river is not yet finished it will not be referred to in more detail).

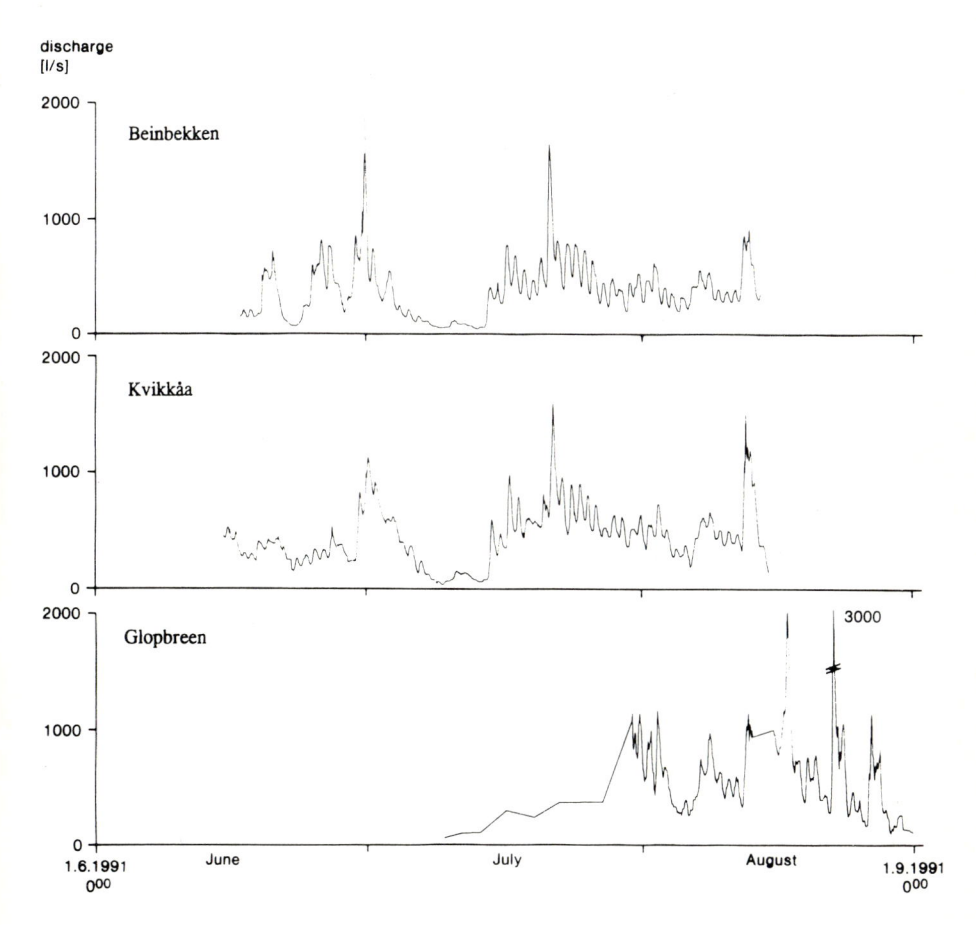

Fig. 3: Discharge 1991 at the gauges of the Beinbekken, the Kvikkåa and the Glopbreen (for the period from the beginning to end of July the hydrograph of the latter is exclusively based on single discharge measurements)

During the summer period in 1991 discharge started in the Beinbekken on June 14th and reached the level of 200 l/s on the 17th. The Kvikkåa broke up two days earlier and increased to 500 l/s, again within three days. Concerning both short and longer period variations, the hydrographs of the Beinbekken and the Kvikkåa run synchronously during the summer period. Differences are noticeable between the absolute values, the total amounts of discharge (Beinbekken: $1.9*10^6$ m^3;

Kvikkåa: $2.4*10^6$ m^3), and the tendency in the hydrographs during the summer. Each hydrograph embodies three periods of higher discharge: at the end of June discharge increased until a peak was reached on June 30th/July 1st with a maximum of 1800 l/s (Beinbekken) and 1100 l/s (Kvikkåa), in the second half of July with a peak on the 21st with 1600 l/s in both rivers and in the middle of August with a maximum on the 12th/13th of 900 l/s and 1500 l/s, respectively.

From the data the different trends of the two hydrographs are visible: highest discharge at the Beinbekken occurs in the first part of the period and the level remains above that of the Kvikkåa for several days. In the advancing discharge period the maximum levels of floods decrease in the Beinbekken. In contrast to this, the Kvikkåa hydrograph shows maximum values in August, too. Both hydrographs confirm the high variability in discharge. This is especially evident in diurnal variations including maximum discharge values late in the evening or around midnight.

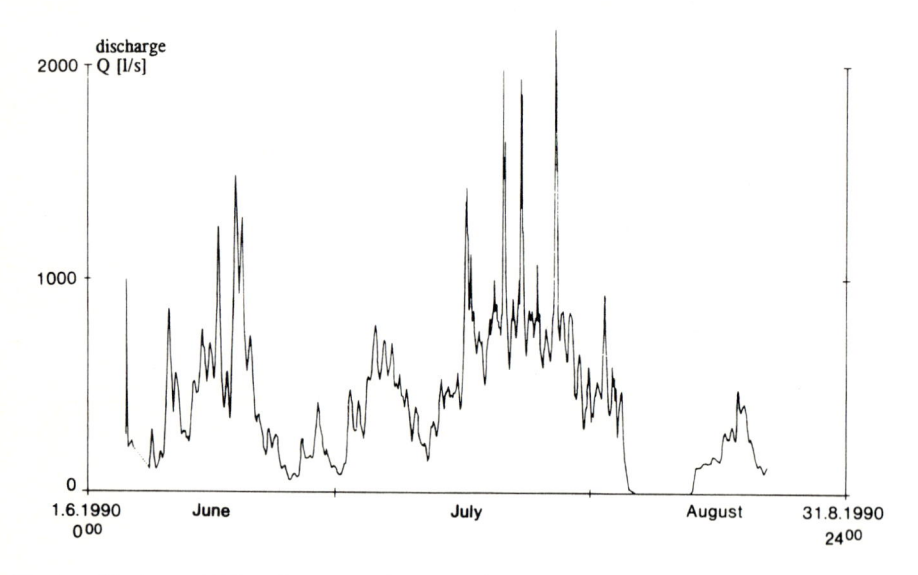

Fig. 4: Discharge 1990 at the gauge of the Kvikkåa

Comparison with the hydrograph of the Kvikkåa in 1990 reveals similar diurnal and longer period variations. Nevertheless, the tendency during the period shows considerable differences (cf. Fig. 4). Breaking up on June 5th, the discharge rapidly reaches a maximum value of 1000 l/s, which is about twice as much as the incipient discharge in 1991 (500 l/s). In June 1990 the discharge level remains higher (maximum value 1500 l/s) than in 1991. In both years, the hydrographs of July include the highest amount of discharge, while in 1990 the absolute height and number of peak discharges were higher. The maximum discharge in 1990 reached 2200 l/s, about 600 l/s (or 37.5 %) more than the maximum in 1991. The tendency in August was also different: the discharge in 1990 temporarily disappeared and altogether it only reached shallow peaks. By contrast, the 1991 hydrograph of the same month remained on a moderate level embodying distinct floods. The reversed pattern of the discharge in the two years results in a slight difference between the individual total amounts of discharge (cf. above).

5. Sediment Transport in the Kvikkåa

5.1 Suspension

As mentioned above (cf. chapter 2), suspension concentration was measured by filtration of samples of 1 l each, taken by hand and by autosampler respectively (CLARK et al. 1988). The problems of correlation between discharge and suspension concentration are well-known for both glacial rivers (GURNELL 1987b) and for non-glacial rivers (WALLING & KLEO 1979). The primary cause of these problems is the availability of sediment in the catchment or in the river bed itself. The empirical correlation between discharge and suspension is shown in Fig. 5. The values for suspension concentration differ considerably for identical discharges. However, the appropriate equation offers a sufficient basis for load calculation (cf. Fig. 5).

The running analysis of sediment origin and mobilization will enable more detailed statistics concerning the correlation of suspension concentration and discharge. Factors controlling the concentration of suspension are the flow velocities and the origins of fine grains (channel erosion, overland flow and glacier melt). The field evidence shows that the suspension concentration is strongly correlated to bed load. This is due to the destruction of the pavement in the channel when the flow velocity is rising. In the consequence, the fine grains underlying the gravels are exposed. Erosion in the channel is assumed to be initiated when the rising flow velocity exceeds a threshold. Therefore a more detailed analysis of suspension concentration versus discharge has to be considered in the context of this threshold.

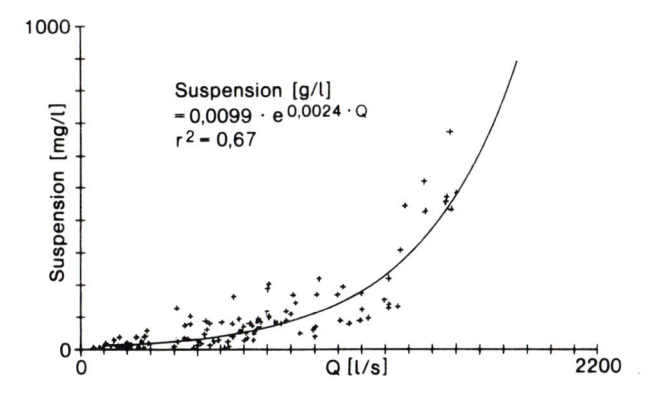

Suspension [g/l] $= 0{,}0099 \cdot e^{0{,}0024 \cdot Q}$
$r^2 = 0{,}67$

Fig. 5: Correlation between discharge and suspension load at the gauge of the Kvikkåa in 1990

Fig. 6 shows the total load at the gauge of the Kvikkåa, calculated on the basis of the equation quoted in Fig. 5. Although the equation is of a preliminary character especially in the extrapolated range, the domination of the floods in suspension load is obvious. The three floods in July project from the graph, having load amounts of 12,000 kg/h, 6,000 kg/h, and 4,000 kg/h. Total load at the gauge in 1990 adds up to 237 tons with the bulk of it (172 tons) deriving from the flood in July mentioned above. For the total load this equals a specific sediment yield of 46 g/m^2 during the measurement campaign.

In addition to the gauge station at the Kvikkåa, the suspension concentrations were also measured at the river's outlet into the fjord below the alluvial fan of approx. 500 m length (cf. Fig. 1). Here, the total suspension load for the 1990 campaign amounts to 177 tons, i. e. a "loss" of 60 tons. This difference is assumed to be the accumulation on the alluvial fan.

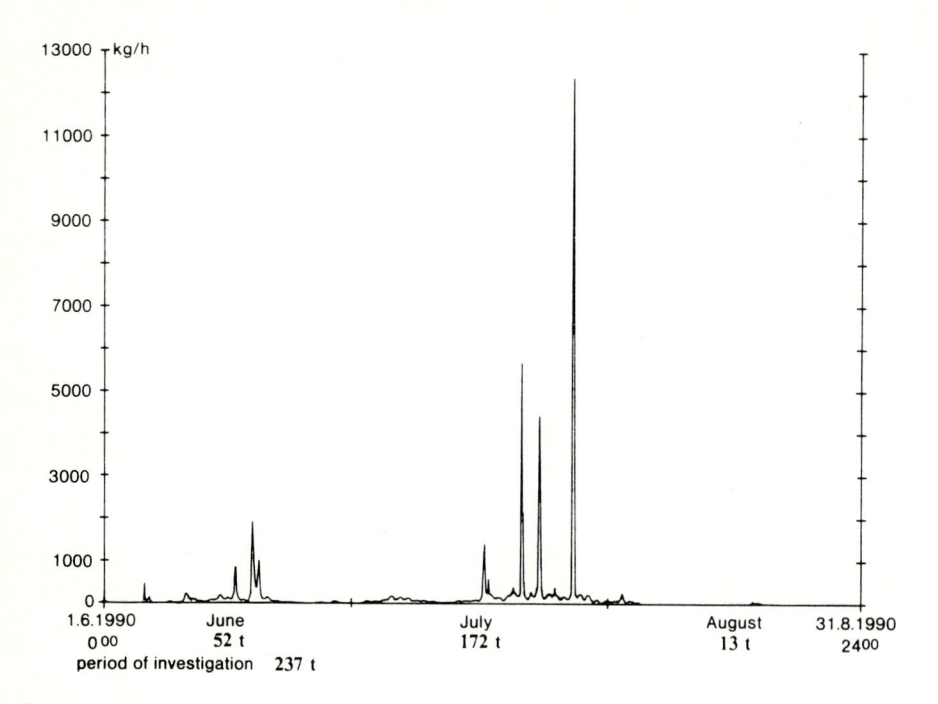

Fig. 6:   Suspension load at the gauge of the Kvikkåa in 1990

5.2   Bed Load

The sediment baskets are installed in the channel before the melting of basal ice. This contrasts with the sediment pits and traps embedded in the gravels of the channel. High discharges required the baskets to be frequently emptied. High discharges also lead to problems at the sediment traps. Increasing velocities remobilize the sediments in the traps whenever a protecting pavement has not developed. As a result the range of measurements is restricted.

Low discharges cause higher values for collected sediment in the traps than in the baskets; on the contrary, at high discharges the baskets collect larger quantities of sediment. This effect results from the 14 mm meshscreen of the baskets: during low discharges, the fine grains can pass through the meshscreen. A calculation of the loads during floods is only possible when the collected sediment in the baskets is removed frequently. The sediment traps allow no emptying during floods owing to their weight.

Discharge and calculated loads correlate significantly, as becomes apparent in Fig. 7, which refers to the flood on July 16th/17th 1990. Similar to the suspension concentration, the initiation of gravel transport is also connected to a threshold in flow velocity.

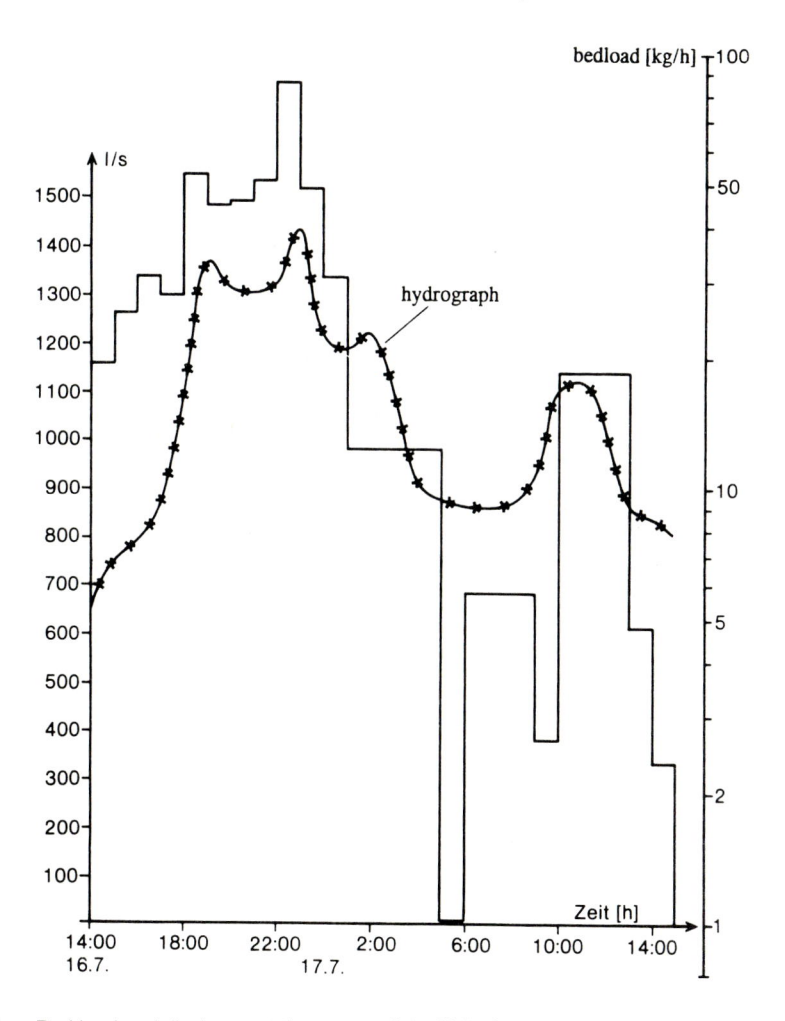

Fig. 7:   Bed load and discharge at the gauge of the Kvikkåa on July 16th/17th 1990

For the flood in July 1990 the yields of the sediment traps are distinctly lower than those collected in the baskets, as verified also in floods during the 1991 campaign. Grain size distribution, however, differs only insignificantly.

Another problem derives from the differentiation of bed load transport in the longitudinal river profile. Therefore, in 1991 gross sediment pits were installed at the gauge of the Kvikkåa and at the outlet into the fjord as well. Bedload collected in the gross pit at the outlet amounts only to 90 kg, whereas that at the gauge adds up to more than 3,000 kg. The sediment traps show the same relation between the sampling sites but reveal lower total amounts. Accumulation on the alluvial fan is evident for both bed load and suspension (cf. chapter 4).

The relation between total yields of suspension and bed load was unexpected (cf. table 1): suspension load dominates clearly, reducing the bed load to a small fraction of the total amounts of transport. Compared to other investigations in glacierized catchments (GURNELL 1987b) the relation between suspension and bed load in the Kvikkåa catchment tends significantly to the dominance of suspension load.

5.3 Measurement of distances of bed load transport

The transport distances of the pebbles embedded in the channel at different sites stayed low after the initiation of the discharge in 1990. Until the end of July 1990 the maximum distance was 62.5 m. The longer distances measured in the middle of the following campaign (July 1991) show that during the initiation of discharge at the beginning of the intensified snowmelt period a distinct transport of pebble occurs, probably primarily caused by slush-flows and water flowing over basal ice. Between June 1990 and July 1991 maximum transport distances are more than 400 m. There is no noticeable influence of weight and shape of the marked pebbles on the distances of transport.

6. Conclusions

The investigations presented lead to the following general results concerning the fluvial system and the sediment transport (cf. table 1).

During break-up periods discharge and sediment transport are highly variable owing to slush-flows. The yield of sediment transport by fluvial activity in this period depends primarily on the existence of basal ice in the channels.

The discharge is characterized by a high variability from year to year but also during one melting period. The total yield of discharge within two periods differs only inconsiderably.

		1990	1991
discharge	total at gauge	$2.8 \ast 10^6$ m^3	$2.4 \ast 10^6$ m^3
	annual runoff total	540 l$\ast$m$^{-2}\ast$y^{-1}	460 l$\ast$m$^{-2}\ast$y^{-1}
	maximum	2200 l/s	1600 l/s
	floods > 1000 l/s	8	4
	break-up date	05.06.	14.06.
suspension	total load at gauge	237 t	124 t
	total load at mouth	177 t	103 t
bed load	total load at gauge (baskets/gross pits)	5/- t	2.2/3 t
	total load at mouth (small/gross pits)	227/- kg	60/90 kg

Table 1: Discharge dynamics and transports in 1990 and 1991 at the Kvikkåa

Kvikkåa, Spitsbergen	24-46	t$\ast$km$^{-2}\ast$y^{-1}
Periglacial catchments		
Sweden	0.8-4.7	t$\ast$km$^{-2}\ast$y^{-1}
Yenisei, USSR	5	t$\ast$km$^{-2}\ast$y^{-1}
Ob, USSR	7	t$\ast$km$^{-2}\ast$y^{-1}
Mecham River, Kanada	22.1	t$\ast$km$^{-2}\ast$y^{-1}
Four large USSR rivers	3-25	t$\ast$km$^{-2}\ast$y^{-1}
Karkevägge, Sweden	53	t$\ast$km$^{-2}\ast$y^{-1}
Colville River, Alaska	80-140	t$\ast$km$^{-2}\ast$y^{-1}
Baffin Island, Canada	400-790	t$\ast$km$^{-2}\ast$y^{-1}
Alpine glacierized catchments		
Tsidjiore Nouve, Switzerland	2700-3300	t$\ast$km$^{-2}\ast$y^{-1}
Bondhusbreen, Norway	580-740	t$\ast$km$^{-2}\ast$y^{-1}
Engabreen, Norway	390-480	t$\ast$km$^{-2}\ast$y^{-1}
Nigardsbreen	260-500	t$\ast$km$^{-2}\ast$y^{-1}
Hilda, Canada	600-800	t$\ast$km$^{-2}\ast$y^{-1}

Table 2: Selected measures of denudation rate in periglacial and glacial catchments (after: CLARK 1988 and GURNELL & CLARK 1987, modified by the authors)

Bed load transport begins at distinct discharge levels. The sediment load is dependent on the availability of solids following destruction of the pavement of the channel bed. This level is estimated at 600-800 l/s.

The suspension load is also connected to floods, but reveals a much higher dynamic. The suspension load during the floods far exceeds those of bed material.

In comparison with other periglacial and glacial catchments the denudation rate remains moderate (cf. Table 2). It is obvious that the characteristics of the Kvikkåa concerning the sediment yield resemble periglacial catchments rather than glacial ones. This is mainly due to the low proportion of glacierized area (35% of catchment area) and to the fact that the two glaciers are presumed to be frozen at the base, which cosiderably limits the amount of suspension.

Obviously, material is transported through steep gorge reaches. On the other hand, the alluvial fan represents an actual accumulation area. Only a small part of the transported material reaches the fjord except for the break-up period of the fluvial activity, when a significant material transport onto the sea ice occurs.

References

BARSCH, D., GUDE, M., MÄUSBACHER, R., SCHUKRAFT, G. & SCHULTE, A. (1992): Untersuchungen zur aktuellen fluvialen Dynamik im Bereich des Liefdefjorden in NW-Spitzbergen.- In: Stuttgarter Geographische Studien, Bd. 117: 217-252.

BARSCH, D., GUDE, M., MÄUSBACHER, R., SCHUKRAFT, G., SCHULTE, A. & STRAUCH, D. (1993, in press): Slush stream phenomena - process and geomorphic impact. - Z. f. Geomorph., Suppl. Bd. 92

BLÜMEL, W.-D. (ed.) (1992): Geowissenschaftliche Spitzbergen-Expedition 1990 und 1991 "Stofftransporte Land-Meer in polaren Geosystemen", Zwischenbericht. - Stuttgarter Geographische Studien, Bd. 117., 416 p

BLÜMEL, W.-D., LESER, H. & STÄBLEIN, G, (1988): Wissenschaftliches Programm der Geowissenschaftlichen Spitzbergen-Expedition 1990 (SPE90) "Stofftransport Land-Meer in polaren Geosystemen". - Materialien und Manuskripte, 15: Bremen, 47 p

CLARK, M. J. (1988): Periglacial Hydrology. - In: CLARK, M. J. (ed.): Advances in Periglacial Geomorphology: 415-462

CLARK, M. J., GURNELL, A. M. & THRELFALL, J. L. (1988): Suspended sediment transport in arctic rivers. - In: Proc. 5th Inter. Permafrost Conf., Trondheim: 458-463

CLARK, M. J. & SEPPÄLÄ, M. (1988): Slushflows in the subarctic environment Kilpisjarvi, Finnish Lappland.- In: Arctic and Alpine Research 20: 97-105.

ERGENZINGER, P. & SCHMIDT, K.- H. (1990): Stochastic elements of bed load transport in a step-pool mountain river.- In: IAHS-Publication 184 (Hydrology of Mountain Regions II): 39-46.

FLÜGEL, W.A. (1981): Hydrologische Studien zum Wasserhaushalt hocharktischer Einzugsgebiete im Bereich des Oobloyah Tales, Nord Ellesmere Island, N.W.T., Kanada.- In: BARSCH, D. & KING, L.(eds.): Ergebnisse der Heidelberg Ellesmeere Island-Expedition.- Heidelberger Geographische Arbeiten 69: 311-382.

GINTZ, D. & SCHMIDT, K.-H. (1991): Grobgeschiebetransport in einem Gebirgsbach als Funktion von Gerinnebettform und Geschiebemorphometrie.- In: Zeitschrift für Geomorphologie, Supplement-Band 89: 63-72.

GOMEZ, B. (1987): Bedload.- In: GURNELL, A.M. & CLARK, M.J. (eds.): Glacio-fluvial sediment transfer: 355-368.

GURNELL, A. M. (1987a): Suspended Sediment.- In: GURNELL, A. M. & CLARK, M.G. (eds.): Glacio-fluvial sediment transfer: 305-345.

GURNELL, A. M. (1987b): Fluvial sediment yield from alpine, glacierized catchments.- In: GURNELL, A. M. & CLARK, M.G. (eds.): Glacio-fluvial sediment transfer: 415-420.

RAPP, A. (1960): Recent development of mountain slopes in Kärkevagge and surroundings, Northern Scandinavia.- In: Geogr. Ann. 42: 138-147

RÖTHLISBERGER, H. & LANG, H. (1987): Glacial hydrology.- In: GURNELL, A.M. & CLARK, M.J. (eds.): Glacio-fluvial sediment transfer: 207-284.

WALLING, D.E. & KLEO, A.H.A. (1979): Sediment yield of rivers in areas of low precipitation: a global view.- In: The Hydrology of Areas of Low Precipitation (Proceeding of the Canberra Symposium, December 1979), International Association of Hydrological Science Publication 128: 479-493.

Acknowledgement

The investigation of fluvial processes was funded by the German Science Foundation. The support of the coordinator of the geoscientific project, Prof. W.-D. Blümel, is gratefully acknowledged as well as that of the assistants, who helped in organisation and field work: A. Fieber (†), Th. Glade, H. Gündra, B. Sandler, and D. Strauch.

General investigations on sediment transport dynamics

HYDRAULICS AND SEDIMENT TRANSPORT DYNAMICS CONTROLLING STEP-POOL FORMATION IN HIGH GRADIENT STREAMS: A FLUME EXPERIMENT

Gordon E. Grant, U.S. Forest Service, Pacific Northwest Research Station,
3200 Jefferson Way, Corvallis, OR, U.S.A.

ABSTRACT: A series of flume experiments was undertaken to determine the domain of flow conditions under which step-pool sequences are formed. This domain can be characterized by the ratio of average shear stress to the shear stress required to move the largest grain sizes (τ/τ_{cr}) and the Froude number (Fr). Step spacing was correlated with the antidune wavelength and steps did not form when particle motion was continuous under high sediment transport rates.

1. INTRODUCTION

Many high gradient, boulder-bed streams are characterized by alternating steep and gentle segments which typically create a staircase structure of steps and pools. This structure is commonly expressed at two distinct scales (Fig. 1). The first is channel-spanning steps composed of large boulders interspersed with small pools less than one channel width in length; the step and its associated pool is termed a step-pool sequence. Multiple step-pool sequences alternating with large pools longer than the channel width are termed channel units (Grant et al, 1990) or swells (Kishi et al, 1987).

Channel units and individual step-pools are important from several perspectives. These structures constitute a major component of flow resistance in mountain streams and dissipate stream energy that might otherwise be available for sediment transport and channel erosion (Ashida et al, 1976; 1986 a,b). Sediment transport and storage in mountain

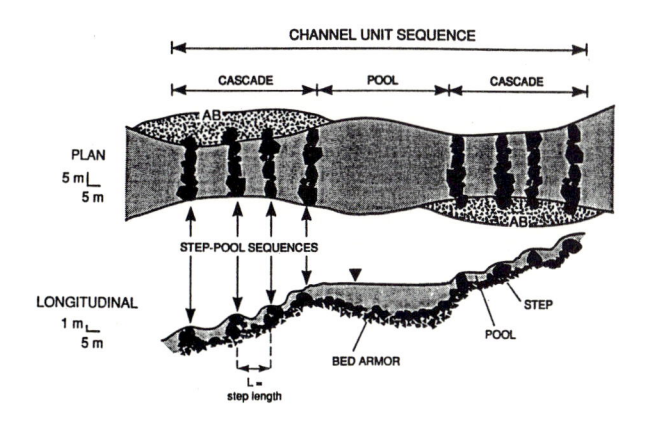

Figure 1: Plan and side view of channel unit and step-pool sequences, showing bed armor and alternate bars (AB).

streams are regulated by the geometry of step-pool sequences; bedload transport during low to moderate discharges involves movement of relatively fine material (gravel and sand) through steep channel units and deposition in pools (Whittaker, 1987 a,b; Sawada et al, 1983). Complex hydraulics generated by alternating steep and gentle gradient units provide diverse habitat for stream organisms, notably fish, which show strong preferences for specific channel units (Gregory et al, 1991). Maintaining channel unit and step-pool morphology is thus important to minimize channel instability, reduce sediment transport, and maintain stream health for aquatic ecosystems.

Little is known, however, about the mechanisms and flow conditions required for step-pool formation. Step-pools generally are stable during most flows but can be reworked or destroyed by large floods and debris flows (Hayward, 1980; Sawada et al, 1983; Ashida et al, 1981). Because recurrence intervals for flows forming channel units are likely to be 20 to 50 years or greater (Grant et al, 1990), field data on step-pool formation are sparse and limited to visual observations of recovery of step-pool topography following debris flow (Sawada et al, 1983). Most research on step-pool morphology has been done in flumes and much of this important work is known only in Japan (Judd and Peterson, 1969; Whittaker and Jaeggi, 1982; Ashida et al, 1984; 1985; 1986 a,b; Hasegawa, 1988).

In this paper we report on a series of flume experiments which examined the basic processes of step-pool formation under a range of flow and sediment transport conditions. These experiments included both clear water flows and runs where sediment was supplied from upstream. The latter is a more realistic representation of natural river systems, since step-pools form during large floods when bedload transport rates are high.

2. METHODS

A series of flume experiments were conducted in fall, 1988 in the 11 m long by 0.5 m wide adjustable-slope flume located at the Public Works Research Institute in Tsukuba, Japan. Water was delivered to the flume by a pump with a maximum capacity of 50 l/sec; discharges were measured at the lower end of the flume through a calibrated V-notch weir. Unimodal bedload mixtures, with maximum grain sizes equal to 64 and 30 mm, were used as bed material (Fig. 2) and supplied at constant rates during some runs by a belt-conveyor feed system located at the upstream end of the flume. When the finer mixture was used, the flume width was reduced to 0.25 m to maintain a constant ratio of maximum grain size diameter to channel width.

Before each run, slope was adjusted, and wetted bed material was added to a uniform depth of 10 cm along the entire length of the flume. Water surface and bed material slope were then allowed to reach equilibrium with the imposed water and (in some runs) sediment discharge. Water depth, bed elevation, and velocity were measured at 7 minute intervals using point gauges and micro velocity meters placed at 0.5 m stations along the flume; three readings equidistant across the flume width were taken at each station. Bedload discharged at the flume outlet was collected for 30-seconds at 1-min intervals throughout the run. The collected bedload was weighed and volumes determined, and the diameter of the intermediate (b) axis of the 10 largest particles caught was measured. After each run, the bed was examined and the location, orientation with respect to flume sidewalls, and percent of channel spanned by each step formed during the run were measured. Beds were photographed and bed material taken for particle size analysis for some runs.

Total boundary shear stress for each station τ_s was calculated as:

$$\tau_s = \rho g d \sin\theta_s \tag{1}$$

where ρ is the density of water (1.0 g/cm^3), g is the gravitational constant, d is the depth of water, and θ_s is the local bed slope averaged over 0.5m up- and downstream from

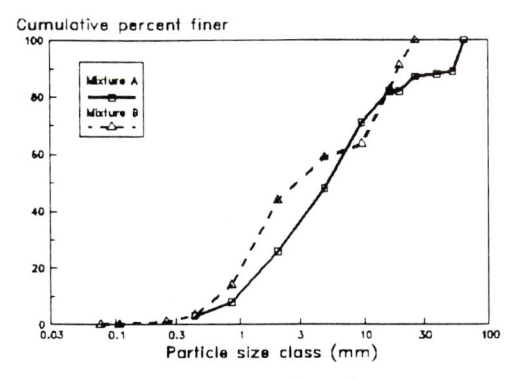

Figure 2: Particle size distributions for the two experimental mixtures.

the station. The Froude number (FR_s) for each station was determined as:

$$FR_s = \frac{v_s}{\sqrt{gd}} \qquad (2)$$

where v_s is the average measured velocity for each station. Average shear stress for the run (τ_r) was determined by averaging all τ_s values and average Froude number for the run (FR_r) was similarly calculated by averaging all FR_s values.

3. EXPERIMENTAL RESULTS

The experiments were carried out in three parts. We conducted preliminary runs to characterize the relationship between shear stress and size of particle entrained for the two mixtures. Next, we varied slope and discharge in a series of clear-water (no sediment added) runs to define the range of flow conditions required to form step-pools. Finally, we fed sediment into the upstream end of the flume to investigate the effects of high sediment transport rates on step-pool formation.

3.1 Incipient motion experiments

A series of incipient motion experiments were conducted with both bedload mixtures to determine critical shear stresses for entraining bed material. Previous field work had determined that steps are composed of the largest bed particles in the stream (Grant et al, 1990), so results were expressed by the average run shear stress τ_r and the largest particles captured during the run (Fig. 3). Results from these experiments were used in later runs to define a dimensionless shear stress τ^*:

$$\tau^* = \frac{\tau_r}{\tau_{cr}} \qquad (3)$$

where τ_{cr} is the critical shear stress for entraining the largest particles in the mixture (330 dynes/cm² for mixture A and 170 dynes/cm² for mixture B).

3.2 Mechanisms and flow conditions for step formation

Step-pools were observed in the bed of the flume following some runs. Steps were distinguished as a linear arrangement of imbricated particles oriented more or less

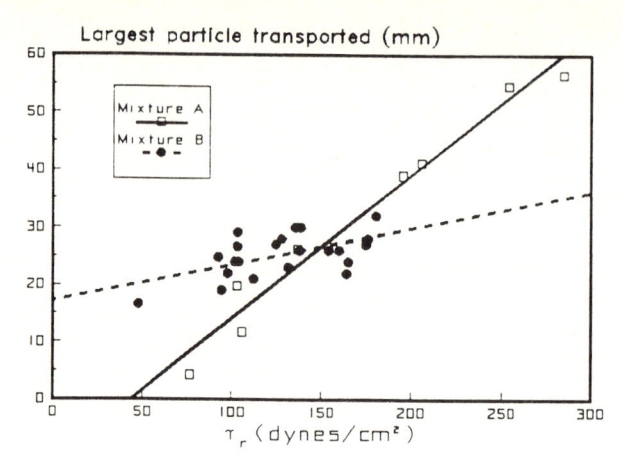

Figure 3: Results from incipient motion experiments for the two mixtures shown in Fig. 2.

perpendicular to flow (Fig. 4). Pools were more difficult to observe but appeared after the flume was drained as pockets of residual water, often partially filled with fines, located just downstream of steps. Reducing flow depth to less than the diameter of the largest grains following a run assisted recognition of steps and pools. Because the height of individual steps was the same as the grain roughness, it was not possible to distinguish steps by longitudinal bed profiles. Instead a visual rating system was employed to evaluate the degree of step development. The rating system went from 1 (very poor or obscure step, no clear imbrication, particles aligned obliquely to flow, no clear pool) to 5 (very well-developed step, particles clearly imbricated, step oriented perpendicular to flow, well-defined pool downstream). Steps with a rating of 3 or greater were termed 'well-developed' steps while steps with a rating of 1 or 2 were termed 'obscure' steps.

In runs where step-pools formed, step formation could be observed through the clear sidewall of the flume. The process we observed is similar to that described by Whittaker and Jaeggi (1983) and Ashida et al. (1984) with some modification. Initially, particle transport rates were high and the entire bed was active with rolling, sliding, and saltating grains. Within 1-2 minutes after the beginning of the run, regular antidunes formed on the water surface, resulting in bed deformation in phase with the water surface. As the bed deformed, individual large particles would intermittently come to rest under or immediately downstream of the antidune crest. These large particles trapped other smaller particles, creating a cluster of imbricated grains. The shallower depth over these clusters caused the formation of a hydraulic jump in the antidune trough. The turbulence associated with this jump scoured the bed immediately downstream of the stalled grains, accentuating the relief of the step. This entire process occurred simultaneously with development of an armor or coarse surface layer over the entire bed. Infilling of the scour hole with fines occurred when the water level dropped at the end of the run reducing the strength of the hydraulic jump.

As predicted by these observations, there was a good correlation ($r^2 = 0.50$) between the interstep spacing (measured from step crest to crest) and the antidune wavelength, defined by the wave number $L = 2\pi v_r^2/g$, where v_r is the mean velocity for the run (Kennedy, 1963) (Fig. 5). Some scatter is probably due to using the flume as opposed to local (station) averages when calculating spacing and velocity. These results suggest that step spacing might be a useful field indicator of paleovelocities required for particle entrainment and step formation.

Figure 4: An example of a step-pool. Note the imbricated particles in the step and fines filling the downstream area (pool). Flow is from right to left. Scale is graduated in centimeters.

The domain of step formation can be characterized by two dimensionless parameters: the ratio of average shear stress to the shear stress required to move the largest grain sizes (τ^*) and the average run Froude number (Fr_r). Flow conditions for formation of well-developed steps were defined reasonably well in our experiments by the range $0.5 < \tau^* \leq 1.0$ and $0.7 < Fr_r \leq 1.0$ (Fig. 6). For comparison, data replotted from Ashida et al (1984) is presented. They show steps forming over the same ranges of τ^* but at Froude numbers approximately twice as high as what we observed (Fig. 6). Reasons for this discrepancy are unclear but may have to do with differences in how the Froude number was calculated as Ashida et al used the shear velocity rather than Equation 2. There may also be differences in criteria used to define steps. Only four of our runs had average Froude numbers greater than unity and we observed steps forming in all of them, so steps may continue to form under supercritical flow conditions, as suggested by Ashida et al. In both studies, there is common agreement that the critical conditions for step-pool formation are: 1) heterogeneous bed mixture; 2) critical shear stress for the bed material exceeded but less than critical shear stress for the maximum grain size; 3) Froude numbers near unity and formation of antidunes.

3.3 Step formation during sediment transport

Several runs were conducted to examine the effect of high sediment transport rates on step formation. We used an initial set of hydraulic conditions that had produced steps in previous clear-water runs. In addition, we fed sediment (mixture B) into the upstream end of the flume at what we calculated to be an equilibrium transport rate, using the method of Bathurst (1987). In the example reported here, the discharge was 4 l/s, initial bed slope was 0.04, and sediment was fed at a rate of 1,140 cm^3/min. The flow conditions corresponded to an initial τ^* = 0.7 and Fr = 0.8, well within the region of step formation (Fig. 6). The experiment was conducted in three parts (Fig. 7). First, steps were allowed to form under a clear water regime (0-20 minutes). After the flume was drained to permit observation of the bed, both water and sediment were fed at constant rates for 130 minutes, interrupted only at 90 minutes to observe the bed. Sediment feed was stopped at 150 minutes and clear water allowed to run for an additional 30 minutes. Sediment outflow and grain size distribution of the bedload exported from the flume were measured throughout the experiment, as was the grain size distribution of the initial and final bed.

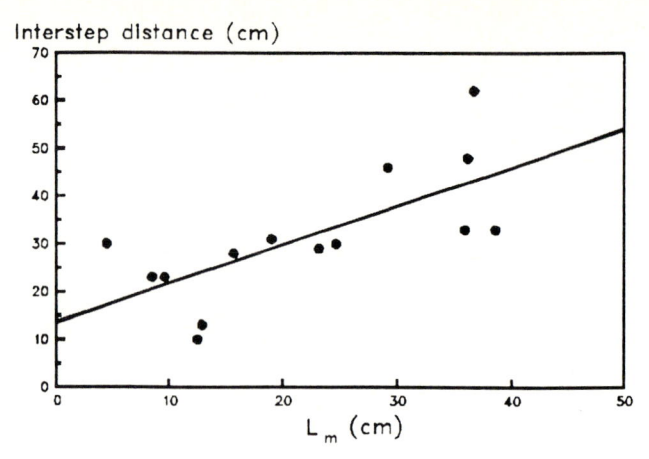

Figure 5: Interstep distance (measured crest-to-crest) as a function of antidune spacing as defined by the wave number $L_m = 2\pi v^2/g$.

Results from this experiment demonstrate the interactions among development of steps, bed armor, and alternate bars. During the initial 20 minute period when no sediment was introduced, well-developed steps formed and the bed armored in the manner previously described (Fig. 7). Alternate bars did not form during this period and steps spanned the entire flume width. Both the sediment transport rate and the maximum size of transported particles were initially quite high but dropped markedly as the bed armored. Armoring of the bed proceeded from the upper to the lower part of the flume, suggesting that absence of upstream supply was a key factor contributing to armor formation, as has been suggested elsewhere (Dietrich et al, 1989).

A different set of bedforms and sediment transport processes was observed during the 130 minute period when sediment was fed into the flume. As it turns out, the sediment input and outflow rates were not in equilibrium during the experiment; average transport rate from 20 - 150 minutes was 515 cm^3/min, roughly half of the feed rate. This resulted in significant aggradation in the upper end of the flume. The slope steepened from 0.037 at 12 minutes to 0.053 at 146 minutes. Because of the increased gradient, Fr increased from 0.78 to 0.93 and τ^* increased from 0.73 to 0.94 during this period. These increases were still within the domain of step formation (Fig. 6), so steps should have formed. Only poorly-formed steps (rating number $\leq$ 2) were observed, however, after either 90 or 150 minutes when the flume was drained. Instead, well-developed alternate bars emerged parallel to the walls along both sides of the flume.

Step formation was inhibited by high sediment transport rates and lack of bed armoring. Several factors contributed to this. High transport rates resulted in frequent grain-to-grain collisions so individual particles never stopped for very long. Ballistics frequently disrupted accumulating clusters of particles so the local conditions for hydraulic jump and step formation occurred only occasionally. A second factor was the longitudinal segregation of bedload into coarse and fine zones. This phenomenon, described by Iseya and Ikeda (1987), was observed in both the size of largest particles in transport (Fig. 7) and the grain size distribution of the bedload (Fig. 8). From 20 - 90 minutes, bedload was predominantly fine sediment while from 90 - 140 minutes, the bedload was significantly coarser than the initial bed material and sediment feed (Fig. 8). Another fine sediment zone was beginning to appear at the flume outlet at the time sediment feed was stopped at 150 minutes (Fig. 8). The mechanism by which this segregation inhibited step formation depended on whether bedload was either fine or coarse sediment. In areas

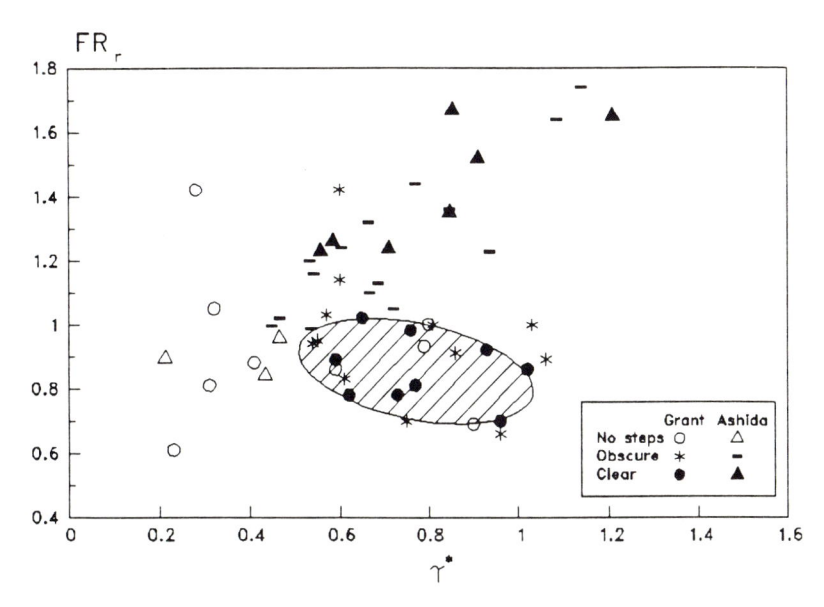

Figure 6: Domain of step formation. Shaded area represents zone of clear steps. Data from Ashida et al (1984) also shown.

of the bed dominated by fine sediment, individual larger grains rarely stopped because they protruded higher in the water column and were thus exposed to higher shear stresses. In addition, fine sediment filled the interstices between larger grains, thereby smoothing the bed and reducing flow resistance (Iseya and Ikeda, 1987; Whiting et al, 1988). In regions dominated by coarse sediment, grain-to-grain collisions, as described above, were more frequent. In neither case were well-developed steps observed.

Steps did reform, however, when water alone was allowed to run for 30 minutes after sediment feed was stopped at 150 minutes (Fig. 7). Step formation was accompanied by formation of a bed armor layer (Fig. 8). Most steps formed adjacent to alternate bars where zones of higher shear stress were concentrated; these steps only spanned the wetted channel and could not be traced onto the surface of the adjacent bar. The constricted width adjacent to bars favored antidune formation which may have promoted forming steps.

4. Discussion and summary

Taken with the earlier work of Whittaker and Jaeggi (1983) and Ashida et al (1984), these experiments provide the basis for understanding the mechanisms and flow conditions for step-pool formation. Widely-sorted bed material is required for two reasons. First, so large grains which protrude into the flow column can produce hydraulic jumps and second, so there is sufficient disparity in entrainment thresholds and bed roughness that larger grains can trap smaller ones. Sorting coefficients for the bed mixtures, defined by the ratio D_{84}/D_{16}, were 17 and 15 for the A and B mixtures, respectively (Fig. 2); sorting coefficients of 20 or greater have been observed in step-pool streams in the field (Grant et al, 1990). Bedforms somewhat analogous to steps, such as transverse ribs, may be produced by a process similar to step formation where bed material is more uniformly sorted (McDonald and Bannerjee, 1971; Mcdonald and Day, 1978; Koster, 1978; Kishi et al, 1987).

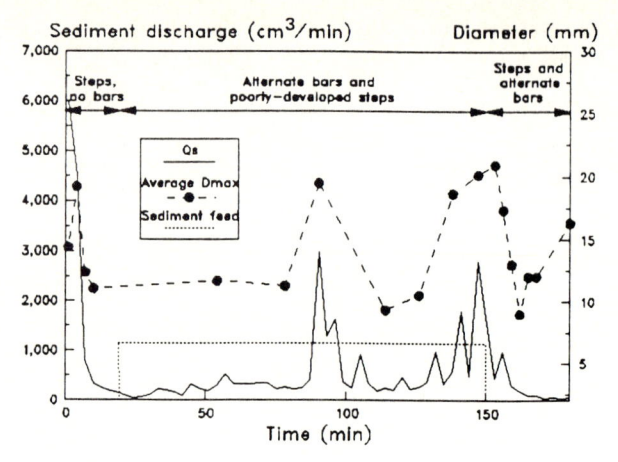

Figure 7: Sediment feed, bedload discharge, and average diameter of largest 10 particles caught at flume outlet during run 22. Bedforms observed during the run are noted.

Flow conditions for step formation require near-critical to supercritical flow conditions over the bed and must be close to but not exceed the entrainment thresholds for the largest (D_{90} or larger) particles. Paleohydraulic calculations from field observations give similar estimates of Froude numbers for these entrainment thresholds (Grant et al, 1990). A further constraint on step-pool formation is that the relative roughness (ratio of particle size to flow depth) be close to unity. Otherwise protruding particles cannot generate hydraulic jumps. This conclusion has also been confirmed by field observations and paleohydraulic calculations (Bowman, 1977; Jarrett and Costa 1986; Grant et al, 1990).

While step-pool formation requires that critical shear stresses be exceeded for most of the bed material, our results also indicate that sediment transport rates cannot be too high. This suggests why step-pools are typically found only in streams where sediment

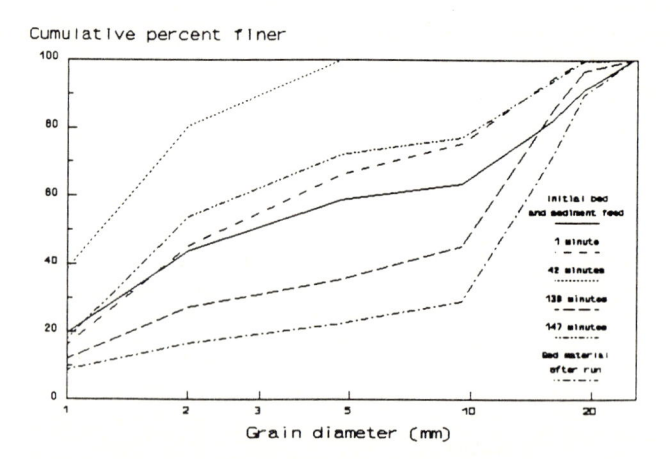

Figure 8: Particle size distributions for bedload and bed material during run 22. Times correspond to those shown in Fig. 7.

supply rates are relatively low and canyons are narrow, such as steeply dissected channels draining areas with competent bedrock. Step-pools are not common in wide alluvial channels with abundant sediment supply -- glacial outwash streams, for example -- both because the bed material tends to be more uniform and sediment transport rates are high (e.g. Fahnestock, 1963).

Factors controlling formation of step-pools appear to be fundamentally different than those creating large pools. While step-pools are features whose horizontal and vertical dimensions are scaled by the depth of flow and coarsest grain sizes, large pools have lengths scaled by the channel width (Fig. 1). These larger features owe their origin to secondary circulation patterns developed around channel curvature or resistant boundary material (Lisle, 1986).

Given the important hydraulic, geomorphic, and ecologic roles that step-pools play in stream and riparian systems, we must recognize conditions which favor step formation. Effective strategies for restoring ecologic functioning in streams following disturbances such as debris flows or large floods may involve promoting formation of step-pool sequences. This may include modifying channel geometry -- channel width, for example -- to create high Froude numbers and shear stresses during high flow events, introducing large particles or resistant boundary material, or in the case of regulated streams, by altering the flow regime to produce the required flow conditions, at least temporarily. Future research should evaluate the effectiveness of these strategies in a range of channel environments.

REFERENCES

Ashida, K., Takahashi, T., and Sawada, T. 1976. Sediment yield and transport on a mountainous small watershed. Bull. Disas. Prev. Res. Inst., Kyoto Univ., v. 26, Part 3, No. 240, p. 119-144.

Ashida, K., Egashira, S., and Ando, N. 1984. Generation and geometric features of step-pool bed forms. Annuals, Disas. Prev. Res. Inst., Kyoto Univ., No. 27 B-2, p. 341-353.

Ashida, K., Egashira, S., Sawada, T., and Nishimoto, N. 1985. Geometric structures of step-pool bed forms in mountain streams. Annuals, Disas. Prev. Res. Inst., Kyoto Univ., No. 28 B-2, p. 325-335.

Ashida, K., Egashira, S., and Nishimoto, N. 1986a. Sediment transport mechanism on step-pool bed form. Annuals, Disas. Prev. Res. Inst., Kyoto Univ., No. 29 B-2, p. 377-390.

Ashida, K., Egashira, S., and Nishino, T. 1986b. Structure and friction law of flow over a step-pool bed form. Annuals, Disas. Prev. Res. Inst., Kyoto Univ., No. 29 B-2, p. 391-403.

Bathurst, J.C. 1987. Critical conditions for bed material movement in steep, boulder-bed streams. Erosion and Sedimentation in the Pacific Rim. I.A.H.S. Publ. No. 165, p. 309-318.

Bowman, D. 1977. Stepped-bed morphology in arid gravelly channels. Geological Society of America Bulletin, v. 88, p. 291-298.

Dietrich, W.E., Kirchner, J.W., Ikeda, H., and Iseya, F. 1989. Sediment supply and the development of the coarse surface layer in gravel-bedded rivers. Nature v. 340 p. 215-217.

Fahnestock, R.K., 1963. Morphology and hydrology of a glacial stream -- White River, Mount Ranier, Washington. U.S. Geological Survey Professional Paper 422-A.

Grant, G.E., Swanson, F.J., and Wolman, M.G. 1990. Pattern and origin of stepped-bed morphology in high-gradient streams, Western Cascades, Oregon. Geological Society of America Bulletin v. 102, p. 340-352.

Gregory, S.V., Swanson, F.J., McKee, W.A., Cummins, K.W., 1991. An ecosystem perspective of riparian zones. Bioscience, v. 41, p. 540-551.

Hasegawa, K. 1988. Morphology and flow of mountain streams. Hydraulic Engineering Series 88-A-8, Hydraulic Committee, Japanese Society of Civil Engineers, p. 1-22. (in Japanese).

Hayward, J.A., 1980. Hydrology and stream sediments in a mountain catchment, Tussock Grasslands and Mountain Lands Institute Special Publication No. 17, Canterbury, New Zealand, 236 p.

Iseya, F., and Ikeda, H., 1987. Pulsations in bedload transport rates induced by a longitudinal sediment sorting: a flume study using sand and gravel mixtures, Geografiska Annaler, v. 69 A-1, p. 15-27.

Jarrett, R.D. and Costa, J.E. 1986. Hydrology, geomorphology, and dam-break modelling of the July 15, 1982 Lawn Lake and Cascade Lake dam failures, Larimer County, Colorado. U.S. Geological Survey Professional Paper 1369. 78 p.

Judd, H.E., and Peterson, D.F., 1969. Hydraulics of large bed element channels, Report No. PRWG17-6, Utah Water Research Laboratory, Utah State University, Logan, Utah, 115 p.

Kennedy, J.F., 1963, Dunes and antidunes in erodible-bed channels, J. Fluid Mechanics, v. 16, p.521.

Kishi, T., Mori, A., Hasegawa, K., and Kuroki, M. 1987. Bed configurations and sediment transports in moutainous rivers. In: Comparative Hydrology of Rivers of Japan: Final Report, Japanese Research Group of Comparative Hydrology, Hokkaido University, Sapporo, Japan, p. 165-176.

Koster, E.H., 1978, Transverse ribs: their characteristics, origin, and paleohydraulic significance, in Miall, A.D., ed., Fluvial Sedimentology, Memoir 5, Canadian Society of Petroleum Geologists, Calgary, Alberta, p. 161-186.

Lisle, T.E. 1986, Stabilization of a gravel channel by large streamside obstructions and bedrock bends, Jacoby Creek, northwestern California. Geological Society of America Bulletin v. 97, p. 999-1011.

McDonald, B.C., and Banerjee, I. 1971. Sediments and bed forms on a braided outwash plain. Can. Journal of Earth. Science, v. 8, p. 1282-1301.

McDonald, B.C. and Day, T.J. 1978. An experimental flume study on the formation of transverse ribs. Current Research, Part A, Geol. Surv. Can., Paper 78-1A, p. 441-451.

Sawada, T., Ashida, K., and Takahashi, T., 1983, Relationship between channel pattern and sediment transport in a steep gravel bed river, Zeitschrift fur Geomorphologie, N.F. Suppl.-Bd 46, p. 55-66.

Whiting, P.J., Dietrich, W.E., Leopold, L.B., Drake, T.G., and Shreve, R.L., 1988, Bedload sheets in heterogeneous sediment, Geology, v. 16, p. 105-108.

Whittaker, J.G, 1987a, Sediment transport in step-pool streams, in Thorne, C.R., Bathurst, J.C., and R.D. Hey, eds., Sediment Transport in Gravel-Bed Rivers, John Wiley and Sons, Ltd. Chichester, England, p. 545-579.

Whittaker, J.G, 1987b, Modelling bed-load transport in steep mountain streams, in Beschta, R.L., Blinn, T., Grant, G.E., Ice, G.G., and Swanson, F.J., eds., Erosion and Sedimentation in the Pacific Rim, International Association of Hydrological Sciences Publication No. 165, p. 319-332.

Whittaker, J.G. and Jaeggi, M.N.R., 1982, Origin of step-pool systems in mountain streams, Journal of the Hydraulics Division, ASCE, v. 108, No. HY6, p. 758-773.

SHORT TERM TEMPORAL VARIATIONS IN BEDLOAD TRANSPORT RATES: SQUAW CREEK, MONTANA, USA AND NAHAL YATIR AND NAHAL ESTEMOA, ISRAEL

Peter Ergenzinger & Carmen de Jong,
B.E.R.G., Institut für Geographische Wissenschaften, Freie Universität Berlin,
Grunewaldstr. 35, 12165 Berlin, Germany

Jonathan Laronne
Department of Geography and Environmental Development, Ben Gurion
University of the Negev, Beer Sheva, P.O. Box 653,
84105 Beer-Sheva, Israel

Ian Reid
Department of Geography, Loughborough University, Loughborough,
LE11 3TU, England

Abstract

Short term temporal variations in bedload transport rates were investigated in two different environments during flood flows, one at Squaw Creek, Montana, a temperate mountain stream with coarse-grained sediment, the other at Nahal Yatir and Nahal Eshtemoa, Israel, both ephemeral channels subject to a semi-arid climate and also with a coarse gravel-bed. The results of the Squaw Creek electro-magnetic bedload measuring station were compared to the Birkbeck type slots in Israel. Bedload transport rates for both measuring stations are comparable since they both take high resolution measurements at minute intervals across the entire channel width. Whereas the Squaw Creek measuring station has the advantage of continuously measuring single particle transport as induced by the transfer of naturally magnetic particles over two detector sills, the Nahal Yatir and Eshtemoa measure bedload on the basis of trap filling rates. The temporal nature of sediment transport in both environments is similiar, as indicated by repeated pulsing. Short term bedload fluctuations are independant of the discharge magnitude but dependant on its duration or change in magnitude.

1. INTRODUCTION

Continous monitoring of bedload transport rates is a widely recognized problem in present-day sedimentary process studies. This is mainly due to the difficulties in development of automatic monitoring technology, and the difficulties in predicting rainfall events in many environments. Until recently, most studies on temporal bedload variations under natural conditions have been confined either to indirect bedload measurements such as acoustic techniques or they have consisted of sporadic measurements by samplers such as the Helley-Smith. None of these methods have been satisfactory enough to enable detailed monitoring of bedload transport at minute and second intervals.

Little is, therefore, known concerning the nature of bedload transport under natural conditions over short time scales (Hoey 1992). A pioneer study (Einstein 1937) showed that single particle transport is an unsteady process. Later studies (Gomez 1983; Gomez et al 1989; Philips and Sutherland 1990; Kuhnle 1992 Hoey 1992) have demonstrated though that bedload transfer is very erratic and of a "pulsed" nature. There remains a lack of understanding of these complex temporal processes. Flume experiments (Iseya and Ikeda, 1987) have confirmed that the pulsed nature of bedload transport results from the storage and release of material from an alternating sequence of congested and free zones. Several ideas have been put forward on the nature of bedload transport but there is still considerable controversy. Motion-picture camera observations (Drake et al 1987, Dinehart 1987) have for example shown that bedload is able to move in sheets. This is mainly the case in finer-grained media where bedload is said to move by bar translation or as a translation of waves (Griffiths and Sutherland 1977, Lekach and Schick, 1983). Automatic detection sites are rare but Ergenzinger and Conrady (1982), Reid et al. (1985) and Bänziger and Bursch (1990) have indicated how discontinuous and erratic bedload transport is under natural circumstances. Others have agreed that bedload should be observed as a discontinuous, pulsed process that is not directly related to discharge. (Bänziger & Bursch 1990, Ergenzinger 1988, De Jong 1993, Bunte, 1992). This is the case in coarser grained material that is incapable of moving as a "carpet" during average flood waves. Jackson and Beshta (1982) suggested that bedload transport occurs as a two-phase process, i.e. alternating phases of sheets and pulses but their experiments were mostly carried out over longer time scales.

Even with some present knowledge on the nature of bedload transport, high resolution studies are lacking both spatially and temporally in coarse-grained environments. This study, therefore, comprises a detailed, comparative study of bedload transport in coarse, cobble-sized and fine-grained gravel material in two contrasting environments. It is our intention to show whether bedload is persistently pulsed at different time scales, including daily, hourly, minute and second intervals and how pulsing is related to discharge.

2. STUDY AREAS

Squaw Creek is a small perennial channel draining a catchment area of 106 km^2 with an average longitudinal slope of 2%, average flood discharges of 7 m^{3}s^{-1} and a mean annual rainfall of 800 mm (Ergenzinger et al. 1994). Flood flows are snow-melt induced and are therefore predictable, usually lasting for 24 hours. The study site extends along an almost straight reach directly beneath a bend and has a mean width of 12 m. The D_{50} of the surface and subsurface are 125 and 22 mm respectively. Two cross channel sills have been installed 30 m apart to identify and continously monitor the input and output of naturally magnetic particles.

Nahal Yatir and Eshtemoa drain the Southern Hebron Mountains into the Northern Negev. The climate is semi-arid with a mean annual precipitation of 286mm. Hydrograph rise times are very rapid and flood events of short duration. Their drainage areas are 19 and 110 km^2 and their widths are 3.5 and 5.5 m respectively. The Nahal Yatir and Eshtemoa reaches are straight as those at Squaw Creek. About 5 bedload transporting events occur annually and the bed is dry throughout most of the year. Their average longitudinal bed slopes are 1.0 and 0.6 % respectively but gradients are very steep in bar areas and very shallow in intervening "flats". The median grain size of the surface and subsurface of the bars is 19 and 18 mm and in the "flats" it is 6 mm for both. The monitored (submerged mass) unit bedload discharges are among the highest recorded with values in the range of 1-5 kg/ms. This typifies the semi-arid environments where unit bedload flux may be hundred-fold of that in temperate regions (Laronne et al 1992, Laronne & Reid, 1993). Such high, unsteady rates of bedload transport occurred during events of small magnitude with average dimensionless shear stress is in the range of 1-3.

3. METHODOLOGY

At Squaw Creek, bedload was measured on the basis of the Faraday principle, where magnetically induced signals of naturally magnetic particles were recorded over two detector sills and sent directly to a PC (Spieker & Ergenzinger 1988, Ergenzinger et al 1994, De Jong & Ergenzinger 1992).

The detector logs consist of two parallel rows of 6 segments, each 2 m long, inserted across the entire river width. Each segment contains magnetically sensitive coils which permit the continuous recording of bedload down to hecto-second intervals (see Fig. 2).

At Nahal Yatir and Eshtemoa, bedload discharge is determined continuously by three slot samplers that are set at intervals across the channel, causing least interference with the bed (Laronne et al 1992). The technique is based on a Birkbeck type pressure pillow system and is fully automated. The slots have a capacity of 0.25-0.4 m^3, which means that the traps usually fill before the cessation of bedload transport. Sampling intervals are on average 0.25 seconds and are recorded either on data loggers. Water surface slope is measured synchronously. At 5 kg/ms (submerged weight) the error is 2.2 and 13.2 % for sample averaging at 60 and 10 seconds respectively. Errors increase linearly as bedload flux decreases, at for instance 3 and 1 kg/ms (submerged weight) the error is 11 and 67 % respectively.

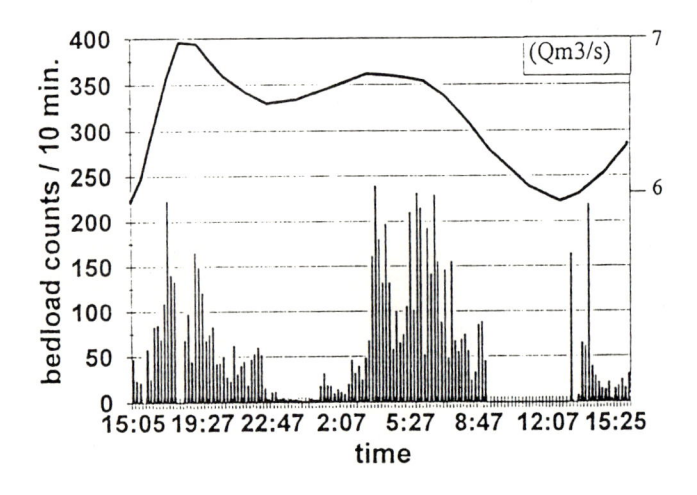

Fig. 1 Nature of bedload transporting event during flood of 5-6th June 1991, Squaw Creek, Montana in main channel, upper sill. Bedload pulses occur independantly of discharge. Notice largest pulse during descending flood limb.

4. SHORT TERM VARIATIONS

For comparison reasons with the Israeli rivers, short term variations in bedload transport were examined for the main channel over the upper sill at Squaw Creek for the 5-6th June 1991 flow event (Fig. 1). The diagram shows bedload counts in 10 seconds averaged over 10 minutes for the whole flood wave. Bedload transport occurs in 2 main pulses during the flood; one during the ascending limb, around 16:00-18:00 hours, the other during the waning stages, between 04:00-08:00. Thus, there is a phase during the rising flood limb when a large pulse of material ensues, followed by a period of lesser activity during peak discharge. A second much larger pulse occurs during the recession. Over hecto-second intervals (Fig. 2), bedload also occurs as single separate pulses. In most cases these are normally distributed. It is obvious that bedload transport is of a pulsed nature over all intervals. This is true not only for the hecto-second intervals (Fig. 2) but also for the 5 minute and hourly intervals (Fig. 1). Hence bedload transport occurs as a discontinuous process over time.

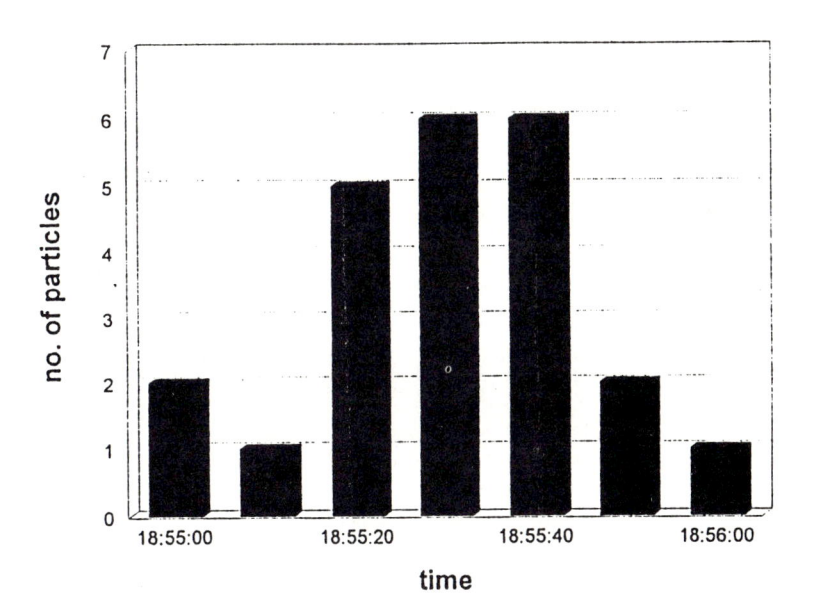

Fig. 2 The character of a single pulse recorded over 1 min. (a summation over 10 sec. intervals) between 18:56 and 18:57, Squaw Creek, 5-6th June 1991.

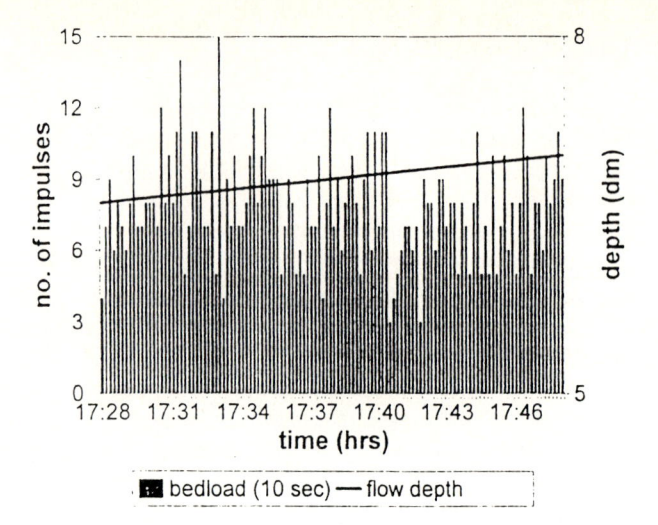

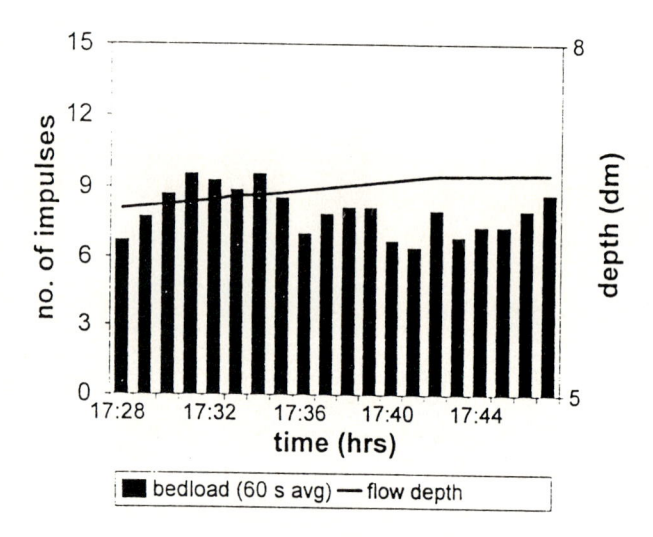

Fig. 3 Nature of bedload from single particle counts during first 20 min. of flood of 5-6th June 1991, Squaw Creek, Montana calculated a) over 10 sec. averages and b) over 1 min. averages.

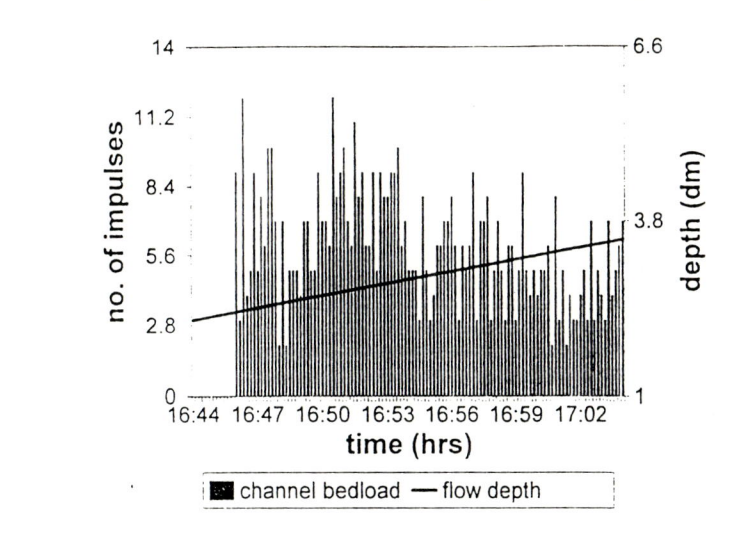

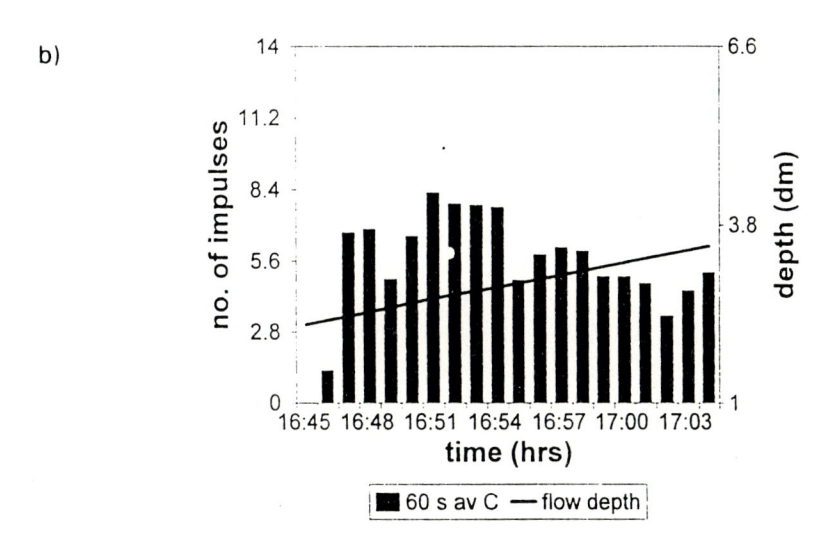

Fig. 4 Bedload transport in Nahal Yatir, Israel, 8th February, 1991: submerged mass unit bedload discharge calculated for first 20 min. of flood a) over 15 sec. averages and b) 1 min. averages.

The most active bedload transporting channel (Fig. 3a&b) was selected for comparison with Nahal Yatir and Eshtemoa. Fig. 3a shows the 10 second average bedload counts for the main channel during the first 20 minutes of the flood. In Fig. 3b the data is averaged over one minute intervals for the same period. The comparisons show that at both intervals, bedload occurs in separate pulses, which occur seemingly independant of flow stage.

At the Yatir and Eshtemoa, bedload transport is also of a pulsed nature over these intervals (Fig. 4a&b). The event of February 8th, 1991 in the Yatir was monitored in detail; the first 20 minutes of the event compared for the same time intervals as Squaw Creek. The first 20 minutes were also calculated for the Nahal Eshtemoa flood on the 26th February, 1992 (Fig. 5a&b). For both of the ephemeral channels, bedload transport occurs in wavy pulses of 5 minute intervals, again independant of stage. Since the slots are usually full after 20 minutes, the final phases of the bedload transport could not be compared. Nevertheless, the beginning of the flood events are comparable to Squaw Creek, since the largest pulses of material occur during the ascending limb rather than during peak flow. There is a divergence between the stage-hydrograph and the bedload flux curve at Squaw Creek. The shorter 30 second pulses, measured at 10 second intervals are of a very irregular nature and do not show clear trends. In contrast the 1 minute averages indicate that there is a gradual increase in bedload flux as stage rises. After an initial period of intensive transport, bedload transport rates remain stable or start to decrease, even though stage is rising.

5. DISCUSSION

The results from Squaw Creek, Nahal Yatir and Eshtemoa, suggest that bedload transport occurs in three different phases (Ergenzinger 1988). The first phase can be attributed to random, low intensity transport, such as would occur during non-flood conditions i.e. before the onset of first flood-induced bedload pulses and at the very end of the flood, when bedload transport is ceasing. During this phase bedload flux is low and erratic. The second phase is wavy periodic transport. This refers to the phenomena observed at 10-20 minute intervals at Squaw Creek and at 3-5 minute intervals at Yatir and Eshtemoa. These correspond to bedload pulses during

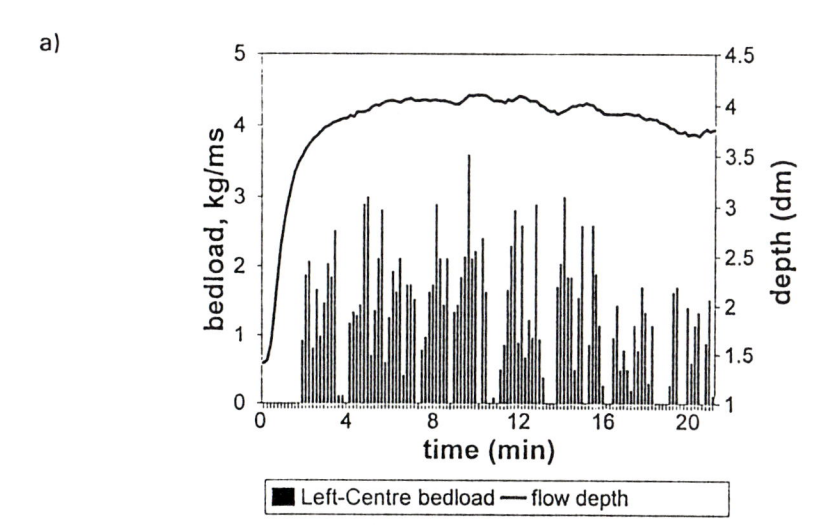

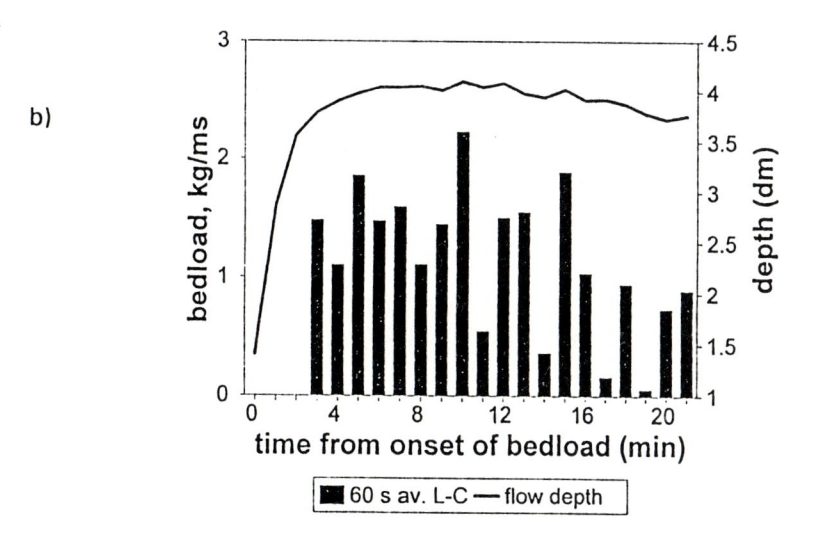

Fig. 5 Bedload transport in Nahal Eshtemoa, Israel, 26th February 1992. Submerged unit bedload discharge calculated for first 20 min. of flood a) over 15 sec. averages and b) 1 min. averages.

active bedload transport. A third phase of intensive bedload transport may be typified by gravel lobes or gravel sheets sustained only over very short periods of time with large amounts of bedload. At Squaw Creek this may last at most for 15-20 minutes (Fig. 3a), and for 5-8 minutes in the ephemeral channels (Fig. 4a & 5a). This phase may involve the mobility of the entire bed.

Comparing the observations from the two types of fluvial environments shows that bedload is shorter and more intense at the Yatir and Eshtemoa, corresponding to the shorter flood durations. Bedload transport continues over longer time intervals at Squaw Creek, where pulse duration is also comparitively longer.

It can be speculated that the sub-second data (Fig.2) may be related to turbulent fluctuations. Other short term fluctuations may be due to the break up of clusters. Thus when considering the sub-second pulses at Squaw Creek, they usually proceed as normally distributed groups of particles. This could mean that a large moving particle is surrounded by a group of smaller particles in its lee. Such minor pulses occur both during intensive and non-intensive bedload phases. Assemblages of this kind will probably not move for very long distances, since there is very extensive interchange between mobile particles and the bed. It is for this reason that the idea of bedload sheets (Dinehart 1992) being transported under normal flood flows cannot be conceptualized with the limited amount of sediment available during most floods. Thus for a bedload sheet to be under way, the entire bed surface would need to be covered by a mobile layer at least two particles in thickness. Since most bedload transporting events at Squaw Creek do not show evidence for this, except during very short intervals of time, the concept of bedload sheets should only apply to very intensive bedload transporting phases. During the remaining less intensive bedload transporting periods, bedload transport will not cover the entire river bed but will move along very localized bedload streets, causing the erratic bedload pulses observed.

The case may be different for the Nahal Eshtemoa and Yatir, where finer grained material or larger volumes of transported material may facilitate thicker layers of bedload movement. As at Squaw Creek, bedload discharge is transport controlled, as indicated by the wavy transport pattern, and often occurs independantly of water stage. Thus the transport phenomena seem to be occurring in at least two time scales, i.e. at 30 second and 3-5 minutes intervals. At the Nahal Eshtemoa, these intervals occur both during the rising flood limb and the following phase of constant water depth i.e. regardless of flow volume. Since the magnitude of the pulses decreases after approximately the first 10 minutes of the flood during both increasing and steady water levels at Nahal Yatir and Eshtemoa (Fig. 4b and 5b) respectively, this would suggest that the amount of bedload in motion is controlled by the amount of sediment available, not by the actual flood strength. At Squaw Creek (Fig. 3b), the intensity of pulsing also decreases after the first 10 minutes on the rising flood limb, only to start increasing again after a further 5 minutes. The large amounts of energy that are required to keep bedload in motion may have caused groups of moving particles to become congested as a sheer result of the intensity of motion within the first phase of the flood. As a result, the release of bedload from temporary storage or because of the exchange of particles would seriously limit the magnitude of the pulses observed. As these zones become decongested a new series of high intensity pulses may result (last five minutes at Squaw Creek, Fig. 3b).

The shorter term second-length pulses are most probably the result of bursts of bed particles from the break-up of clusters or single particle protuberances. This may explain why their transport nature is so irregular and discontinuous. The longer term, 3-5 minute pulses may demonstrate layers of bedload in motion, perhaps resulting from the propagation of coarse-grained dunes. Such relatively longer term movements (over approx. 5 minutes duration) most probably occur in the form of gravel sheet advances. The temporary storage and exchange of particles with the bed suggests that for both Squaw Creek and Nahal Yatir and Eshtemoa, the influence of the local hydraulics are probably much more significant than the discharge intensity. Since the structure and arrangement of the bed are significantly influenced by the nature of the local hydraulics, bedload pulses should be considered to be a result of these local turbulences.

6. CONCLUSIONS

These joint studies have shown that bedload transport is a very erratic independent process that occurs in a pulsed manner at short time scales. A certain periodicity of pulses may be discerned at both Squaw Creek and Nahal Yatir and Eshtemoa in relation to the duration of discharge. One mechanism alone cannot be responsible for the bedload pulses at these different time scales. As shown at Nahal Yatir and Eshtemoa, pit-slot sampling would have limitations during large and prolonged flood events, since the duration of sampling has to decrease rapidly to ensure the upkeep of representative measurements. At Squaw Creek, the continuous sampling technique helps to identify the character of the entire bedload transporting event. Although bedload pulses may become larger with prolongation of the flood, the high intensity and short duration pulses remain unpredictable. More flood measurements in other areas of the world are necessary to help explain this pulsed transport process. Speculations have been made over the reasons for the fluctuations at different scales but these require further investigations.

Since the irregular nature of bedload transport cannot be explained by the nature of the stage/discharge curve alone, investigations into the interrelationships between the water surface gradient and river bed morphology as well as roughness conditions are a necessity in future studies.

7. REFERENCES

BÄNZIGER, R. & BURSCH, H. (1990). Acoustic sensors for bedload transport in a mountain torrent. In *Hydrology in Mountain Regions*, 193 (pp. 207-214). Lausanne: International Association of Hydraulic Sciences.

BUNTE, K. (1992). Particle number grain-size composition of bedload in a mountain stream. In P. Billi, R. D. Hey, C. R. Thorne & P. Tacconi (Eds.), *Dynamics of Gravel-Bed Rivers* (pp. 55-72). Chichester: John Wiley and Sons.

DE JONG, C., & ERGENZINGER, P. (1992). Unsteady flow, bedload transport and bed geometry responses in steep mountain torrents. In P. Larsen & N. Eisenhauer (Ed.), *5th Intenational Symposium on River Sedimentation. Sediment Management*, Vol. 1 (pp. 185-192.). Karlsruhe: International Society of River Sedimentation.

DE JONG, C. (in press) Temporal and spatial interactions between river bed roughness, geometry, bedload transport and flow hydraulics in mountain streams - examples from Squaw Creek, Montana (USA) and Lainbach/Schmiedlaine, Upper Bavaria (Germany). *Berliner Geographische Abbhandlungen,* Ph.D. thesis, Free University of Berlin.

DINEHART, R. L. (1992). Evolution of coarse gravel bed forms: field measurements at flood stage. *Water Resources Research*, 28(10), 2667-2689.

DRAKE, T. G., SCHREVE, R. L., DIETRICH, W. E., WHITING, P. J., & LEOPOLD, L. B. (1987). Bedload transport of fine gravel observed by motion-picture photography. *Journal of Fluid Mechanics*, 192, 193-217.

EINSTEIN, H. A. (1937). Der Geschiebetransport als Wahrscheinlichkeitsproblem. *Mitteilungen der Versuchsanstalt für Wasserbau an der ETH Zürich.*

ERGENZINGER, P. (1988). The nature of coarse material bedload transport. In *Sediment Budgets,* 174 (pp. 207-216), Porto Allegro: International Association of Hydraulic Sciences

ERGENZINGER, P., DE JONG, C., & CHRISTALLER, G. (1994). Interrelationships between bedload transfer and river bed adjustment in mountain rivers. In M. Kirkby (Ed.), *Theory in Geomorphology.* Leeds: J. Wiley & Sons.

ERGENZINGER, P., & CONRADY, J. (1982). A new tracer technique for measuring bedload in natural channels. *Catena,* 9, 77-80.

GOMEZ, B. (1983). Temporal variations in bedload transport rates: the effect of progressive bed armouring. *Earth Surface Processes & Landforms,* 8, 41-54.

GOMEZ, B., NAFF, R. L., & HUBBELL, D. W. (1989). Temporal variation in bedload transport rates associated with the migration of bedforms. *Earth Surface Processes & Landforms,* 14, 135-156.

GRIFFITHS, G. A., & SUTHERLAND, A. J. (1977). Bedload transport by translation of waves. *American Association of Civil Engineers J. Hydraulics Division,* 103 (HY11), 1279-1291.

HOEY, T. (1992). Temporal variations in bedload transport rates and sediment storage in gravel-bed rivers. *Progress in Physical Geography,* 16(3), 319-338.

ISEYA, F., & IKEDA, H. (1987). Pulsations in bedload transport rates induced by a longitudinal sediment sorting. *Geografiska Annaler,* 69A(1), 15-27.

JACKSON, W. L., & BESHTA, R. L. (1982). A model of two-phase bedload transport in an Oregon coast range stream. *Earth Surface Processes & Landfroms,* 7, 517-527.

KUHNLE, R. A. (1992). Fractional transport rates of bedload on Goodwin Creek. In P. Billi, R. D. Hey, C. R. Thorne, & P. Tacconi (Eds.), *Dynamics of Gravel-Bed Rivers* (pp. 141-155). Chichester: John Wiley and Sons.

LARONNE, J. B., REID, I., YITSCHAK, Y., & FROSTICK, L. E. (1992). Recording bedload discharge in a semiarid channel, Nahal Yatir, Israel. In J. Bogen, D. E. Walling, & T. Day (Ed.), *International Symposium on Erosion and Sediment Transport Monitoring Programmes in River Basins,* 210 (pp. 79-86). Oslo: International Association of Hydraulic Sciences.

LARONNE, J.B. & REID, I. (in press). Very high rates of bedload sediment transport by ephemeral desert rivers. *Nature.*

LEKACH, J., & SCHICK, A. P. (1983). Evidence for transport of bedload in waves: analysis of fluvial sediment samples in a small upland stream channel. *Catena,* 10, 267-279.

PHILLIPS, B. C., & SUTHERLAND, A. J. (1990). Temporal lag effect in bedload sediment transport. *International Association of Hydraulic Research, Journal of Hydraulic Research,* 28, 5-23.

REID, I., FROSTICK, L. E., & LAYMAN, J. T. (1985). The incidence and nature of bedload transport during flood flows in coarse-grained alluvial channels. *Earth Surface Processes & Landforms.,* 10(1), 33-44.

SPIEKER, R., & ERGENZINGER, P. (1990). New developments in measuring bed load by the magnetic tracer technique. In D. E. Walling, A. Yair, & Berkowicz (Ed.), *Erosion, transport and deposition processes,* 189, IGU (pp. 169-178). Jerusalem:

MEASURING SYSTEMS TO DETERMINE THE VELOCITY FIELD IN AND CLOSE TO THE ROUGHNESS SUBLAYER

Karin Hammann de Salazar and Andreas Dittrich
Institute of Hydraulic Structures and Agricultural Engineering, University of Karlsruhe
Kaiserstr. 12, D-76128 Karlsruhe

ABSTRACT

River beds are often characterized by marked surface textures. To extend the applicability of the model suggested by *Shields (1936)* to natural river beds with marked surface textures, research was carried out to develop a physical model that allows the stability and instability of river beds to be determined more precisely than previous ones (see *Dittrich (1992)*). The model is based on the exact knowledge of the turbulence characteristics of the liquid phase in combination with parameters describing the solid phase. For a good description of the velocity field it is not only necessary to know the mean velocity but also the deviations of the mean values up to the fourth moment close to the roughness elements. Until now, the only method to measure the turbulence in the roughness sublayer is Laser Doppler Anemometry. As this method attains its range of validity close to the roughness elements, three different measuring systems were tested. The received data show system dependent relationships. Thus, it is important to consider errors and failures that can be related to the utilized LDA-system. A description of the three different LDA-systems as well as a list of errors and failures and a short summary of main results will be the content of this paper.

1 INTRODUCTION

One of the important problems in engineer practice is the estimation of river bed stability. Fig. 1 shows the balance of forces acting on a particle of the river bed according to the theory of *Shields (1936)*. The parameters result from the assumption of equilibrium between the resisting force R and the shear force F_T. Theoretically as well as practically it has been shown that the balance of forces has to be extended by a further component, the so-called lift force L. For the explanation of its origin and reaction two models have been developed. The one sees the reason of this force in the pressure difference between the bottom and top of a particle (*Raudkivi (1976)*). The second, supported by *Grass (1971)*, sees the reason for the lift force in the flow induced turbulence that can be characterized by bursts. These bursts are intermittent series of different events that transport high momentum fluid towards the bed by sweeps and low momentum fluid away from the bed by ejections. Sweeps are often assumed to be responsible for incipient motion of bed particles.

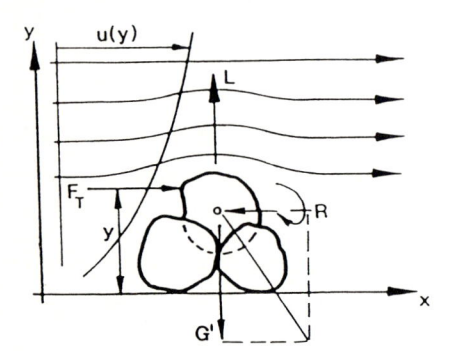

Fig. 1: Forces acting on a particle

As a consequence of the existence of sweeps and ejections the instantaneous streamwise velocity u follows a characteristic probability distribution that is far away from being symmetric or even Gaussian. Fig. 2 shows that the shape of this distribution is indeed important for the understanding of the processes causing river bed instability. The model given by Fig. 2 was suggested by *Grass (1971)*. Instead of the mean values $\bar{u}$ and $\bar{u}_K$ assumed by *Shields* (with u_K : the critical velocity to lift a single grain), the two probability distributions of u and u_K are required to determine river bed stability. Even if the mean values differ remarkably from each other, the two probability distributions may overlap and cause incipient motion. Thus, it is necessary to know the exact shape of the velocity distribution close to the top of the particles.

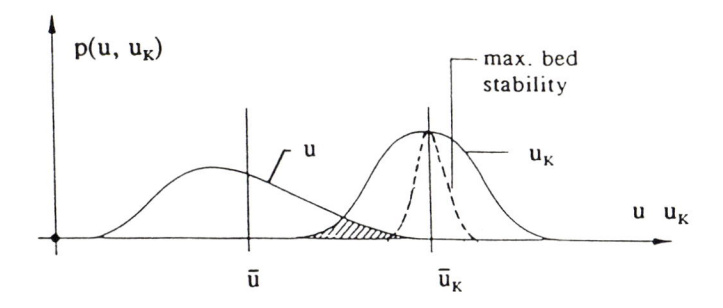

Fig. 2: Probability distributions of velocity u and critical value u_K to lift a single grain

To describe the shape of the velocity distribution u the moments of the mean value, the standard deviation or rms-value, the skewness and flatness have to be known. Over smooth surfaces the distributions of the four parameters over depth are known. However, over rough surfaces, especially in the vicinity of the roughness elements, there is not yet enough information available. Therefore, the distributions of these parameters in the roughness sublayer very close to the elements of a rough bed have to be determined. As this is a region which is not easily accessible, it is important to ensure that the data are free from influences of the chosen measuring system. Apart from theoretical considerations and the presentation of main results this paper concerns the determination of the system specific errors and the estimation of their influences on the four moments.

2 THEORY

2.1 Logarithmic wall law

In a fully developed two-dimensional open channel flow over rough surfaces the distribution of the velocity $\bar{u}$ close to the boundary has been shown to follow the logarithmic wall law, in the way (see e.g. *Grass (1971)*)

$$\frac{\bar{u}}{u^*} = 2{,}5 \cdot \ln\left(\frac{y}{k_s}\right) + 8{,}5 \tag{1}$$

where k_s is the eqivalent sand roughness, y the vertical distance from the boundary, u^* the shear velocity and $\bar{u}$ the local mean velocity. By fitting the experimental results to this law the parameters characterizing the examined flume bed u^*, k_s and the shift of the origin of the velocity profile y' can be obtained. A detailed description of this procedure can be found e.g. by *Sumer and Deigaard (1981)*.

2.2 The statistical moments

The four moments describing the u-distribution in a sufficiently accurate way are defined as follows:

$$\text{mean}\quad \bar{u} = \frac{1}{N}\sum_{i=1}^{N} u_i \tag{2}$$

$$\text{rms}\quad u' = \sqrt{\frac{1}{N-1}\sum_{i=1}^{N}(u_i-\bar{u})^2} \tag{3}$$

$$\text{skewness}\quad S = \frac{1}{N-1}\frac{\sum_{i=1}^{N}(u_i-\bar{u})^3}{u'^3} \tag{4}$$

$$\text{flatness}\quad F = \frac{1}{N-1}\frac{\sum_{i=1}^{N}(u_i-\bar{u})^4}{u'^4} \tag{5}$$

As $\bar{u}$, u', S and F are statistical values of a totality of samples, the moments are determined by u_i and N, where u_i is one sample of the instantaneous streamwise velocity and N the number of samples.

2.3 Distribution of the statistical moments for smooth-bed surfaces

The distributions of the four moments over depth for smooth bed surfaces are given in the article of *Durst et al. (1987)*. Fig. 3 shows the distribution of the second, third and fourth moment over the dimensionless depth y^+ ($=y \cdot u^*/v$). As illustrated by Fig. 3 very high values of the skewness and flatness are found in the vicinity of smooth bed surfaces compared with the values that characterize the Gaussian distribution : $S = 0$ and $F = 3$. These circumstances indicate that knowledge of these moments is of great importance close to the elements in the roughness sublayer.

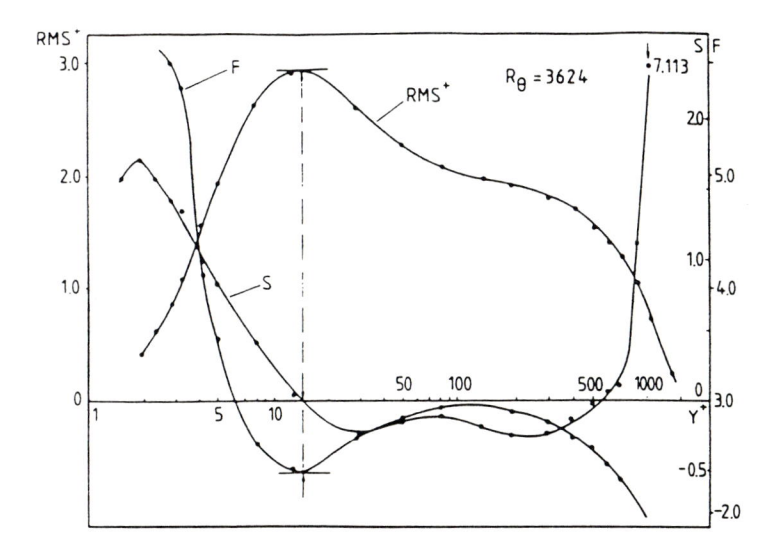

Fig. 3: u_{rms}/u^*, skewness S and flatness F for smooth-bed surfaces as a function of y^+

2.4 Roughness Sublayer

A so called roughness sublayer has been identified by various authors below the logarithmic region in the case of rough surfaces. In general the deviation of the time-averaged velocity from the logarithmic wall law is used to define the roughness sublayer (*Nakagawa et al. (1988)*). Alternatively, the zone where the skewness depends on the roughness was chosen by *Raupach (1981)* to define the thickness of this sublayer.

3 EXPERIMENTAL PROCEDURE

3.1 Flume

The experiments were conducted in a 13,2 m long and 30 cm wide tilting flume of the Theodor-Rehbock-Laboratory at the University of Karlsruhe. The flume is made of steel plates and the 2-m long measuring section has glass-sided walls. A magnetic flow meter was used to determine flow rate and a point gauge to measure flow depth. For more technical data see *Dittrich (1992)*. The flume bed was covered with glass spheres in-line which were fixed on steel plates. The spheres had a diameter of 9 mm and an absolute height above the steel plates of 8,3 mm.

3.2 Roughness Textures

In the course of experiments the bed of the 2-m-long measuring section was covered with four different artificial roughness textures two of them composed of glass spheres and two of PVC cubes. Diameter and absolute height of the spheres in the measuring section were the same as those along the flume bed. The PVC cubes had a height of 8 mm. The textures used were characterized by different shapes, arrangements and roughness densities. As illustrated in Fig. 4, the roughness density c_K is defined as the ratio of the upstream projected area of a roughness element to the floor area occupied by the element. The main characteristics of the four textures are summarized in table 1.

c_K = A'/A = upstream projected area/floor area

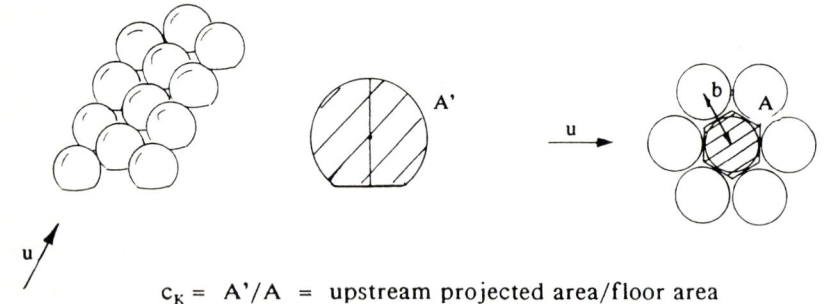

Fig. 4: Definition sketch of c_K

table 1: characteristic properties of the four textures

Texture	shape	arrangement	roughness density c_K
1	spheres	in-line	0,70
2	spheres	staggered	0,70
3	cubes	staggered	0,28
4	cubes	staggered	0,58

3.3 LDA Systems

Laser doppler anemometry was used to measure the horizontal velocity components in the roughness sublayer. To estimate and eliminate the errors arising from the measuring system three different laser doppler anemometer units were chosen and the results obtained with each system compared. In tables 2 and 3 the optical parameters of the three systems and the signal processors including some notes concerning their handling are summarized. Photo 1 shows the measuring system 3.

The *optical set-up* of the three systems was similar and the main differences were found in the power of the laser and the dimensions of the measuring volume. The laser of system 1 was weak and did not permit any adaptability of its power to the instantaneous conditions while system 2 and 3 had more powerful laser that could be controlled and thus adapted to the instantaneous conditions. The length of the measuring volume l was shortest for system 2 and longest for system 1 (see table 2).

A great difference existed between the three *signal processors*. The working principle of processor 1 is similar to a counter and provided no possibilities to control the processing like the voltage of the photomultiplier and the signal-to-noise ratio. The signal processing of system 1 is based on the presupposition that noise disturbs a LDA-signal and changes its period. Thus, only signals that change their period within a given range are taken as

table 2: optical parameters of the three units

Measuring system	Measuring system 1	Measuring system 2	Measuring system 3
Laser	Helium-Neon (Spectra-Physics), 35 mW	Argon-Ion (Spectra-Physics), 0-100 mW	Argon-Ion (Coherent Innova 300), 0-4 W
Wavelength λ of laserlight	$\lambda = 632,8$ nm	$\lambda = 514,5$ nm	$\lambda = 514,5$ nm
Used beamseparators, probes etc.	9180 Frequency Shift, Translator 9259, Coupling 9262, 9272 Probe(ϕ14mm)	60X40 Transmitter 60X24 Manipulators 60X14 Probe (ϕ14mm)	TRCF1 Multicolor Beam Separator, 9811-12 Probe (ϕ25mm)
Beam Spacing D/ Focal lenght f (water)/ Beam Diameter d	$D = 8$ mm $f = 80,57$ mm $d = 0,30$ mm	$D = 8$ mm $f = 67,09$ mm $d = 0,27$ mm	$D = 15$ mm $f = 147,57$ mm $d = 0,53$ mm
Beam half angle ϕ	$\phi = 2,84°$	$\phi = 3,41°$	$\phi = 2,91°$
Measuring volume in water: Length l / Width b	$l = 4,37$ mm $b = 216,7$ µm	$l = 2,74$ mm $b = 163,1$ µm	$l = 3,59$ mm $b = 182,6$ µm
Fringe spacing d_f	$d_f = 5,39$ µm	$d_f = 3,23$ µm	$d_f = 3,78$ µm
Number of fringes N	$N = 40$	$N = 50$	$N = 48$

table 3: signal processing and notes concerning handling of the three units

Measuring system	Measuring system 1	Measuring system 2	Measuring system 3
Photomultiplier voltage adjustable?	not adjustable	adjustable between 0-2000 V	adjustable between 1000-2000 V
Processor	IFA550 (Counter)	FVA (Correlator)	IFA750 (Correlator)
Frequency Range / Data Rate	1 kHz - 15 MHz 650 kHz	-6 MHz - +30 MHz 100 kHz	300 Hz - 90 MHz 125 kHz
Used Bandwidth	30 kHz - 300 kHz Shift 100 kHz	-70 kHz - 280 kHz	30 kHz - 300 kHz Shift 100 kHz
Typical Data Rate: upper/lower measuring range	upper measuring range: 100 - 250 Hz lower measuring range: 15 - 50 Hz	upper measuring range: 100 - 400 Hz lower measuring range: 10 - 50 Hz	upper measuring range >500 Hz lower measuring range: 15 - 100 Hz
Signal-to-Noise-Ratio adjustable?	not adjustable	adjustable	adjustable

correct, and frequency is determined from the period of the correct signal. The processors of systems 2 and 3 are both based on the autocorrelation principle of a signal and provided possiblities to control the processing in the way mentioned above. The autocorrelation of a signal means that the incoming LDA signal is shifted by a given time τ and correlated with the original signal. If the signal is correct the phase shift Φ between the two signals is constant and the correlator gives an analog output that depends linearly upon this phase shift. If the incoming signal is noisy no constant phase shift will exist between the two signals and the output of the correlator will be zero. The frequency of the signal is evaluated from the phase shift of the signal by $f = \Phi/\tau$.

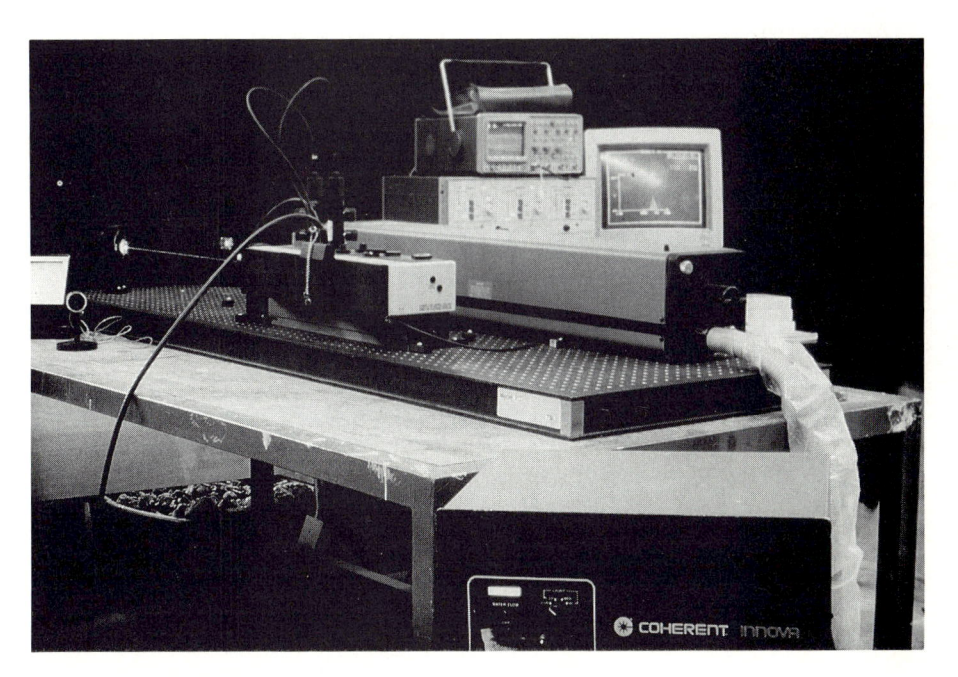

photo 1: View of measuring system 3

3.4 The measurements

For each texture three representative locations had been chosen to measure the horizontal velocity field. The first location was situated above the elements, the second behind and the third between the elements. At each measuring point three measurements with 10,000 samples were carried out. To be able to measure between the elements the probe of the LDA-system was fixed on a point gauge and moved vertically through the water (see photo 2). The position of the probe could be determined within an accuracy of ±0,05 mm.

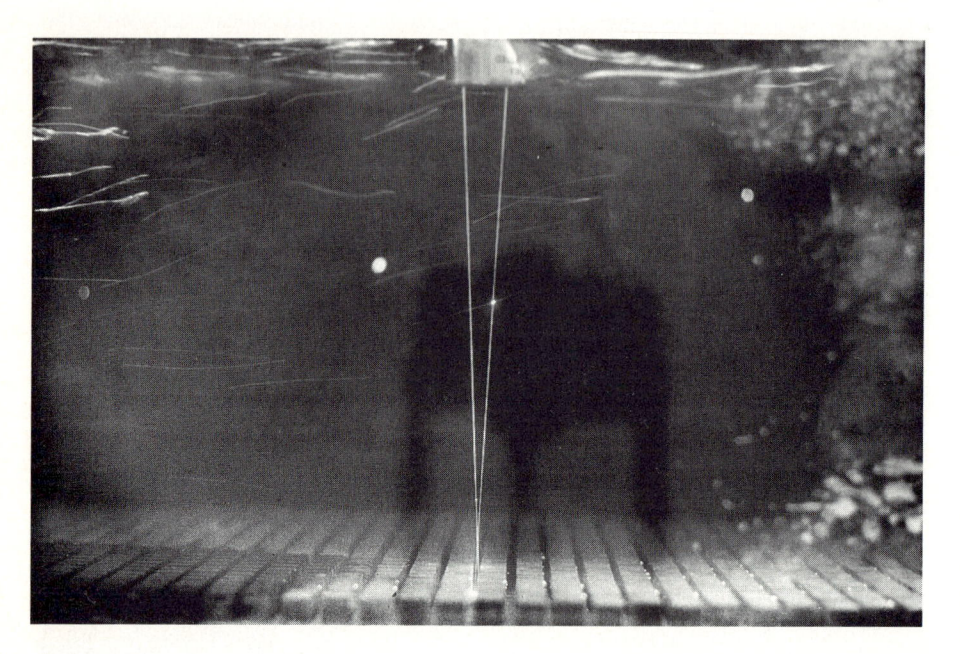

photo 2: Orientation of the probe

4. EXPERIMENTAL RESULTS

In Fig. 5 a semi-logarithmic plot of the mean velocities versus y/k_s is presented. The linear regression analysis for the log-law was carried out in the region $y/h < 0,25$. For all profiles the regression correlation coefficients were very high and varied between 0,995 and 0,999. The mean velocities were found to follow the logarithmic wall law almost down to the roughness elements (see Fig. 5). As illustrated in Fig. 5 the mean velocity do not depend upon shape, arrangement or roughness density. The depth dependent separation point of the mean velocities from the logarithmic wall law was used, as mentioned before, to define the upper boundary of the roughness sublayer. In Fig. 6 the thickness of this sublayer is plotted as a function of the dimensionless roughness k_s/k (with k = geometric roughness height). As indicated by Fig. 6, the thickness increases linearly with increasing k_s/k.

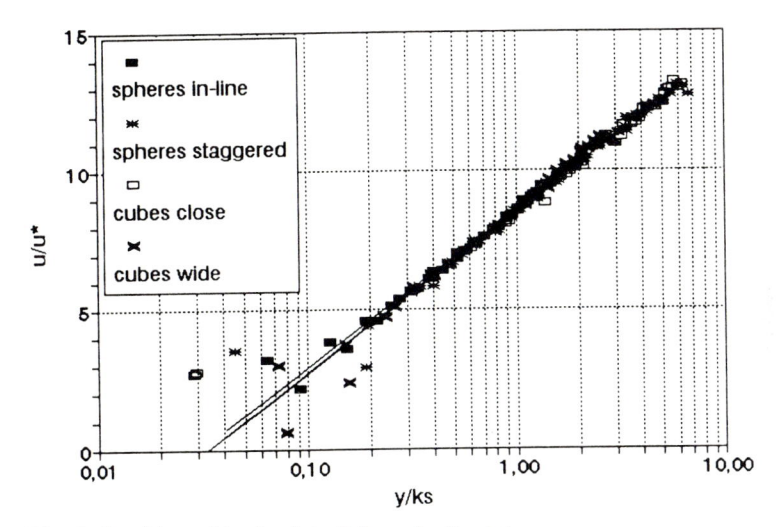

Fig. 5: Semi-logarithmic plot of the velocity data

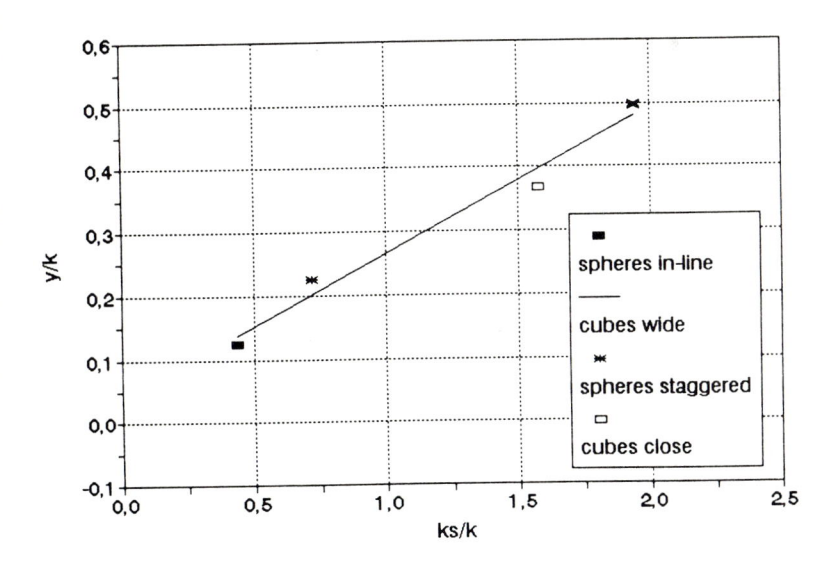

Fig. 6: Thickness of the roughness sublayer

In Fig. 7 the distributions of the rms-values, the skewness and the flatness for the wide cubes are plotted versus the dimensionless depth y^+. The distributions of the rms-values as well as those of the skewness and the flatness are similar for the four roughness textures and do not show any significant dependance on shape or arrangement of the roughness elements. Thus, Fig. 7 is representative for the four roughness textures.

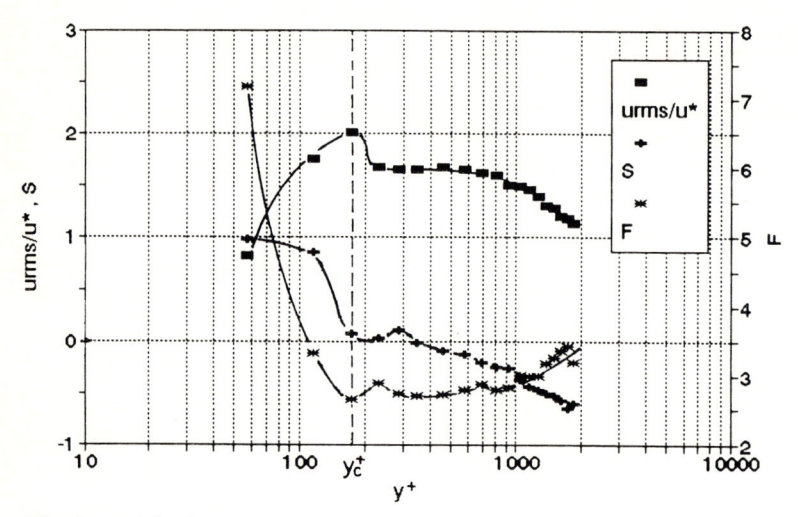

Fig. 7: u_{rms}/u^*, skewness S and flatness F for the wide cubes as a function of y^+

The distributions of the *rms-values* are charaterized by maximal values of 2,4 to 2,8 at the upper limit of the roughness sublayer behind the elements. Above the roughness sublayer the rms-values decrease with increasing y^+. The distributions of the *skewness* obtain maximal values of 1 to 1,5 within the roughness sublayer behind the roughness elements. Above the roughness sublayer the values of the skewness decrease to negative values with increasing y^+. The distributions of the *flatness* reach maximum values of 5,5 to 7,5 within the roughness sublayer behind the roughness elements at the same y^+-location as the maximum of the skewness factor S. The values of F reach a minimum value of about 2,6 at the upper limit of the roughness sublayer and increase above this sublayer with increasing y^+. The maxima of the distributions of the rms-values as well as those of S and F are less apparent between the elements.

By comparing the results of Fig. 7 and Fig. 3 the similarity of the distributions over rough and smooth bed surfaces is obvious. As in the case of smooth surfaces, the minimum of the flatness factor is correlated with the maximum of the rms-values and the change of sign of the skewness factor (see Fig. 3 and Fig. 7). These characteristic points of u_{rms}, S and F fall together at the location marked with y^+_c. The distributions of the rms-values, the skewness and the flatness are therefore universal for rough and smooth surfaces. The only significant difference between the distributions over the four rough and the smooth bed surfaces can be found in the value of y^+_c. In the case of the rough surfaces y^+_c indicates the upper limit of the roughness sublayer and increases with increasing roughness. Thus, the effect of roughness is reflected in the value of y^+_c. In the case of smooth surfaces y^+_c is constant with a value of ≈ 13 (see Fig. 3).

Summarizing the results from above, the conclusion can be drawn that the burst character of the outer flow field is similar over rough and smooth surfaces. Although, no direct dependance on shape, arrangement or roughness density could be identified, the influence of roughness is reflected in the location of the characteristic point y^+_c (see Fig. 7).

5 DISCUSSION OF THE RESULTS WITH REGARD TO SYSTEM SPECIFIC INFLUENCES

Laser doppler anemometry is a measuring method with a lot of advantages. However, under difficult measuring conditions, as for example close to rough boundaries, the method can reach its limits of applicability. Especially three parameters are important for the validity and quality of signals in the vicinity of rough bed surfaces:

- the length and orientation of the measuring volume
- the processing of the signals and
- the power of the laser.

In Fig. 8 the orientation of the measuring volume with respect to the direction of flow is shown. As the probe is moved vertically through the water column, the measuring volume is oriented parallel to the gradient of the mean velocity profile and thus the length of the volume plays an important role.

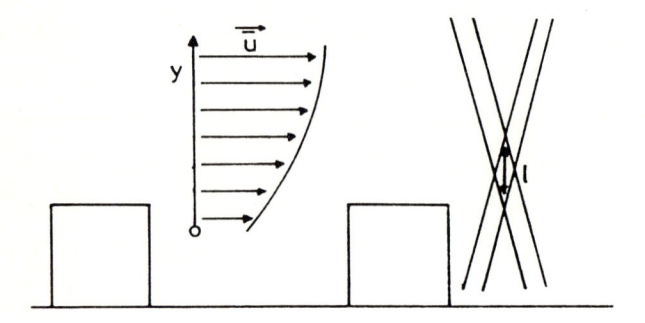

Fig. 8: Orientation of the measuring volume

The processing of signals or more precisely the algorithms decide on the reliabilty of a signal. Thus, a good adaptation to the special flow conditions by adjusting the voltage of the photomultiplier and the signal-to-noise-ratio can be helpful for the processing to distinguish valid signals from noise. Not only strong noise can change the results in an unacceptable way, but also week noise can influence e.g. the shape of the velocity distributions and therefore the values of skewness S and flatness F. The power of the laser determines the intensity of the scattered light and as a consequence the number of valid signals. This fact is especially important in the roughness sublayer as the particle density is extremely low in this region and therefore the processing of the signals is difficult to handle.

The main charateristics of the three measuring systems concerning the above mentioned parameters are summarized in tab. 2 and 3.

5.1 System specific influences causing unacceptable results

As the measuring volume is oriented perpendicular to the roughness boundary, a contact between roughness elements and measuring volume can not always be avoided. When a contact occurs, the produced signal will transfer the information of the velocity at the roughness elements (which is zero) to the processor. As the touched area of the elements is of large extension, its capability to scatter light is greater than the one of the few particles in the fluid and therefore the probability distributions will be characterized by a peak of zero velocities. In Fig. 9 a) and b) two probability distributions influenced by the scattered light and measured with system 1 close to the surface of the elements are shown. In the case of Fig. 9 b) the distance from the surface is greater than in the case of Fig. 9 a). A reduction of this effect can only be achieved by using a smaller measuring volume. Therefore this problem was greatest in the case of system 1 with the longest measuring volume and almost negligible in the case of system 2 with the shortest measuring volume. When working with system 3 the effect was slightly stronger than with system 2.

In locations of the roughness sublayer with very low particle flux, the power of the laser and the processing of the signals become very important. If the laser is not powerful enough the intensity of the scattered light will be too weak to produce a useful signal. On the other hand the laser should not be too strong in this region because light will be scattered from the surface of the roughness elements. Therefore, the laser should not only be powerful but also adjustable and adaptable to the specific conditions.

The task of the electronic processor, firstly, is to distinguish between valid signals and noise and, secondly, to eliminate the noise and process the valid signals. Thus, the above mentioned adaptation of voltage of the photomultiplier and the signal-to-noise-ratio are important properties of a good processing unit. Nevertheless, the most important property of the processor is its algorithm to distinguish between noise and valid signals. A system dependent shape of the probability distributions could be observed as a consequence of the differences between the algorithms of the three systems. In Fig. 10 a) and b) two examples of probabilty distributions determined 3 mm below the edge of the roughness elements are given. In Fig. 10 a) the probability distribution of the velocity u measured with system 1 is shown and in Fig. 10 b) the distribution measured with system 3. Laser and processor of system 1 are obviously not able to overcome the difficult conditions in the roughness sublayer while laser and processor of system 2 and 3 can manage them.

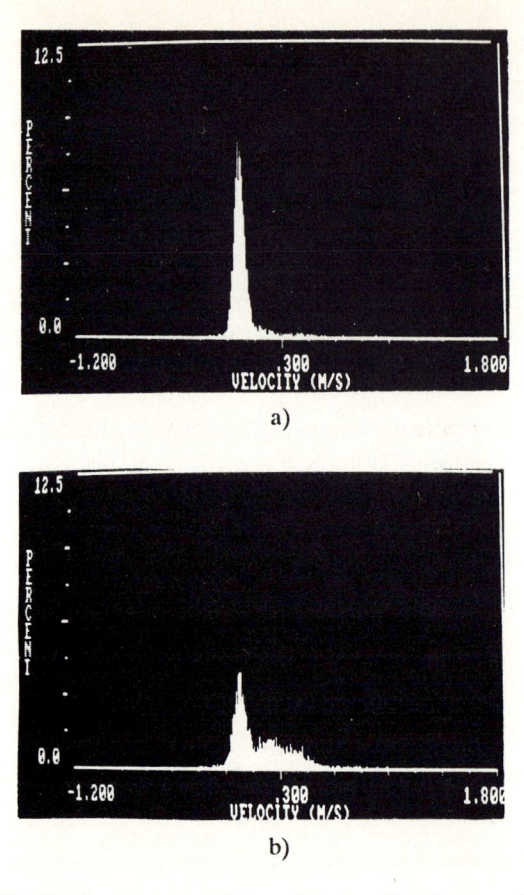

Fig. 9: Velocity distributions measured at a distance of a) 1mm and b) 2 mm from the surface of the elements

Due to the low particle flux the processor of system 1 takes signals arising from cables and contacts as correct and produces flat and skewed distributions. As skewness and flatness depend on the third and fourth order of the deviations of the false velocity signals from the mean velocity they are influenced by noise to a greater extent than the first two moments. Thus, the higher moments measured with system 1 within the roughness sublayer exceeded all bounds while the values measured with the other two systems did not exceed the physically senseful limits of 2,0 to 2,5 for S and 8 to 10 for F. In Fig. 11 a typical probability distribution measured with system 1 below the edge of the roughness elements is given. Fig. 12 shows the effect of such distributions on the flatness F. As can be seen

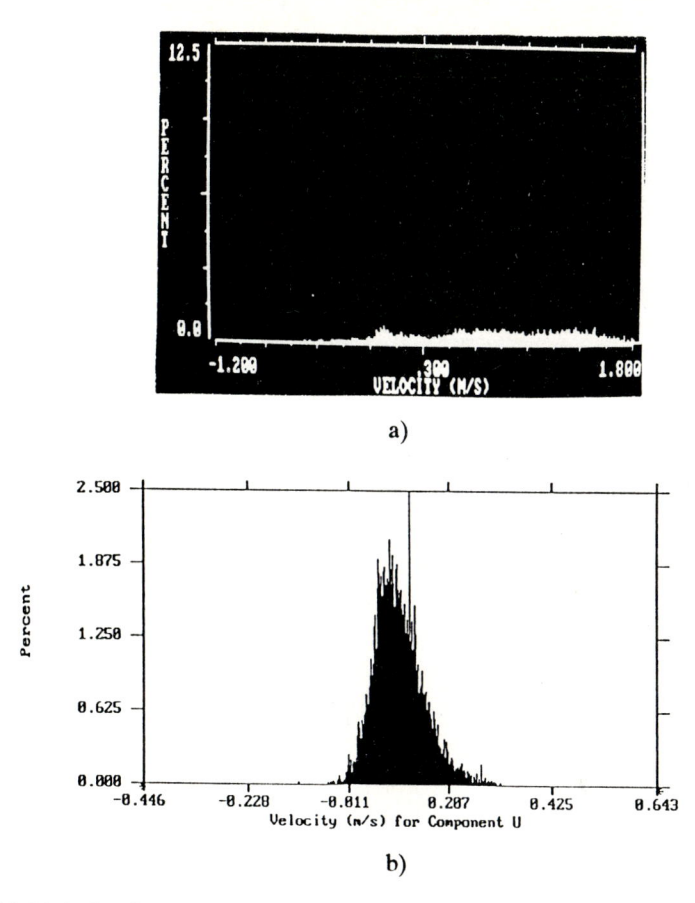

Fig. 10: Velocity distributions between the roughness elements determined a) with measuring system 1 and b) with measuring system 3

from a comparison of Fig. 12 with the distribution of F determined by system 2, shown in Fig. 16 b), much too high values of up to 50 are obtained for the flatness F by the use of system 1.

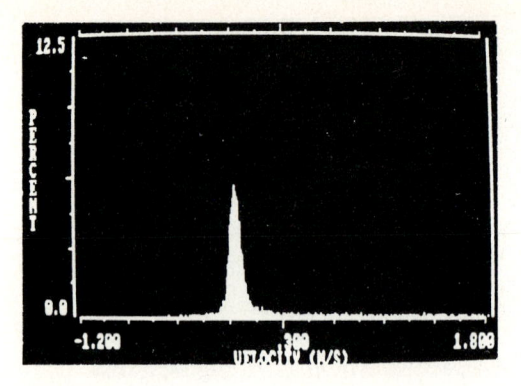

Fig. 11: Velocity distribution in the roughness sublayer determined with measuring system 1

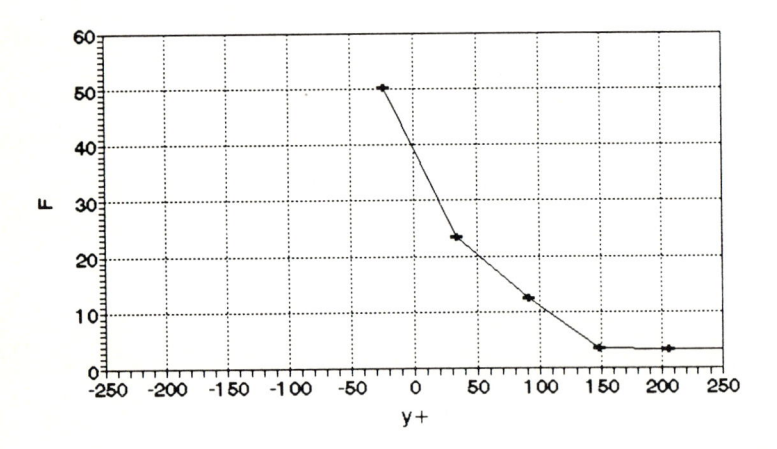

Fig. 12: Flatness F determined with measuring system 1 in the roughness sublayer as a function of y^+

Summarizing the results from the above, it can be stated that the measuring system 1 was not able to overcome the difficult measuring conditions in the roughness sublayer whereas with systems 2 and 3 most of the problems could be solved.

5.2 System dependent influences on valid signals

In the following, the results from the measurements over the staggered PVC-cubes with $c_K = 0,28$ will be discussed with respect to system dependent influences on the four moments. However, mainly the results obtained with system 2 and 3 can be compared due to the lack of valid measurements conducted with system 1 in and close to the roughness sublayer.

In Fig. 13 the $\bar{u}/u^*$-data of the location behind the cubes, measured in the roughness sublayer, are plotted versus y^+. As can be seen from Fig. 13, only a small difference of the mean velocity-data obtained with system 2 and system 3 could be observed at this location near the edge of the cube ($-200 < y^+ < +200$). The reason for this deviation results from the different length of the two measuring volumes (system 2: $l = 2,74$ mm; system 3: $l = 3,59$ mm). As a consequence of the great difference between the particle flux above and below the edge of the cube the upper part of the measuring volume will be emphasized and the real measuring point displaced towards the outer flow field. Therefore, the distribution of $\bar{u}/u^*$ determined by system 3 will be displaced by a small amount towards the bed with respect to the distribution of system 2.

For the u_{rms}-data the effect of the length l close to the roughness elements can also be recognized. As can be seen from Fig. 14, the location of the maximum in the distribution is slightly displaced towards the bed. However, as illustrated in Fig. 15, the length of the measuring volume has another effect on the u_{rms}-values: the distributions are shifted to higher values with increasing length of the measuring volume. Although the slope of the distributions is not affected, the values obtained with system 3 are shifted by 0,4 units and the values of system 1 are shifted by 0,45 units with respect to the values obtained with system 2. If the orientation of the measuring volume and thus of the integration length parallel to the gradient of the mean velocity (see Fig. 8) is kept in mind, the increase of the u_{rms}-values with increasing length of the measuring volume is logical. Concerning the higher moments, the effect of the length of the measuring volume is only evident in the vicinity of the roughness elements. In Fig. 16 a) and b) the plots of the S- and F-values

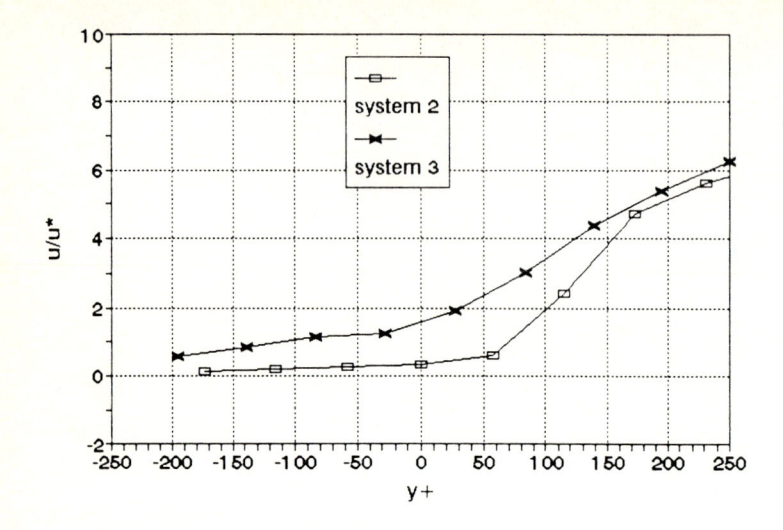

Fig. 13: $\bar{u}/u^*$ measured with systems 2 and 3 as a function of y^+

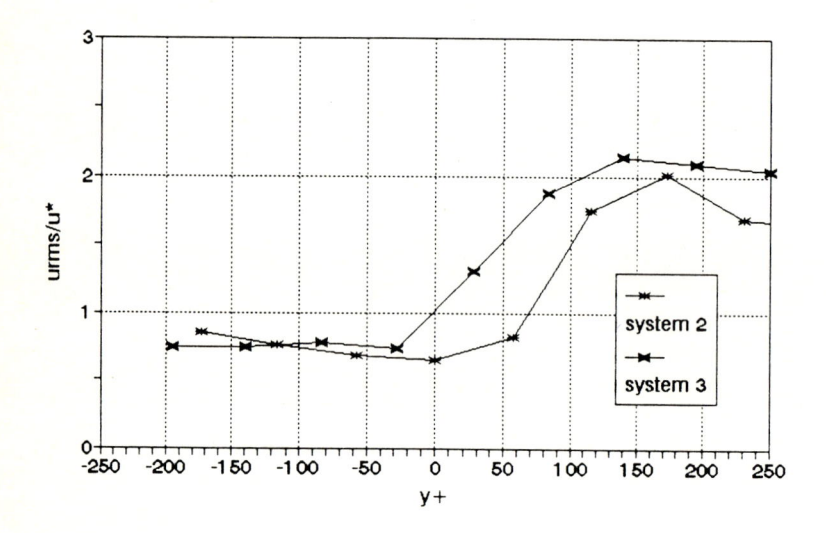

Fig. 14: u_{rms}/u^* measured with systems 2 and 3 as a function of y^+

versus y^+, obtained from the measurements behind the cubes, indicate that again the effect is relatively small.

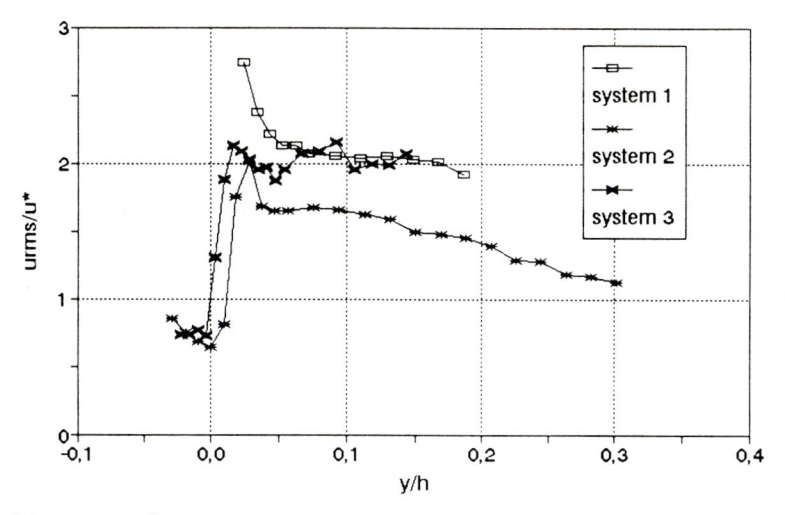

Fig. 15: u_{rms}/u^* measured with systems 1, 2 and 3 as a function of y/h

It was mentioned above that skewness and flatness in the roughness sublayer are influenced by the signal processing. With the processor of system 1, noise could not be eliminated in all cases, even in the outer flow field. As a consequence skewness and flatness determined with system 1 differ from those of system 2 and 3 over the whole measured water depth y/h. Especially the flatness, which depends on the fourth order of the derivations from the mean value, is affected to a high extent by the processor. As can be seen from Fig. 17, the F-distribution determined with system 1 is stretched and shifted towards higher values compared to system 2 and 3.

6 CONCLUSIONS

(i) The distributions of all statistical moments with depth y^+ are similar for the four rough and the smooth bed surfaces. Thus, the burst character of the flow field is similar over rough and smooth bed surfaces.

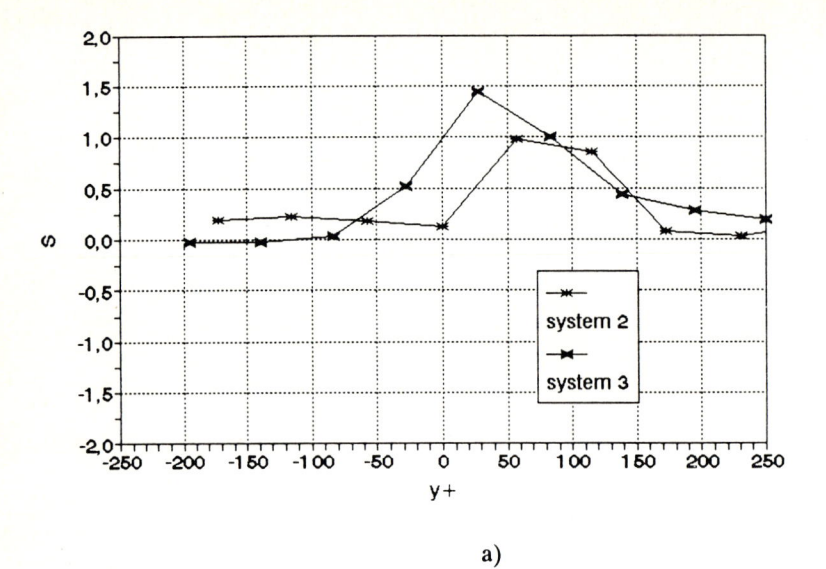

Fig. 16: a) Skewness S and b) Flatness F measured with systems 2 and 3 as a function of y^+

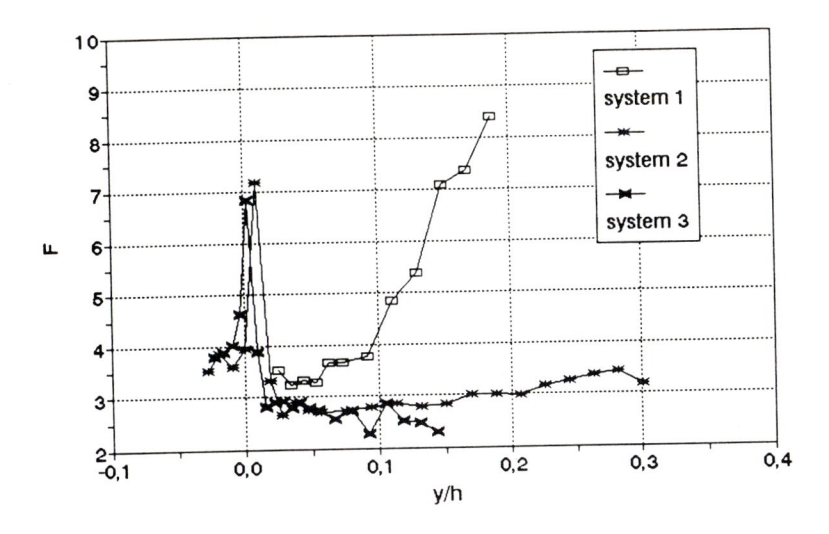

Fig. 17: Flatness F measured with systems 1, 2 and 3 as a function of y/h

(ii) The value of the point y^+_c, where the minimum of the flatness, the maximum of the u_{rms}-values and the change of sign of the skewness fall together, reflects the effect of roughness on the flow field. The value of y^+_c increases with increasing roughness. No direct dependence on shape, arrangement or roughness densitiy of the elements could be identified.

(iii) Although the slope of the u_{rms}-distributions are not affected by the length of the measuring volume, the u_{rms}-values are shifted to higher values with increasing length of the measuring volume over the whole measured water depth y/h. Within the roughness sublayer the distributions of all moments are slightly displaced towards the bed with increasing length of the measuring volume.

(iv) The distributions of skewness and flatness in the roughness sublayer and in the outer flow field are significantly influenced by the signal processing. Within the roughness sublayer the signal processing can even cause unacceptable results.

(v) The power of the laser and its adjustment are important to obtain useful signals but the power has no influence on the quality of the results.

REFERENCES

Dittrich, A., 1992: Untersuchungen zum Stabilitätsverhalten natürlicher Gerinnesohlen; Mitteilungen des Instituts für Wasserbau und Kulturtechnik, Karlsruhe, Heft 182

Durst, F., Jovanovic, J., Kanevce, Lj., 1987: Probability density distribution in turbulent wall boundary-layer flows; Turbulent Shear Flows 5, Springer-Verlag Berlin Heidelberg, pp. 197-220

Grass, A., 1970: Initial instability of fine bed sand; ASCE, Journal of the Hydraulics Division, Vol. 96, HY 3, pp. 619-632

Grass, A., 1971: Structural features of turbulent flow over smooth and rough boundaries; J. Fluid Mech., Vol. 50, Part 2, pp. 233-255

Nakagawa, H., Tsujimoto, T., Shimizu, Y., 1988: Velocity profiles of flow over rough permable bed; Proc. 6th Congress IAHR, Kyoto, Japan, 20-22 July, pp. 449-456

Raudkivi, A., 1976: Loose boundary hydraulics; 2nd Edition, Pergamon Press, Frankfurt, New York

Raupach, M.R., 1981: Conditional statistics of Reynolds stress in rough-wall and smooth-wall turbulent boundary layers; J. Fluid Mech., Vol. 108, pp. 363-382

Shields, A., 1936: Anwendung der Ähnlichkeitsmechanik und Turbulenzforschung auf die Geschiebebewegung; Mitteilungen der Preußischen Versuchsanstalt für Wasser- und Schiffsbau, Heft 26, Berlin

Sumer, B. M., Deigaard, R., 1981: Particle motions near the bottom in turbulent flow in an open channel; J. Fluid Mech., Vol. 109, Part 2, pp. 311-337

AN ATTEMPT AT DETERMINATION OF INCIPIENT BED LOAD MOTION IN MOUNTAIN STREAMS

Alicja Michalik, Wojciech Bartnik
Department of Hydraulic Engineering, Agricultural University
Al. Mickiewicza 24/28, 30-059 Cracow, Poland

ABSTRACT

Radioactive tracer measurements carried out in three Polish mountain rivers have allowed to determine the critical dimensionless shear stress, f_i, required to set in motion a given size d_i of particles. The different functions, $f_m/f_i = \varphi(d_i/d_m)$ (f_m - the critical dimensionless shear stress relative to the mean grain diameter, d_m), have revealed that extrapolation of results from some rivers studies is very difficult. An attempt at determination of incipient bed load motion, on the basis of the grain size distribution has been made. The measured grain size distributions before and after flood could be compared with the radio tracer measurements results. The Gessler (1970) function has been applied to determine the armour layer motion beginning. The critical shear stress values were obtained for the all particle diameters at which their motion had begun, f_m = 0.034 - 0.040 corresponding to d_m = 0.07 - 0.14 m. The standard deviation of the grain size distribution was variable δ = 1.8 - 3.2 .

The presented investigations concern the floods, when the maximum water flow was Q_{max1}= 5.50 m³/s, Q_{max2}= 2.75 m³/s, Q_{max3}= 5.75 m³/s, Q_{max4}= 5.90 m³/s. The amounts of the transported bed material during these floods were: G_1= 6.25 T, G_2= 2.70 T, G_3= 57.4 T, G_4= 181.1 T.

It was concluded that the proposed method can be used in many similar cases if the incipient motion of bed materials is too difficult to measure, but only when the changes of the river bottom caused by flood are known.

INTRODUCTION

In mountain streams, where there is an armour at the bottom, the bed material will be transported, if the drag force of the flow is sufficient to break down an armouring layer and to remove the particular grains. The bed material is nonuniform and its characteristics can be presented by the grain size distribution.

Determination of the hydraulic conditions at which bed material begins to move is much more difficult than measuring continuous bed load transport. The initiation of motion is basically different when the bed material is nonuniform instead of being uniform, and therefore the Shields beginning of motion criterion cannot be used in mountain streams.

With reference to other studies, the best information for coarse-grains entrainment can be obtained from radioactive- tracer studies in rivers. This paper presents:
- the radioactive-tracer method applied in the investigations of the threshold hydraulic conditions necessary to initiate motion of particular bed material fractions
- the possibility of determination the critical shear stress at which grain motion begins, if the grain size distribution before and after flood is known.

THE MEASUREMENT OF INCIPIENT BED LOAD MOTION

The direct measurements of transport beginning have been carried out for some bed material fractions in three Carpathian rivers (Michalik, 1990). About 30 native grains of different diameter have been traced with radioisotope tantalum-182, injected to a small cored hole in each grain, which was filled with silicon resin. In Fig. 1, some grains prepared for measuring are shown.

Figure 1. Native grains traced by radioisotope

The traced samples have been situated in the river bed in such points of a cross-section to depths less than critical for a given fractions. Every traced grain has been introduced in place of non-traced similar grain. The detectors of radiation have been fixed above the samples in this manner not to disturb the bed load motion. Their position in a cross-section of the Raba river is presented in Fig. 2.

Figure 2. Position of the detectors in measuring cross-section of the Raba river.

All detectors have been connected with the electronic device and computer system, placed in mobile laboratory shown in Fig. 3.

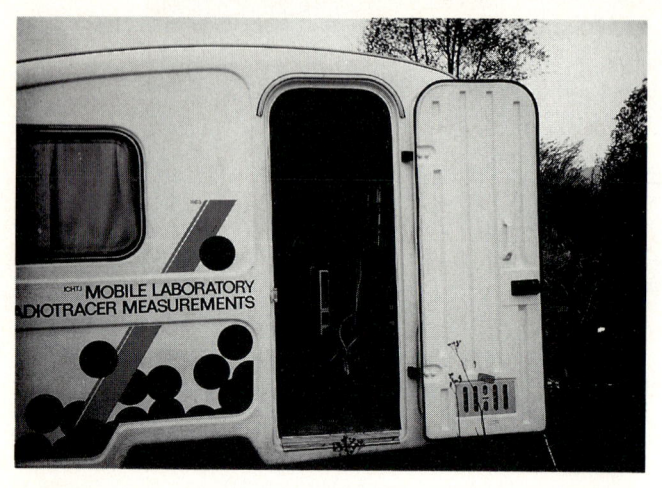

Figure 3. Mobile laboratory with the electronic device

The measurement has relied upon continuous record of radiation intensity and simultaneous observations at a monitor screen. Moreover, water stage was recorded at the same time which enabled determination of water depth at which motion began. In Fig. 4, recording for the fraction diameter, d_i = 0.066 m is presented. This curve shows the relation between radioactivity rate, in impulses per second, and time (since moment of the sample introducing). Point on time axis noted: 11^{44}, determines the beginning of this fraction motion; at this moment the intensity of the radiation drops to the value of natural background. In accordance with that point, the water stage and water surface slope is found. These parameters allow the computation of the critical shear stress.

Measurements have been carried out in three rivers: the Wisloka river, the Dunajec river and the Raba river. The Wisloka and the Dunajec rivers were investigated nearby the their mouths. Their bed material was different. The most important parameters were the following:

	the Wisloka river	the Dunajec river
water-surface slope, S	0.002 - 0.005	0.002 - 0.004
width of river channel, B(m)	30.0	65.0
depth of flow h(m)	1.0 - 3.1	1.0 - 4.7
mean grain diameter d_m(m)	0.0031	0.0247
standard deviation of grain size distribution σ	4.00	2.77

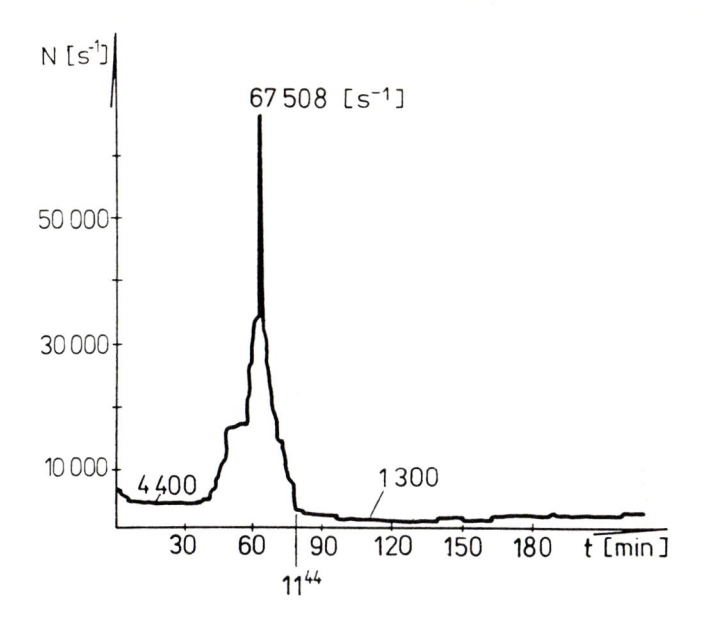

Figure 4. Recording of variation of radioactivity rate with time for the fraction d_i = 0.066 m - measured in 1988, in the Raba river

Results of measurements for five fractions were the same, in other words, the conditions of the beginning of bed load motion in these rivers were similar. The following function of dimensionless shear stress has been obtained:

$$f_m/f_i = (d_i/d_m)^{0.44} \qquad (1)$$

where: f_m is critical dimensionless shear stress of mean grain diameter d_m, and f_i - critical dimensionless shear stress of fractional diameter d_i.

In the Raba river seven fractions have been observed. The characteristic data: S = 0.002 - 0.004, B = 60.0 m, d_m = 0.072 m, = 3.8. The measuring cross-sections were located in upper part of the Raba river, and the following relation has been obtained:

$$f_m/f_i = (d_i/d_m)^{0.94} \qquad (2)$$

In spite of different hydraulic conditions, the critical dimensionless shear stress, f_m, has the same value in all investigated rivers, $f_m = 0.030$. Results of measurements are shown in Fig. 5 and compared with the standard Shields curve 1, corrected Shields curve 2 by Gessler (1970) and with the data of measurements presented by Komar (1989). This figure demonstrates that the Shields diagram cannot be used for nonuniform bed material in natural streams. Dimensionless shear stresses in all compared rivers are less than Shields values and they diminish if the Reynolds number Re_* increases. Every fraction diameter has own dimensionless shear stress.

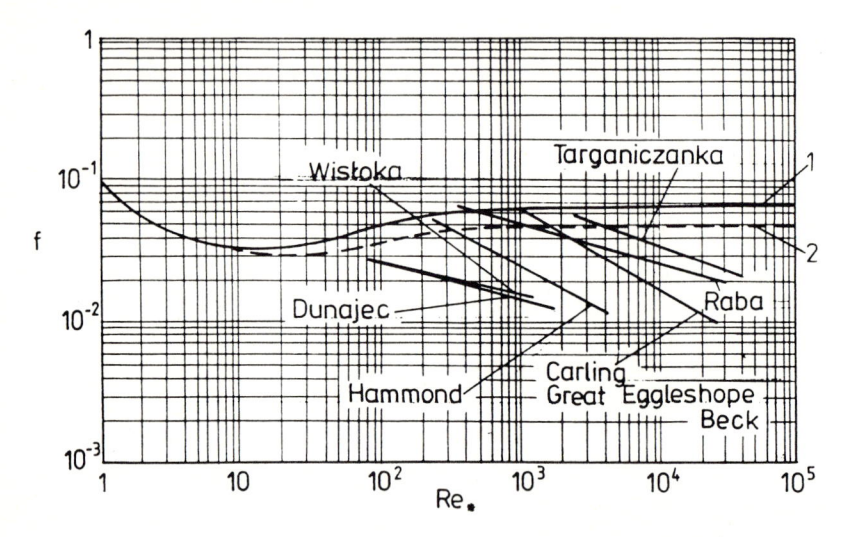

Figure 5. The standard Shields curve 1, corrected Shields curve 2, results of the investigations (the Wisloka river, the Dunajec river, the Raba river, the Targaniczanka river) and the data presented by Komar (1989)

The functions (1) and (2) are used to compute the bed load transport rate by the Meyer-Peter and Mueller formula, previously modified by Gladki et al. (1981).

METHOD OF INTERPRETATION OF THE GRAIN-SIZE DISTRIBUTION

In the Targaniczanka river, a small Carpathian stream, ($S = 0.016$, $B = 6.0$ m, $h = 0.60$ m, $d_m = 0.07 - 0.014$ m, $\sigma = 1.8 - 3.2$)

the radioactive tracer method has been used to measure bed load transport during 3 floods. The obtained results have given not only the amount of the transported bed material, but the grain size removed by flood as well. In each case, the thickness of the transported layer of the bottom has been determined (Gladki et al. 1981). Results of these measurements were used to find verification of the bed load transport formulas. The best agreement between measured and calculated data has been obtained by the Meyer-Peter and Mueller formula, modified by introducing the F.Wang (1977) relations.

For six years, the grain-size distribution in observed cross-sections has been investigated before and after floods. Eight of these data are presented in Fig. 6 and corresponding with them changes in the bottom are shown in Fig. 7,(flood number in parenthesis). The grain size distribution after one flood is considered also as the grain size distrubution before the next flood. The changes in the bottom (Fig. 7) have been low for 5 years, and the armouring progress was visible. The last flood only has caused significant change.

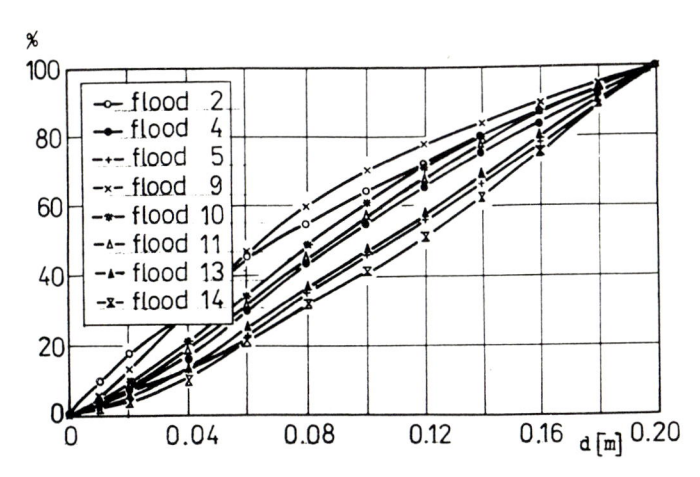

Figure 6. The grain size distributions measured before and after floods, in the Targaniczanka river

Gessler (1970) considered the probability of the removed of grains by transport and determined the resulting mean diameter of grains forming the armouring layer. Practically, this method allows to predict

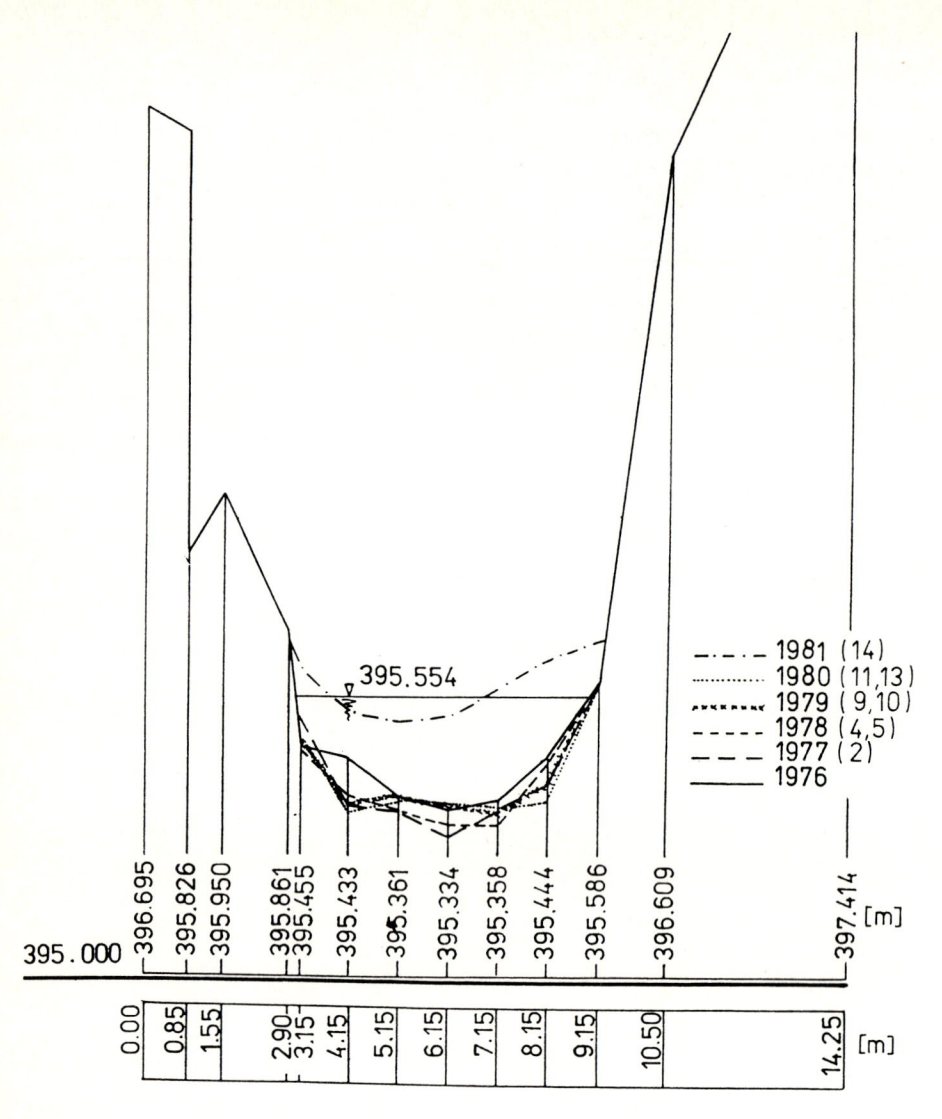

Figure 7. Investigated cross-section in the Targaniczanka river

the grain size distribution. The probability that a grain size d_i will not be removed depends on the standard deviation, , and the critical shear stress for grain size d_i (Simons and Senturk 1977).

Application of the Gessler method to the Targaniczanka river bed material has corresponded well with the measured grain size distributions. In this situation, an attempt at inversion of the question has been undertaken. Knowing the measured grain size distribution before and after flood it was possible to determine the critical shear stress.

Computation has been made for 10 floods and the received relations are as follow:

$$f_m = 0.039 \, \delta^{0.26} \quad \text{if} \quad d_i/d_m < 0.6$$
$$f_m = 0.028 \, \delta^{0.26} \quad \text{if} \quad d_i/d_m \gg 0.6 \tag{3}$$

The critical dimensionless shear stress for a given grain size d_i can be determine by use the functions (3).

The obtained data have been used to compute the bed load transport by the modified Meyer-Peter and Mueller formula. Results of the bed load transport calculation made for 3 floods could be compared with relative results of radioisotope measurements. This comparison is shown in Table 1.

Table 1

Comparison between computed and measured weight of bed material transported by flood

flood N°	G_T calculated (kN)	G_T measured (kN)
10	89.0	62.5
12	29.0	27.0
13	592.0	574.0

CONCLUSION

The direct measurements of the initiation of bed load motion in natural conditions is the best way to study the character of an individual fraction movement in nonuniform river bed material.

However, while still looking for other possibilities of determination the critical shear stress for any grain size, the presented method of interpretation of the grain size distribution can be tentatively recommended.

NOTATION

B - width of river channel
d_i - diameter of the ith fraction of bed material
d_m - mean grain size
$f_i = \dfrac{\gamma h S}{(\gamma_s - \gamma) d_i}$ - critical dimensionless shear stress of the ith fraction
f_m - critical dimensionless shear stress of mean diameter, d_m
G_T - weight of bed material transported by flood
h - depth of flow
S - water - surface slope
γ - specific gravity of water
γ_s - specific gravity of sediment
$\delta = \sqrt{d_{84}/d_{16}}$ - standard deviation of grain size distribution

REFERENCES

Gessler J. (1970) Self stabilizing tendencies of alluvial channels: J.Water Ways and Harbors Division, 235-249

Gładki H.,Michalik A. and Bartnik W. (1981) Measurements of bed load transport in mountain streams using the radioactive tracers method: Proc. of Workshop IAHR, Rapperswil, 45.1

Komar P.D. (1989) Flow - competence evaluations of the hydraulic parameters of floods: an assessment of the technique: "Floods - Hydrological, Sedimentological and Geomorphological Implications"ed. by K. Beven and P. Carling, UK, 107-134

Michalik A. (1990) Investigations of bed load transport in Carpathian rivers, Habilitation thesis N° 138, ed. by Zeszyty Naukowe, Agricultural University in Cracow, (in Polish), 1-115

Simons D.B., Sentürk F. (1977) Sediment transport technology, ed. Water Resources Publications, Fort Collins

Wang F.Y. (1977) Bed load transport in open channels: Proc. of Symp. IAHR Baden-Baden, A-9,63

Orientation

The orientation distributions of sedimentary surfaces of different age are shown in Figure 7. More than 50,000 gravel were measured. In order to avoid ambiguous results only particles with ratios of long to short axis of more than 1.5 were selected. A particle's orientation was calculated as the deviation from the flow direction at the sample site of the particle. A marked bi-modality is noticeable in the orientation distribution for the recent surface. This bi-modality is reduced with increasing age of surface. Finally, on the oldest surface an orientation normal to the direction of flow prevails. It is postulated that the changes in direction are related to solifluction processes.

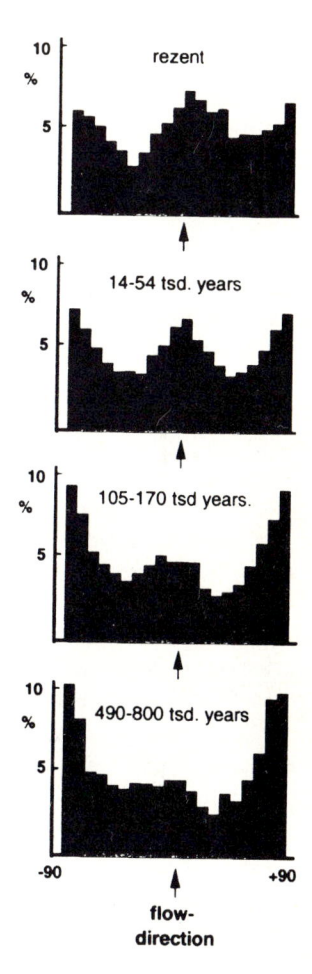

Figure 7: Orientation distributions from sediment surfaces of different age on the Hanapauh fan.

3.3. Schmiedlaine

In the Schmiedlaine, a small mountain tributary of the Loisach drainage system in Upper Bavaria, specialized studies on river bed dynamics concentrating particularilly on changes in roughness conditions were carried out. Some preliminary results dedicated to clusters, which are widespread micro-roughness elements covering more than 50% of the river bed (de Jong 1992) and their influence on their open-bed surroundings are introduced. In these analyses, all clusters were considered and compared to the entire surrounding gravel bar. This consituted a total sample of approximately 5,000 particles. The technique has allowed an improved and detailed characterization of the grain constituents of roughness elements.

Grain Size

Two methods for determining grain size distributions were applied, that calculated from the particle area falling into a particular size class covered in the photo, which is directly comparable to the mechanically sieved distribution and that from the b-axis by number method derived from the ellipsoid approximation. In order to compare the traditional grain size distributions obtained from mechanical sieving and those by the photo-sieving technique, a test area of 150 m^2 representing an entire gravel bar was defined. An area of 2 m^2 was sampled on the surface of the gravel bar for the mechanically sieved sediments (Figure 9). For comparative purposes, only material coarser than 12 mm could be considered. The weight of the surface material was kg. As anticipated, the surface material is typically armoured and coarse. In contrast to photo-sieving, the mechanically sieved samples are far more restricted in sample area and this is one reason why the resulting distribution is less representative. The remarkable efficiency and accuracy of the photo-sieving technique is illustrated by its comparison with the mechanical distributions in

Figure 8a. Not only does the photo-sieving technique allow an area 25 times larger in size to be covered in the same time, it also results in a smoother distribution with negligible noise even when dealing with very small samples. In Figure 8b there is a considerable shift from a D_{50} of 45 mm for the surrounding material to 70 mm for the clusters and this tendency is also true for the D_{95}, where there is a shift from 104 mm for the surrounding material to 180 mm for the clusters. When considering the validity of sampling procedures for rivers and river sections, the photo-sieving technique is without doubt more successful.

Grain area

An additonal interpretation of the digitised images is possible using the area covered by every grain on the photo. The photos thereby offer the possibility of a new approach towards detailed spatial roughness interpretation. Since the particles determine surface roughness and this has a major influence on flow hydraulics, it is useful to determine the surface area of distinctive b-axis classes. For this analysis the D_{50} by area changes from grain sizes of 90 mm for open-bed material to 110 mm for clusters but the related D_{95} for both lie around 225 mm (Figure 8c). This procedure reveals that very few coarse grains cover most of the river bed and demonstrates that very coarse material has a dominant influence on the hydraulic conditions of the river bed. Traditional analyses by Wolman sampling cause the maximum to shift to the finer sizes whereas in the area distribution the coarser cobbles dominate. The difference between clusters and open-bed is very pronounced according to the b-axis grain size distribution yet there is less difference if we compare size by area only.

Figure 8: Comparison of a) surface grain size distributions of Schmiedlaine by mechanical sieving and photo-sieving, b) cluster and open-bed distributions using b-axes from photo-sieving and c) clustier and open-bed distributions using surface areas from photo-sieving. ▷▷

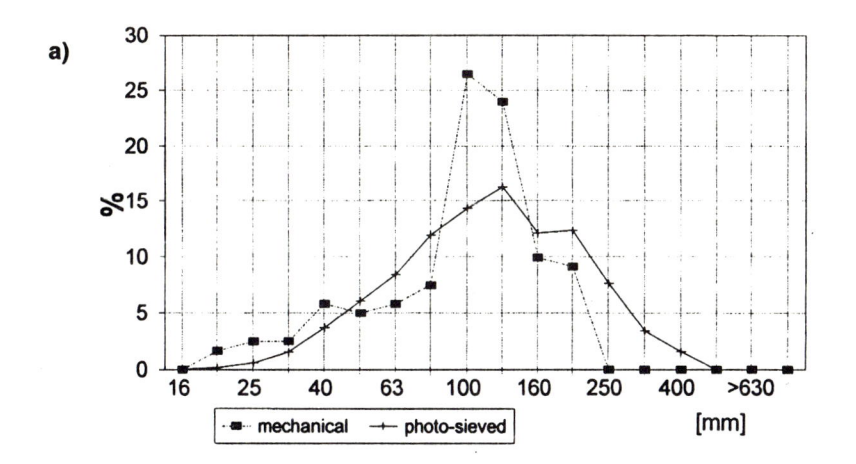

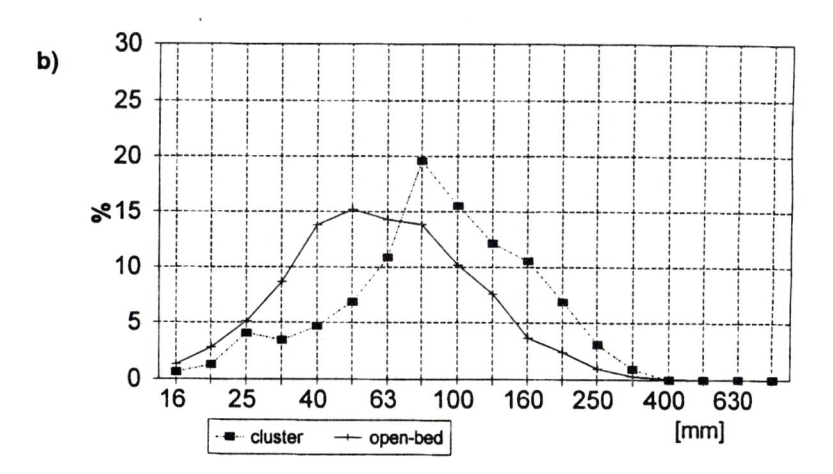

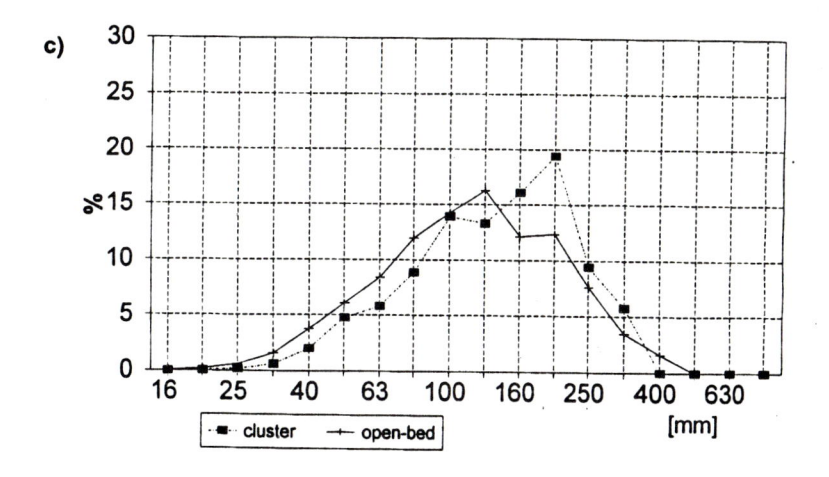

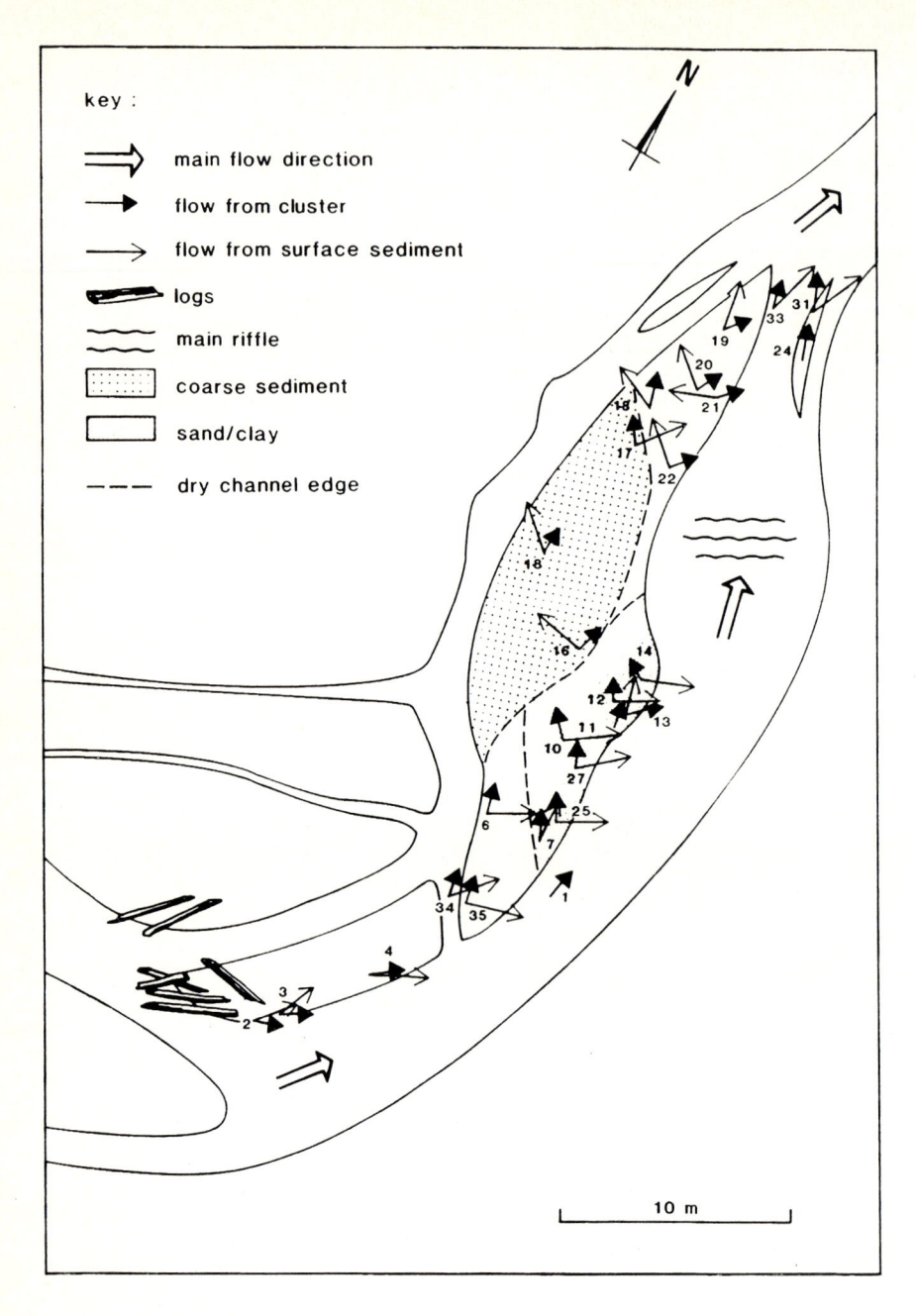

Figure 9: Orientations of clustered and open-bed material in a comparative analysis of the Schmiedlaine.

Orientations

In the Schmiedlaine the orientation of the cobbly surface material is quite dispersed. In contrast, clusters show clearer trends parallel to the main flow directions (Figure 9). The orientation of open-bed (surface) material seems to reflect high stage flow dynamics during an event that completely covered the bar and channel. The flow arrows over the upper bar indicate that flow followed the shortest route directly parallel to the valley shape but that flow was deflected onto the opposite channel wall over the lower bar. In contrast, the clusters thread their way parallel to the least energy flow routes and were probably deposited during the final flood stage (de Jong and Ergenzinger in prep). Johansson (p. 247, 1976) demonstrated that the orientation of pebbles is controlled by low velocity conditions but becomes increasingly dispersed during high velocity conditions. Concerning orientations, Fig. 9 illustrates that in the case of reconstructing flow dynamics at a micro-scale, the micro-bedfroms are very efficient indicators of the main flow directions. The chi-squared test, stating that there is no relationship between single cluster particle orientations and the main flow direction could be rejected at the 97,5 % confidence interval. This reaffirms the usefulness of the technique for reconstructing paleo-flow directions from isolated sedimentary bodies.

4. CONCLUSION

It has been demonstrated that present day sedimentary studies need to be investigated at increasingly higher spatial and temporal resolutions which in turn require more precise and time-efficient analytical techniques. In order to maintain a balance between sample size and grain size distributions in different high gradient, coarse-grained localities, the efficiency of the photo-sieving technique was demonstrated. In paleo-fluvial studies, transitions in flow, grain size distributions as well as particle form and roundness can be related in detail to geological time periods. Rounding increases significantly with fluvial transport distance in contrast to the negligible results obtained from sphericity. In turn, detailed spatial process studies of roughness have indicated that new perspectives on grain size, grain area and orientation have to be taken into account with respect to more realistic roughness modelling in the future.

force-free particles : the decrease of clay-particle network volume and the global increase of solid volume concentration and consequently the decrease of the water available for double-layers.

Beyond a certain concentration limit (cf Figure 6) we can expect the effects of direct interaction between force-free particles to prevail. In the case of a Newtonian interstitial fluid, interaction was referred to as friction. When the interstitial fluid already includes solid elements such as clay particles, the nature of the prevailing direct interaction beyond a concentration limit is not quite clear. However, this percolation threshold is theoretically justified. Coussot (1992)(a) also proved experimentally that above a certain concentration of solid particles there is a sudden rapid increase of the yield stress of the suspension. As this percolation broadly corresponds to solid particle crowding we can assume (Coussot (1992)(a) and Wang Y. (1989)) that this rapid increase of yield stress is due to a sudden increase of the volume of frictional type interactions between force-free particles.

B.8. High force-free particle concentration in a clay water-mixture (cf Figure 7)
When force-free particle concentration is too high the suspension can no longer stand large, continuous, incompressible deformation without a failure occurring through the sample. In fact this phenomenon is clearly explained by the inability of a crowded, solid packing to change its configuration via an incompressible shear. The critical, force-free, particle concentration beyond which this phenomenon appears is lower than with an interstitial fluid such as water.

B.9. High force-free particles concentration in a low concentrated clay water mixture (cf Figure 8)
If we return to the mixture obtained in the previous paragraph and decrease its clay concentration, we may get a non-fracturing yield stress fluid (because then the viscosity of the interstitial fluid is low). But in some cases, when the direct interactions between force-free particles are large, the suspension behaviour may be unstable (minimum in flow curve) (Coussot (1992)(a)).

323

— Clay particle
● Force-free particle
⬭ Double-layer

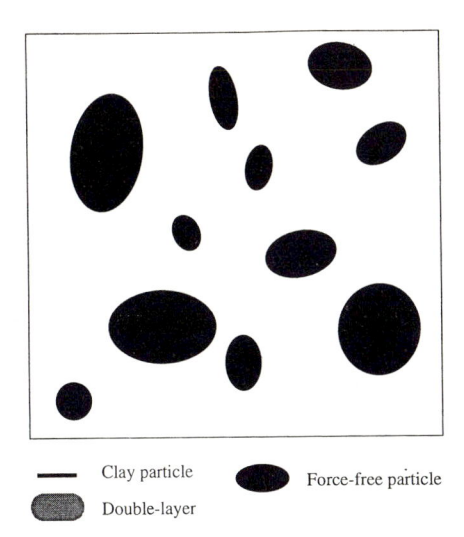

Figure 1 : Few force-free particles in water.

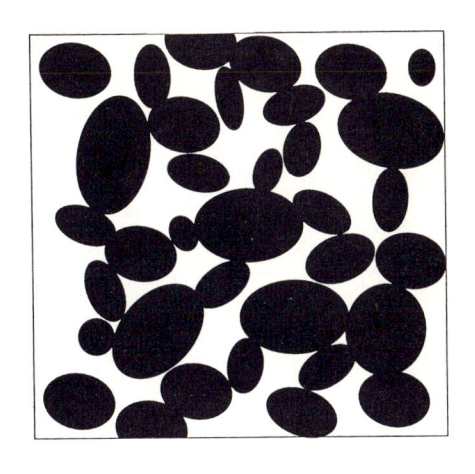

Figure 2 : Many force-free particles in water ; percolation concentration overcome.

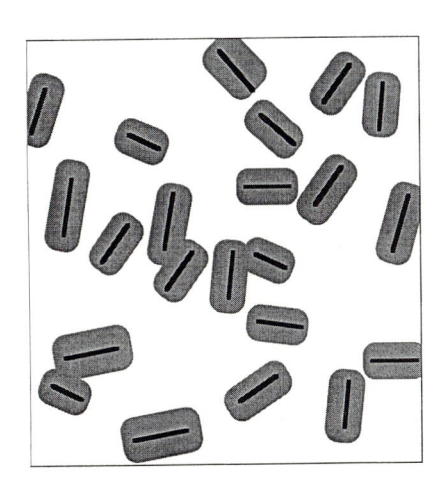

Figure 3 : Few clay particles in water.

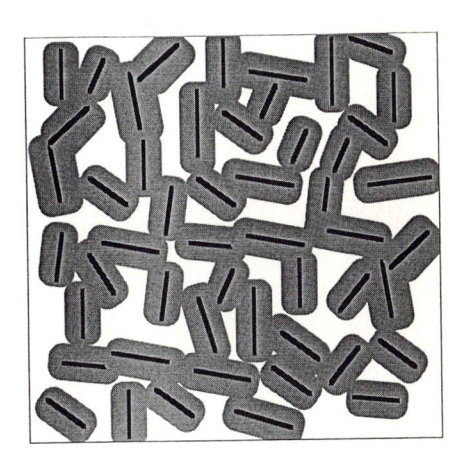

Figure 4 : Many clay particles in water ; percolation concentration overcome.

Figure 5 : Few force-free particles in clay-water mixture.

Figure 6 : Many force-free particles in clay-water mixture ; percolation concentration overcome.

Figure 7 : Many force-free particles in clay-water mixture ; crowding concentration.

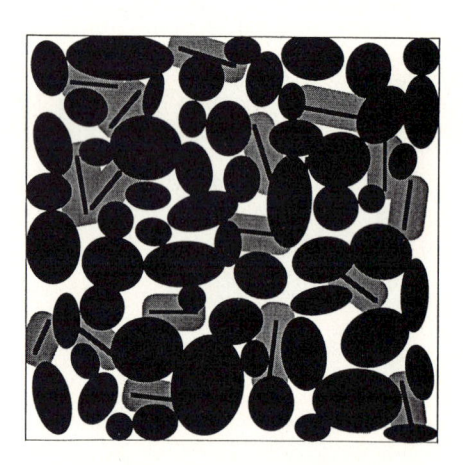

Figure 8 : Many force-free particles in a low concentrated clay-water mixture ; below crowding concentration.

Conclusion

We have tried to review some of the theoretical bases for the behaviour of various clay and force-free particle-water mixtures. We hope this will contribute to a better understanding of the behaviour of various natural mixtures and especially to their classification. In order to complete this review many additional studies such as accurate and consistent experimental work are still necessary. We also need new theoretical models that should be grounded on physical assumptions or observations concerning the evolution of microstructural interactions during shear.

References

Bagnold, R.A., (1954), "Experiments on a gravity-free dispersion of large solid spheres in a newtonian fluid under shear", Proceedings of the Royal Society, A225, pp.49-63.

Barnes, H.A., Jomba, A.I., Lips, A., Merrington, A., and Woodcock, L.V., (1991), "Recent developments in dense suspension rheology", Powder Technology, 65, pp.343-370.

Batchelor, G.K., (1970), "The stress system in a suspension of force free particles", Journal of Fluid Mechanics, 41, pp.545-570.

Chen, C.-L., (1988), "General solutions for viscoplastic debris flow", Journal of Hydraulic Engineering, 114 (3), pp.259-282.

Chen, C.-L., (1991), "Rheological model for ring-shear type debris flows", Fifth Federal Interagency Sedimentation Conference, Las Vegas, Nevada, Subcommittee on Sedimentation, Interagency advisory committee on water data.

Coussot, P., (1992)(a), "Rhéologie des laves torrentielles - Etude de dispersions et suspensions concentrées", Thèse de l'Institut National Polytechnique de Grenoble, France, 420p.

Coussot, P., Leonov, A.I., and Piau, J.-M., (1992)(b), "Rheology of concentrated dispersed systems in low molecular weight matrix", Journal of Non-Newtonian Fluid Mechanics, 46, pp.179-217.

Coussot, P., Leonov, A.I., and Piau, J.-M., (1992)(c), "Rheological modelling and peculiar properties of some debris flows", Proceedings of the International Symposium on Erosion, Debris Flow and Environment in Mountain Regions, Chengdu, China, I.A.H.S. Publication, (209), pp.207-216.

Davies, T.R.H., (1986), " Large debris flows : A macro-viscous phenomenon", Acta Mechanica, 63, pp.161-178.

Einstein, A., (1956), in Investigation of the brownian movement, Dover, New York, p.49, [English translation of Ann. Physik, 19, p.286 (1906), and 34, p.591, (1911)].

Fei Xiangjun, (1982), "Viscosity of the fluid with hyperconcentration coefficient rigidity", Journal of Hydraulic Engineering, 3, pp57-63 (in Chinese).

Iverson, R.M., and Denlinger, R.P., (1987), "The physics of debris flows - a conceptual assessment", Proceedings of the Corvallis Symposium on Erosion and Sedimentation in the Pacific Rim, I.A.H.S. Publication, (165), pp.155-165.

Johnson, A.M., (1970), *Physical Processes in Geology*, Freeman Cooper and Co, 577p.

Kamal, M.R., and Mutel, A., (1985), "Rheological properties of suspensions in newtonian and non-newtonian fluids", Journal of Polymer Engineering, 5, N°4, pp.293-382.

Kytomaa, H.K., and Prasad, D., (1993), "Transition from quasi-static to rate dependent shearing of concentrated suspensions", Powders & Grains, Thornton (ed.), Balkema, Rotterdam, pp.281-287.

M'Ewen, M.B., and Mould, D.L., (1957), "The gelation of montmorillonite. Part II : The nature of interparticle forces in sols of Wyoming bentonite", Transactions of the Faraday Society, 53, pp.548-564.

Major, J.J., and Pierson, T.C., (1992), "Debris flow rheology : experimental analysis of fine-grained slurries", Water Resources Research, 28, N°3, pp.841-857.
Melton, I.E., and Rand, B., (1977), "Particle interactions in aqueous kaolinite suspensions, Part I - Effect of pH and electrolyte upon the mode of particle interaction in homoionic sodium kaolinite suspensions", Journal of Colloid and Interface Science, 60, N°2, pp308-336.
Migniot, C., (1989), "Tassement et rhéologie des vases", La Houille Blanche, N°2, pp.95-112.
Naik, B., (1983), "Mechanics of mudflow treated as the flow of a Bingham fluid", Ph. D. thesis, Washington State University, USA, 166p.
O'Brien J.S., Julien P.Y., (1988), "Laboratory analysis of mudflow properties", Journal of Hydraulic Engineering, 114 (8), pp.877-887.
Phillips, C.J., and Davies, T.R.H., (1989), "Debris flow material rheology - Direct measurement", Proceedings of International Symposium on Erosion and Volcanic Debris Flow Technology, Jogykarta, Indonesia.
Phillips, C.J., and Davies, T.R.H., (1991), "Determining rheological parameters of debris flow material", Geomorphology, 4, pp101-110.
Takahashi, T., (1978), "Mechanical characteristics of debris flow", Journal of Hydraulics Division, 104 (HY8), pp.1153-1169.
Takahashi, T., (1980), Debris flow on prismatic open channel. Journal of the Hydraulics Division, 106 (HY3), pp.381-396.
Takahashi, T., (1981), "Debris flow", Annual Review of Fluid Mechanics, 13, pp.57-77.
Takahashi, T., (1991), *Debris flow*, International Association for Hydraulic Research, Monograph Series, A.A. Balkema, Rotterdam, Netherlands, 168p.
Utracki, L.A., (1988), "The rheology of two-phase flows", in *Rheometrical Measurement*, Edited by A.A. Collyer and D.W. Clegg, Elsevier Applied Science, Chapter 15, pp.479-594.
Wang, Y., (1989), "An approach to rheological models for debris flow research", Proceedings of the Fourth International Symposium on River Sedimentation, Beijing, China, pp.714-721.
Wang, Z., Larsen, P., and Xiang, W., (1992), "Rheological properties of sediment suspensions and their implications", submitted to Journal of Hydraulic Research, I.A.H.R.

QUANTIFICATION OF TEXTURAL PARTICLE CHARACTERISTICS BY IMAGE ANALYSIS OF SEDIMENT SURFACES - EXAMPLES FROM ACTIVE AND PALEO-SURFACES IN STEEP, COARSE-GRAINED MOUNTAIN ENVIRONMENTS

Michael Diepenbroek
Alfred Wegener Institut für Polar- & Meeresforschung
Am Handelshafen 12, 27570 Bremerhaven, Germany

Carmen de Jong
Institut für Geographische Wissenschaften, Grunewaldstr. 35
12165 Berlin, Germany

Abstract

Fundamental properties of sedimentary particles including size, shape and orientation were quantified in three scale- and process-differentiated studies using Fourier Analysis of particle outlines. The shape analyses refer to a large scale study at the River LaVerde, a mountainous fiumare in southern Calabria. A large-scale paleo-surface, the Hanapauh Fan, Death Valley, USA was investigated under geological time scales and contrasted to a small active gravel bar in the Schmiedlaine, S. Germany. Grain size distributions were calculated from digitized photos of sedimentary surfaces using the photo-sieving methodology, which has been upgraded by transforming the particle outlines into a series of Fourier coefficients. A particle's nominal diameter is calculated from its projected area. Other size parameters, as the main axes, are derived from the particle's best approximating ellipse, which is correlated with the second harmonic of the Fourier spectrum. The orientation of a particle corresponds to the orientation of this ellipse. The ellipse's orientation is calculated from the phase angle of the second harmonic. The ellipse corresponds to the particle's form and thus can be used for the calculation of sphericity. Having high quality digitizations of the particles contour lines it is also possible to measure roundness. Roundness data are derived from the complete harmonic spectrum, whereby the aspect of sphericity is eliminated by again using the best approximated ellipse.

1. INTRODUCTION

Size, shape and orientation are fundamental properties of sedimentary particles reflecting hydrodynamics and sedimentary processes during erosion, transport and sedimentation of particles. The quantification of these properties has not proved simple. Grain size analysis of sediments containing cobbles and boulders is almost impossible and measuring their orientation is laborious. The quantification of particle shape has also produced extreme difficulties in the past, which is mainly due to the complexity of the shape information. In particular the differentiation of shape into roundness and sphericity has been very contradictory e.g. they were considered to be independant variables by Wadell (1933) and Selley (p. 39, 1988) but dependant by Krumbein (1941) and Pettijohn (p. 61, 1975). This - together with practical shortcomings in measuring procedures - explains why investigations in this field are so rare.

More recent approaches try to cope with the problems by image analysis techniques. One approach is known as the photo sieving technique (Ibbeken & Schleyer, 1986), which aims in particular on the grain size analysis of coarse clastic sediment surfaces. This paper briefly outlines extensions and improvements of the photo sieving technique, a major upgrade being the quantification of the particles contour lines by a Fourier analysis. The Fourier series hold the information on size, orientation and shape of the particles (Figure 1). A detailed description of the Fourier technique for particle outlines is given by Schwarcz & Shane (1969), Ehrlich & Weinberg (1970) and Diepenbroek (1993).

The aims of the field and laboratory studies presented in this paper were to carry out a scale and processed differentiated analysis of the sedimentary properties of coarse-

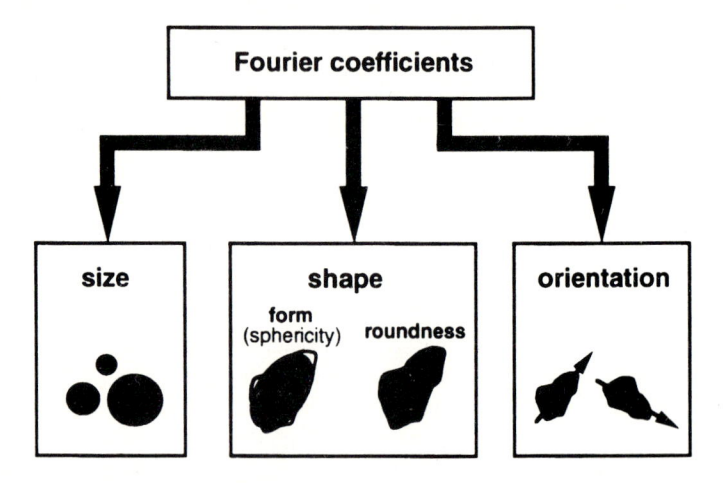

Figure 1: The particle´s outline is transformed into a series of Fourier coefficients. The coefficients hold the information on size, orientation and shape.

grained fluvial surfaces by use of the newly developed photo-sieving technique. Three field sites were considered, the LaVerde River in S. Calabria, Italy, in the large-scale process context, the Hanapauh alluvial fan in Death Valley, California, used as a large-scale, paleo-process example area and an actively reworked gravel bar in the Schmiedlaine, S. Germany for testing the technique in a small-scale, active process area.

2. METHODOLOGY

2.1. Field techniques

In order to carry out the photo-sieving technique, photographs are taken normal to the sediment surface at a defined scale using a frame to which the camera is attached. Oblique photographs are also possible. In this case the image has to be calibrated in order to guarantee a constant scale. Pictures have a limited resolution. So, if the desired range of grain sizes can not be accessed by a single photograph, the surface has to be photographed at different scales, ensuring an overlap between the different size ranges. This is also advantageous to optimize the time necessary for "sieving", because different surface areas are needed to obtain representative amounts of grains for the different size classes.

For shape analysis it is necessary to know the spatial orientation of the particles and to have high quality contour lines. This can be achieved by photographing the particles separately. A 16 mm camera in single picture mode is very suitable for this purpose.

2.2. Image analysis and description of particle outlines by Fourier series

Figure 2 illustrates the laboratory set-up of the image analysis system. In each photograph, the particles contour lines are digitized with a stylo or a similar tool on a digitizer. Alternatively the photographs are scanned and projected on a screen in which case the contour lines can be digitized with a mouse. Thin sections or samples of fine particles can be processed directly with help of a microscope. For samples of loose grains the procedure was automated. About 500 to 600 particles per hour can be processed.

Schwarcz & Shane (1969) and Ehrlich & Weinberg (1970) were the first ones to propose a closed-form Fourier Analysis for the description of a particle´s outline. For the calculation of the Fourier series it is first necessary to find the centroid of the outline. For that purpose an iterative algorithm is used (Diepenbroek 1993, p. 21). Then 64 points at equal angles are choosen from the outline. The coordinates of these points are used as input for the calculation of the first 24 harmonics, which in turn serve as an information pool for the derivation of textural parameters.

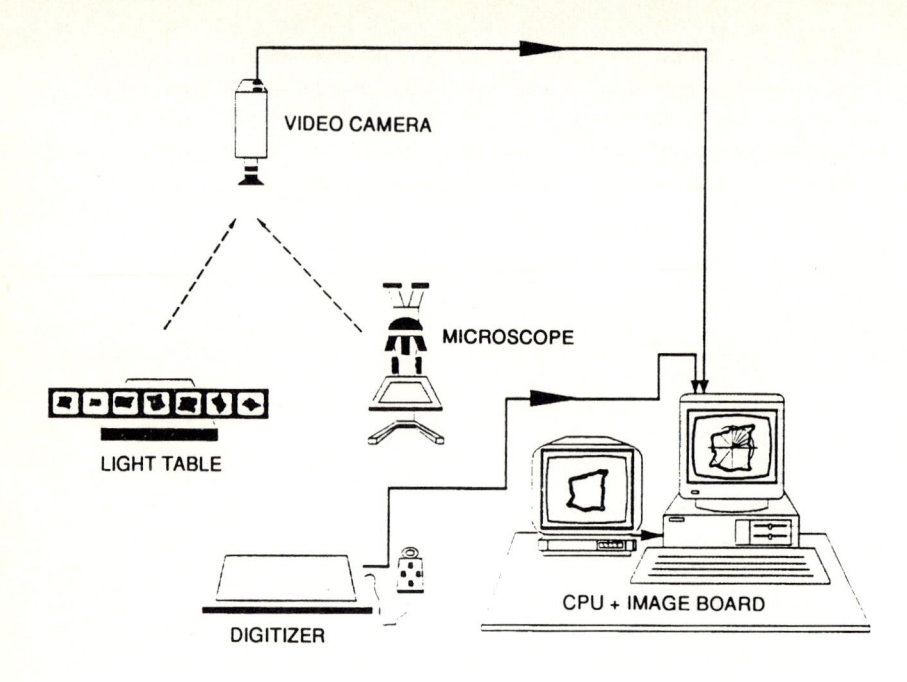

Figure 2: Hardware components for the image processing of clastic sediments.

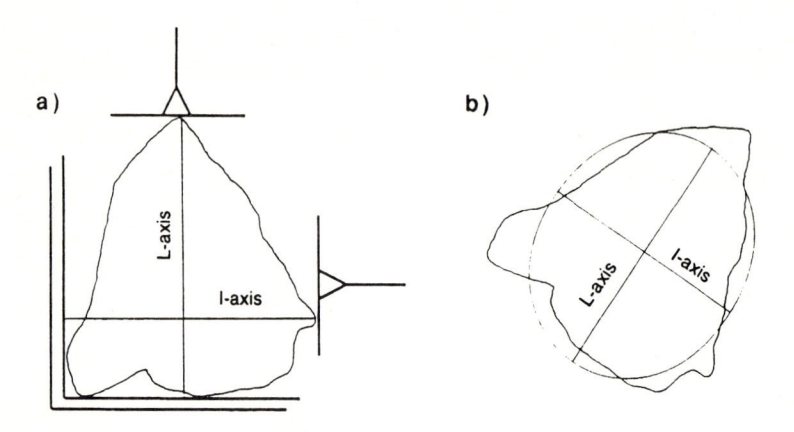

Figure 3: Measurement of a particle's main axes. a) mechanical device. b) ellipse-approximation.

2.2.1. Size

Several commonly used size measures can be choosen. A particle´s nominal diameter is calculated from its mean radius, which is the zeroth harmonic of the particle´s amplitude spectrum. Other size parameters, as the main axes, are derived from the particle´s best approximating ellipse (Figure 3b), which is correlated with the second harmonic of the Fourier spectrum. In order to get a grain size distribution, sizes are classified. The percentage of a specific grain size class is either determined by the number of particles within that class (Wolman sampling) or refers to the total length of the particles a- or b-axes within that class. A grain size distribution which is comparable with those resulting from mechanical sieving can be obtained by relating the class percentages to the area percentages which these classes have in the digitized picture. Thus area is equated with volume or weight (Diepenbroek et al. in prep.).

2.2.2. Orientation

The orientation of a particle is defined by the direction of its long axis. Conventional procedures to measure the orientation are laborious and often the direction of the long axes cannot be determined exactly. With the procedure presented here it is assumed that orientation corresponds to the orientation of the particle´s best approximated ellipse. The ellipse´s orientation is calculated from the phase angle of the second harmonic of the particle´s Fourier series.

2.2.3. Shape

Since Wentworth (1919) many approaches towards the characterization of shape have been undertaken. Due to the many processes affecting particle shape, the information stored in the particle outline is complex. Main aspects are roundness and sphericity. The conventional methods have however failed to separate sphericity and roundness as distinctive shape aspects.

Roundness

Particle roundness is primarily the effect of abrasional forces, which are active during bedload transport. Present concepts on roundness are deficient and a generally accepted method for quantifying this parameter is not available. Therefore, it was necessary to develop a new concept for the evaluation of roundness. The main criteria recognized are that:

- all curvatures on a particle have to be recognized (surface texture excepted).
- the relative position of corners and edges is important.
- curvatures and the position of curvatures have to be related to an ultimate shape, which is assumed to be ellipsoidal.

This concept has been realized using the complete harmonic spectrum from the Fourier series (Diepenbroek et al. 1992).

Sphericity

Sphericity is highly correlated with the dynamical behavior of particles during transport. It is calculated from a particle's main axes. Conventional methods for measuring the main axes use mechanical devices. Figure 3a illustrates the principle of such devices. It is obvious, that the measurement is influenced by protruding corners and edges, that is, it is influenced by the roundness of the particle. In Figure 3b the particle is approximated to an ellipse, using the second harmonic of the Fourier series. This guarantees the independence of roundness and sphericity. The ellipse-approximation is consistent with the concepts on sphericity as given by Krumbein (1941) and Sneed and Folk (1958). Two projections are necessary to get the three main axes of a particle: the maximum projection plane and one projection at right angle to it.

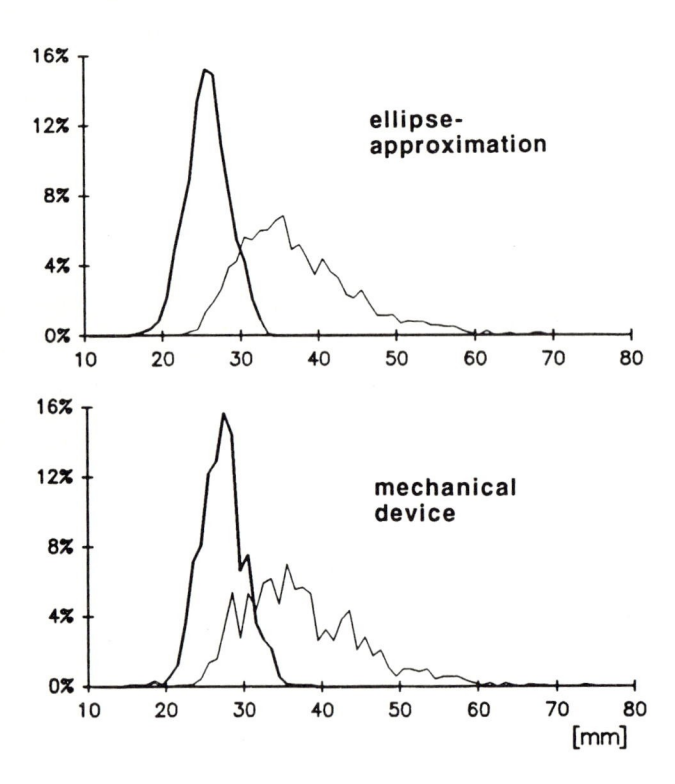

Figure 4: Main axis distributions for Calabrian gravels. The thick lines show the distributions of the intermediate axes, the thin lines represent the long axes.

3. RESULTS AND DISCUSSION

3.1. River LaVerde

The potential of the new approaches of textural particle characterization using image analysis were first demonstrated by Ibbeken & Schleyer (1991) with examples from Southern Calabria. Although shape analysis will be treated only briefly in this article one main result that is noteworthy from the existing investigations at the River LaVerde is that the sphericity of gravel populations is not altered by the abrasional forces during transport (Bartholomä, 1992) thus confirming the independence of roundness and sphericity. Sphericity was measured both with a mechanical device and through image analysis. The comparison between methods clearly demonstrates that the main axes distribution and consequently also the sphericity properties of the gravel population are better represented with the ellipse-approximation, which is due to the fact, that the influence of roundness on the measurement was eliminated (Figure 4).

The effect of abrasion can be demonstrated through changes in particle roundness which occur after only 2 km of fluvial transport (Figure 5). These changes are even more remarkable along the beach.

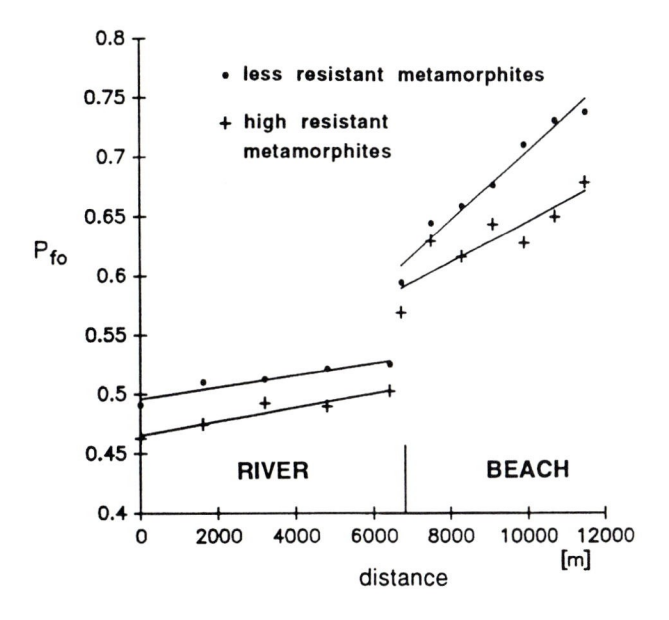

Figure 5: Mean roundness values for gravel populations from Calabria / Italy plotted against distance of transport. 'P_{fo}' is the Fourier roundness. Note the difference in roundness gradients between the fluvial and the coastal section. Remarkable is also the jump in roundness from the river mouth to the coast.

3.2. Hanapauh fan

One of the objectives of the study were to compare the efficiency of photo-sieving against mechanical sieving. A further objective was to investigate the orientation of particles on different fan surfaces and in the main channel. The particular tectonic setting and geomorphological development of Death Valley caused the formation of four distinct fan surfaces of different age. The stages of evolution are differentiated according to topography and differing rock varnishes. The older surfaces are between 14 to 800 thousand years of age (Dorn, 1988).

Grain size

Figure 6a illustrates the grain size distribution for the main channel on the fan obtained from mechanical sieving. Figure 6b shows the distribution resulting from the photo-sieving in the same area. The latter is obviously more representative, since it takes into account the coarse tail end of the distribution. The photo-sieving technique is very time efficient, in fact, when compared to the mechanical technique it is about 12 times faster. In addition most of the photo-sieving analysis can be carried out in the laboratory.

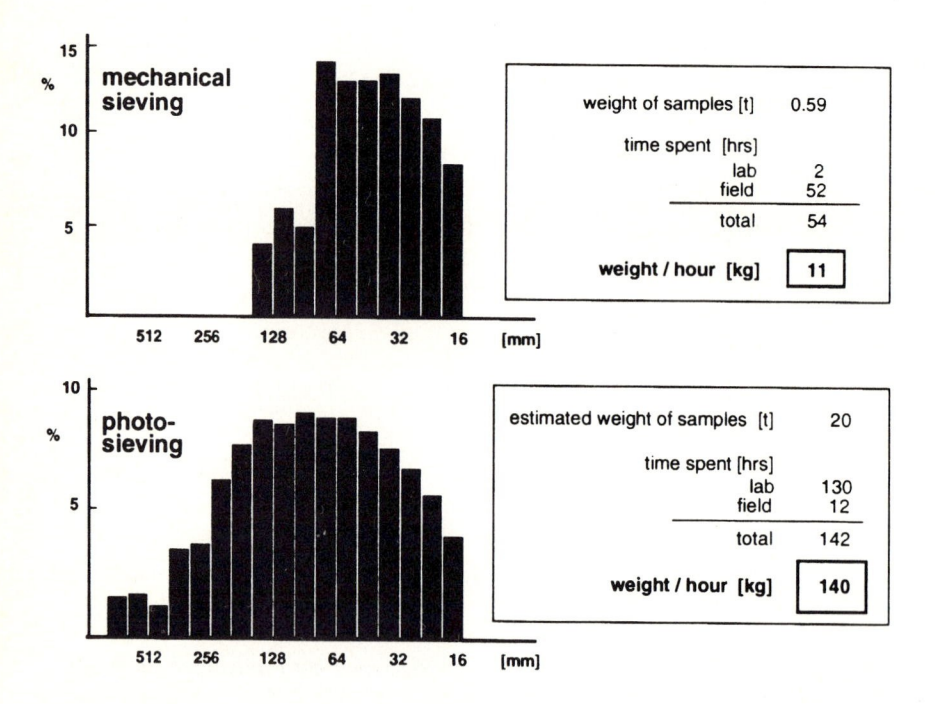

Figure 6: Grain size distributions of main channel sediments > 16 mm from the Hanapauh fan.

REFERENCES

BARTHOLOMÄ, A. 1992, Texturelle und kompositionelle Reifung von Fluß- und Küstenschottern, Kalabrien, Süditalien, unpubl. diss., FB Geowiss., FU-Berlin

DIEPENBROEK, M., 1993, The characterization of grain shape using Fourier-Analysis, diss., Reports on Polar Research, v. 122

DIEPENBROEK, M., BARTHOLOMÄ, A. & IBBEKEN, H. 1992, How round is round? A new approach to the topic ´roundness´ by Fourier grain shape analysis: Sedimentology, v. 39, p. 411-422

DIEPENBROEK, M., IBBEKEN, H. & WARNKE, D. in prep., Textural characteristics of the Hanapauh fan, Death Valley, California

DORN, R.I. 1988, A rock varnish interpretation of alluvial-fan developement in Death Valley, California, National Geographic Research, v. 4, p. 56-73

EHRLICH, R. & WEINBERG, B. 1970, An exact method for characterization of grain shape, Jour. Sed. Petrology, v. 40, No. 1, p. 205-212

IBBEKEN, H. & SCHLEYER, R. 1986, Photo-sieving: A method for grain-size analysis of coarse-grained, unconsolidated bedding surfaces. Earth Surface Prozesses and Landforms, v. 11, p. 59-77

IBBEKEN, H. & SCHLEYER, R. 1991, Source and sediment, a case study of provenance and mass balance at an active plate margin (Calabria, Southern Italy), Berlin, Heidelberg, Springer, 286 p.

JONG, DE C. 1992 A re-appraisal of the siginificance of obstacle clasts in cluster bedform dispersal. Earth Surface Processes and Landforms, Vol. 17, p.

JONG, DE C. & ERGENZINGER,P. 1992 (in prep) Spatial interactions between river bed roughness and flow hydraulics. Earth Surface Processes and Landforms.

KRUMBEIN, W.C. 1941, Measurement and geological significance of shape and roundness of sedimentary particles, Jour. Sed. Petrology, v. 11, No.2, p. 61-72.

PETTIJOHN, F. J. 1975, Sedimentary Rocks, Harper & Row, New York, 628 p.

SCHWARCZ, H.P. & SHANE, K.C., 1969, Measurement of particle shape by Fourier-analysis, Sedimentology, v. 13, p. 213-231

SELLEY, R, C 1988 Applied Sedimentology. (Publ.) Harcourt Brace Jovanovich 446p.

SNEED, E.D. & FOLK, R.L. 1958, Pebbles in the Lower Colorado river, Texas; a study in partical morphogenesis, Journal of Geology, v. 66, No.2, p. 114-150

WENTWORTH, C.K., 1919, A laboratory and field study of cobble abrasion, Journal of Geology, v. 40, No.7, p. 507-521

WADELL, H. 1933, Spericity and roundness of rock particles, Journal of Geology, v 41, p. 310-331

SOME CONSIDERATIONS ON DEBRIS FLOW RHEOLOGY

Philippe Coussot*, Jean-Michel Piau

Laboratoire de Rhéologie,
B.P. 53X, 38041 Grenoble Cedex, France.

Abstract
Our knowledge of the rheology of debris flows is unsatisfactory : a number of different models have been proposed but they lack experimental justification, or are theoretically unsatisfactory, or both. A correct approach to rheology must be based on a description of the evolution of the microstructure. In this paper we present a brief review of the fundamental rheological characteristics of various natural suspensions of clay and coarse particles in water. We define the physical boundary conditions which may be useful for dividing the suspensions in terms of clearly different qualitative behaviour.

Introduction
Debris flow dynamics is an acute problem when predicting either the extent of runout of such flows in inhabited zones or their effect on the environment. We consider that in this field the first step is to acquire a better understanding of the rheology of water-debris mixtures constituting debris flows. Indeed, when the constitutive equations of any material are put into motion equations, it becomes possible at least theoretically to determine any material flow, as soon as the boundary and initial conditions are known. In the case of a highly viscous and laminar flow it is generally possible to obtain a simplified momentum equation that describes the motion with very few parameters (such as the viscosity). Then the constitutive equation of the fluid remains as a key parameter. Debris flow rheology has already been studied by different authors all over the world. Various approaches have been proposed. Iverson & Denlinger (1987) provided some interesting physical critical comments about them and a description of the resulting mathematical problems. Later in the first part we will propose a more simple and pragmatic critical review of the main previous models, by

---

* Present address : CEMAGREF, Domaine Universitaire, B.P. 76,
      38402 St-Martin-d'Hères, France.

distinguishing the phenomenological and physical approaches. In the second part we shall examine the qualitative behaviour of different natural suspensions. Some physical limit conditions between behaviour types are established by assuming the different suspensions have been prepared by adding more and more solid material.

A. Previous work

A.1. Phenomenological approach

Johnson (1970), considering a debris flow as a mixture which is halfway between a soil (following a Coulomb criterion) and a Newtonian fluid, used a Bingham model :

$$\tau = \tau_c + \eta \dot{\gamma} \tag{1}$$

where τ is the shear stress, τ_c the yield stress, η the viscosity and $\dot{\gamma}$ the shear rate. A Bingham model was also considered by several authors (O'Brien & Julien (1988), Fei (1982), Wang Z. et al. (1992), Naik (1983)) as the best model for describing mud-mixture behaviour. These authors based their theories on the results of rheometrical experiments with clay-water mixtures or mixtures of water with fine fraction debris flows. Even if this simple Bingham model clearly conveys the main characteristics of debris flow behaviour, we need experimental or theoretical verifications of complete coarse debris flow sample behaviour and greater detail about the range of validity of this model.

Using a similar approach, Wang Y. (1989) and Major & Pierson (1992) used a Herschel-Bulkley model to describe the behaviour of some mixtures of water and fine fraction debris flows :

$$\tau = \tau_c + K \dot{\gamma}^n \tag{2}$$

where K and n are parameters of the fluid. Additionally Wang Y. (1989) suggested that the yield stress of such suspensions should be regarded as the sum of a term originating in matrix cohesion (very fine sediment-water mixture) and a term originating in friction between coarser particles. In order to obtain an idea of the behaviour of suspensions as close as possible to natural debris flow samples, Major & Pierson (1992) tested mud-mixtures (diameter up to 2 mm) in a large coaxial cylinder rheometer. They fitted a Herschel-Bulkley model to their results but obtained a wide range of variation for parameter "n" depending on material and solid concentration. Phillips & Davies (1989, 1991) built also a special, large (2 m diameter) cone-plate rheometer. They were thus able to test very coarse suspensions (diameter up to 12 cm) but they obtained results which fluctuated considerably. This phenomenon is, according to the above authors, explained by grain collisions and structural packing changes within the fluid.

A.2. Physical approach

Takahashi (1978, 1980, 1981), suggested applying the model that Bagnold (1954) developed for suspensions of particles in a Newtonian fluid to debris flow. This model distinguishes two regimes:

<u>Macroviscous regime</u> : $N < 40$ $\quad\quad \tau = 2.25\, \lambda^{1.5}\, \eta\, \dot\gamma$ (3)
(In this case viscous dissipation in the interstitial fluid prevails)

<u>Inertial regime</u> : $\quad\quad N > 450 \quad\quad \tau = a\, \rho_s (\lambda \sigma)^2\, \eta\, \dot\gamma^2$ (4)
(In this case transfer momentum through interparticle collisions prevails)

where a is an empirical coefficient, $N = \dfrac{\sqrt{\lambda}}{\eta}\, \rho_s\, D^2\, \dot\gamma$, ρ_s the particle density, D the particle diameter, η the viscosity of the interstitial fluid, and $\lambda = \dfrac{1}{\sqrt[3]{\dfrac{C_m}{C_v}} - 1}$ where C_v is the solid volumic concentration and C_m the maximum solid packing concentration. On the subject of debris flow motion initiation, Takahashi (1991) also proposed a yield criterion which is used specifically for the static case. Although all the underlying ideas in Takahashi's model are justified, we think its main defect is that it does not include both fluid yield stress and additional energy dissipation caused by motion in a comprehensive rheological law, which would at least make it possible to predict start and stop flow conditions more precisely. In addition we should point out that in the inertial regime it may be essential to take account not only of collisions but also turbulence within the interstitial fluid and momentum transfer due to fluctuating particle motions around their average motion. However it is doubtful whether, except for some peculiar, uniform, coarse, granular debris flows, a natural debris flow can flow in an inertial regime. Indeed, generally speaking, shear rate is not high enough or grain size distribution is uneven and the solid concentration is so high that the mean free path of particles is not large enough for collisions to prevail. Under these conditions, if we follow Bagnold's theory, we find that debris flow behaviour is generally in the macroviscous regime (Davies (1986)). However, in this case, a viscous model taking into account only interstitial fluid energy dissipation is too simple, especially with high solid concentrations when particle friction or water-clay flocculent structure can exist. Finally we think that, even if one assumes, as Takahashi (1991) did, that it is possible to use a very different Bagnold coefficient "a" in equation (4) to make experimental results fit with theoretical predictions, the question is still open as to whether it is reliable or not to transfer these results, obtained from particular laboratory experiments, to other specific field processes. The same question may be asked with regard to the use of Chen's (1988 (a,b), 1991) generalized viscoplastic model for natural flow prediction.

A.3. Comments

Because of the various suspensions that may be obtained with the various complex components of debris flows, it is doubtful whether a single, simple rheological model can be used to describe debris flow behaviour. Phillips & Davies (1991) came to the same conclusion. Under these conditions it could be useful to try to characterize, at least qualitatively, the very different behaviour which can be observed with these natural mixtures and which depend on clay and coarse particle distribution. What we intend to develop here is a first step to such a classification based on well-known rheological results and some recent, new results.

B. Natural suspension behaviour

We shall consider here mixtures of water with clay, silt, sand and rock particles. These mixtures can be considered as suspensions in the sense that they can be obtained by suspending various particles in water. The coarsest particles interact either hydrodynamically or directly when they are in close contact (friction, collision). We may call them force-free particles. Clay particles may also interact through bounded water in ionized double-layers. The nature of the interactions depends very much on the pH of the suspension, on electrolyte concentration and on clay type. We will not describe here the possible effects of these parameters on clay-water mixture behaviour. Such an attempt is made in Wang et al (1992). In order to simplify the following developments we will assume it is possible to sort out debris flow particles into two groups. The first group will include particles for which the clay type interaction prevails and the second one will include particles for which the coarsest particle type interaction prevails. We will also assume solid particle density is equal to water density. The reality is obviously different and there are a number of other possible reasons for explaining support of coarse particles in suspensions (brownian diffusion, yield stress of the interstitial fluid, inertia, etc). Here we will use this unrealistic assumption because we intend to review the various possible behaviours of suspension from a conceptual point of view and leaving aside all consideration of two-phase flows (here this term refers to flow with significant, mean, relative velocities between two types of particles, which may result from settling, segregation, etc). Later on we will consider various suspensions obtained by increasing solid particle concentration. Actually, one may conceive that in some cases this approach might be related to the different stages of formation of some debris flows in the field.

B.1. Pure water
Pure water behaviour is Newtonian, but at high shear rates (the transition limit is given by a critical Reynolds number) an additional term appears in the stress tensor which originates in the fluctuating motions of particles around their mean motion.

B.2. Few force-free particles in water (cf Figure 1)
When solid concentration is low enough, shearing motion of the interstitial fluid (water) is slightly disturbed. Then theoretical calculations of total energy dissipation may be carried out for simple cases such as spheres or ellipsoids (cf Einstein (1956), Batchelor (1970)). The resulting behaviour is Newtonian with a viscosity slightly higher than interstitial fluid viscosity.

B.3. More force-free particles in water
When the concentration of force-free particles increases, interaction between particles can have a greater influence on behaviour. We can expect hydrodynamic interaction (lubricating process or not), friction, and collisions to take place. Even in simple shear the strain field already appears fairly complex within this simple suspension. However, for spheres, as long as the direct interaction (friction, collisions) is negligible and if the configuration is statistically isotropic and constant for different shear rates, the theory tells us that the behaviour of the suspension will be Newtonian.

Unfortunately the range of solid concentration for which such hypotheses are consistent is not well known yet and there are only semi-empirical predictions of the viscosity of such suspensions (Kamal & Mutel (1985), Utracki (1988)).

B.4. Percolating concentration of force-free particles (cf Figure 2)
When the solid concentration is high enough, direct interaction is very likely to prevail. Collisions will also be numerous at rather high shear rates. Following our introductory remarks we will not consider this case later on. Then, beyond a critical (percolation) concentration (which depends on the components of the initial fluid), a continuous network of frictional particle contacts takes place throughout the sample. In order to break this network by initiating frictional, relative motions of particles and thus initiate flow, it is necessary to impose a minimum shear stress value which is called yield stress (for yield stress of this suspension type, cf : Barnes et al. (1991), Kytomaa & Prasad (1993), Coussot (1992)(a)). Obviously the solid concentration must not be too close to a maximum packing concentration (not well-defined because it depends on the configuration) or otherwise shear will be accompanied by dilatancy.

B.5. Few clay particles in water (cf Figure 3)
The behaviour of double-layers (surrounding particles and caused by the slight diffusion of exchangeable cations in water) is not well known. However, at a low enough clay double-layer concentration and at low enough shear rates the behaviour of the suspension is likely to be Newtonian. Viscosity is given by formulas formally identical to those obtained with force-free particle suspensions (cf B.2.).

B.6. More clay particles in water
Clearly, when the double-layer concentration is such that double-layer interaction is not negligible, no simple microstructural model exists. Indeed suspension behaviour depends on the complex and varied interaction which can take place. For example time effects may be important. However, because of repulsion and attraction between particles, above a certain concentration, a continuous network of links exists throughout the sample (cf Figure 4) (M'Ewen & Mould (1957), Melton & Rand (1977)). In this case it is necessary to impose a certain minimum shear stress (yield stress) to break this network and then initiate a flow.

In the general case, for example when interactions between particles are weak (Coussot (1992)(a)), the complete behaviour of these kinds of suspensions has not yet been explained or described, but generally a Herschel-Bulkley model can be used in a wide range of shear rates. In the case of strong interaction a complete microstructural model has been proposed by Coussot et al (1992)(b). This model, which was established to describe more globally the rheology of concentrated dispersed systems in low molecular weight matrix, takes into account the evolution of the number of broken bonds within the material. The thixotropic behaviour which is obtained can predict quite well the peculiar characteristics (instability, minimum in flow curve) of some bentonite-water mixtures (Coussot et al (1992)(b)) and could also explain some peculiar characteristics of debris flows (Coussot et al (1992)(c)).

B.7. Clay-water mixtures with few force-free solid particles (cf Figure 5)

As far as we know, no physical theory exists for predicting the behaviour of suspensions of force-free particles in yield stress fluids. For instance, assuming that this interstitial fluid follows a Bingham model and taking into account the considerations corresponding to force-free particle suspensions in Newtonian fluid, we could expect at least an increase of the Bingham viscosity. Unfortunately this reasoning is quite inaccurate. Indeed there is no simple analogy between Bingham viscosity of suspensions and Newtonian fluid viscosity. We can only assimilate the total viscous dissipations in all cases, including the viscous dissipations derived from the yield stress term in the total shear stress. Besides, as a rule, when a simple Bingham model is used to describe the rheology of a suspension, the yield stress parameter corresponds to the minimum stress necessary to break the structure. But when this structure is broken there is a complex process of clay particle aggregate reformation and rupture. As a result, viscous dissipations originate in both these phenomena and not simply in classical hydrodynamic interactions.

It has been observed (Migniot (1989), Major & Pierson (1992), Coussot (1992)(a)) that, when the concentration of force-free particles added to the clay-water mixture is not too high, the difference between the behaviour of the suspension and the behaviour of the interstitial fluid is negligible. At least, the almost stable value of yield stress might be explained by the simultaneous action of two opposite effects originating in the addition of

Lecture Notes in Earth Sciences

Vol. 1: Sedimentary and Evolutionary Cycles. Edited by U. Bayer and A. Seilacher. VI, 465 pages. 1985. (out of print).

Vol. 2: U. Bayer, Pattern Recognition Problems in Geology and Paleontology. VII, 229 pages. 1985. (out of print).

Vol. 3: Th. Aigner, Storm Depositional Systems. VIII, 174 pages. 1985.

Vol. 4: Aspects of Fluvial Sedimentation in the Lower Triassic Buntsandstein of Europe. Edited by D. Mader. VIII, 626 pages. 1985. (out of print).

Vol. 5: Paleogeothermics. Edited by G. Buntebarth and L. Stegena. II, 234 pages. 1986.

Vol. 6: W. Ricken, Diagenetic Bedding. X, 210 pages. 1986.

Vol. 7: Mathematical and Numerical Techniques in Physical Geodesy. Edited by H. Sünkel. IX, 548 pages. 1986.

Vol. 8: Global Bio-Events. Edited by O. H. Walliser. IX, 442 pages. 1986.

Vol. 9: G. Gerdes, W. E. Krumbein, Biolaminated Deposits. IX, 183 pages. 1987.

Vol. 10: T.M. Peryt (Ed.), The Zechstein Facies in Europe. V, 272 pages. 1987.

Vol. 11: L. Landner (Ed.), Contamination of the Environment. Proceedings, 1986. VII, 190 pages.1987.

Vol. 12: S. Turner (Ed.), Applied Geodesy. VIII, 393 pages. 1987.

Vol. 13: T. M. Peryt (Ed.), Evaporite Basins. V, 188 pages. 1987.

Vol. 14: N. Cristescu, H. I. Ene (Eds.), Rock and Soil Rheology. VIII, 289 pages. 1988.

Vol. 15: V. H. Jacobshagen (Ed.), The Atlas System of Morocco. VI, 499 pages. 1988.

Vol. 16: H. Wanner, U. Siegenthaler (Eds.), Long and Short Term Variability of Climate. VII, 175 pages. 1988.

Vol. 17: H. Bahlburg, Ch. Breitkreuz, P. Giese (Eds.), The Southern Central Andes. VIII, 261 pages. 1988.

Vol. 18: N.M.S. Rock, Numerical Geology. XI, 427 pages. 1988.

Vol. 19: E. Groten, R. Strauß (Eds.), GPS-Techniques Applied to Geodesy and Surveying. XVII, 532 pages. 1988.

Vol. 20: P. Baccini (Ed.), The Landfill. IX, 439 pages. 1989.

Vol. 21: U. Förstner, Contaminated Sediments. V, 157 pages. 1989.

Vol. 22: I. I. Mueller, S. Zerbini (Eds.), The Interdisciplinary Role of Space Geodesy. XV, 300 pages. 1989.

Vol. 23: K. B. Föllmi, Evolution of the Mid-Cretaceous Triad. VII, 153 pages. 1989.

Vol. 24: B. Knipping, Basalt Intrusions in Evaporites. VI, 132 pages. 1989.

Vol. 25: F. Sansò, R. Rummel (Eds.), Theory of Satellite Geodesy and Gravity Field Theory. XII, 491 pages. 1989.

Vol. 26: R. D. Stoll, Sediment Acoustics. V, 155 pages. 1989.

Vol. 27: G.-P. Merkler, H. Militzer, H. Hötzl, H. Armbruster, J. Brauns (Eds.), Detection of Subsurface Flow Phenomena. IX, 514 pages. 1989.

Vol. 28: V. Mosbrugger, The Tree Habit in Land Plants. V, 161 pages. 1990.

Vol. 29: F. K. Brunner, C. Rizos (Eds.), Developments in Four-Dimensional Geodesy. X, 264 pages. 1990.

Vol. 30: E. G. Kauffman, O.H. Walliser (Eds.), Extinction Events in Earth History. VI, 432 pages. 1990.

Vol. 31: K.-R. Koch, Bayesian Inference with Geodetic Applications. IX, 198 pages. 1990.

Vol. 32: B. Lehmann, Metallogeny of Tin. VIII, 211 pages. 1990.

Vol. 33: B. Allard, H. Borén, A. Grimvall (Eds.), Humic Substances in the Aquatic and Terrestrial Environment. VIII, 514 pages. 1991.

Vol. 34: R. Stein, Accumulation of Organic Carbon in Marine Sediments. XIII, 217 pages. 1991.

Vol. 35: L. Håkanson, Ecometric and Dynamic Modelling. VI, 158 pages. 1991.

Vol. 36: D. Shangguan, Cellular Growth of Crystals. XV, 209 pages. 1991.

Vol. 37: A. Armanini, G. Di Silvio (Eds.), Fluvial Hydraulics of Mountain Regions. X, 468 pages. 1991.

Vol. 38: W. Smykatz-Kloss, S. St. J. Warne, Thermal Analysis in the Geosciences. XII, 379 pages. 1991.

Vol. 39: S.-E. Hjelt, Pragmatic Inversion of Geophysical Data. IX, 262 pages. 1992.

Vol. 40: S. W. Petters, Regional Geology of Africa. XXIII, 722 pages. 1991.

Vol. 41: R. Pflug, J. W. Harbaugh (Eds.), Computer Graphics in Geology. XVII, 298 pages. 1992.

Vol. 42: A. Cendrero, G. Lüttig, F. Chr. Wolff (Eds.), Planning the Use of the Earth's Surface. IX, 556 pages. 1992.

Vol. 43: N. Clauer, S. Chaudhuri (Eds.), Isotopic Signatures and Sedimentary Records. VIII, 529 pages. 1992.

Vol. 44: D. A. Edwards, Turbidity Currents: Dynamics, Deposits and Reversals. XIII, 175 pages. 1993.

Vol. 45: A. G. Herrmann, B. Knipping, Waste Disposal and Evaporites. XII, 193 pages. 1993.

Vol. 46: G. Galli, Temporal and Spatial Patterns in Carbonate Platforms. IX, 325 pages. 1993.

Vol. 47: R. L. Littke, Deposition, Diagenesis and Weathering of Organic Matter-Rich Sediments. IX, 216 pages. 1993.

Vol. 48: B. R. Roberts, Water Management in Desert Environments. XVII, 337 pages. 1993.

Vol. 49: J. F. W. Negendank, B. Zolitschka (Eds.), Paleolimnology of European Maar Lakes. IX, 513 pages. 1993.

Vol. 50: R. Rummel, F. Sansò (Eds.), Satellite Altimetry in Geodesy and Oceanography. XII, 479 pages. 1993.

Vol. 51: W. Ricken, Sedimentation as a Three-Component System. XII, 211 pages. 1993.

Vol. 52: P. Ergenzinger, K.-H. Schmidt (Eds.), Dynamics and Geomorphology of Mountain Rivers. VIII, 326 pages. 1994.

DUE DATE			
DEC 1 1 1995			
DEC 1 2 1995			
	201-6503		Printed in USA

Lecture Notes in Earth Sciences

This series reports new developments in research and teaching in the entire field of earth sciences - quickly, informally, and at a high level. The timeliness of a manuscript is more important than its form, which may be unfinished or tentative. The type of material considered for publication includes

– drafts of original papers or monographs,

– technical reports of high quality and broad interest,

– advanced-level lectures,

– reports of meetings, provided they are of exceptional interest and focused on a single topic.

Publication of Lecture Notes is intended as a service to the computer science community in that the publisher Springer-Verlag offers global distribution of documents which would otherwise have a restricted readership. Once published and copyrighted they can be cited in the scientific literature.

Manuscripts

Lecture Notes are printed by photo-offset from the master copy delivered in camera-ready form. Manuscripts should be no less than 100 and preferably no more than 500 pages of text. Authors of monographs and editors of proceedings volumes receive 50 free copies of their book. Manuscripts should be printed with a laser or other high-resolution printer onto white paper of reasonable quality. To ensure that the final photo-reduced pages are easily readable, please use one of the following formats:

Font size (points)	Printing area (cm)	(inches)	Final size (%)
10	12.2 x 19.3	4.8 x 7.6	100
12	15.3 x 24.2	6.0 x 9.5	80

On request the publisher will supply a leaflet with more detailed technical instructions or a T_EX macro package for the preparation of manuscripts.

Manuscripts should be sent to one of the series editors or directly to:

Springer-Verlag, Geosciences Editorial, Tiergartenstr. 17,
D-69121 Heidelberg, Germany

ISBN 3-540-57569-3
ISBN 0-387-57569-3